THE LIBRARY
ST. MARY'S COLLEGE OF MARYLAND
ST. MARY'S CITY, MARYLAND 20686

GRAPH THEORY AND COMPUTING

CONTRIBUTORS

CLAUDE BERGE

N. G. de BRUIJN

D. D. COWAN

J. R. FIKSEL

SOLOMON W. GOLOMB

C. C. GOTLIEB

FRANK HARARY

B. R. HEAP

A. HOLLIGER

JOEL D. ISAACSON

L. O. JAMES

C. A. KING

D. E. KNUTH

W. KUICH

W. F. LUNNON

GEORGE MARBLE

DAVID W. MATULA

JOHN F. MEYER

RONALD C. READ

S. O. RICE

DONALD J. ROSE

P. ROSENSTIEHL

B. ROY

ALLEN J. SCHWENK

R. G. STANTON

W. T. TUTTE

W. A. WALKER

GRAPH THEORY AND COMPUTING

Edited by RONALD C. READ

Department of Combinatorics and Optimization
University of Waterloo
Waterloo, Ontario, Canada

ACADEMIC PRESS   New York and London   1972

COPYRIGHT © 1972, BY ACADEMIC PRESS, INC.
ALL RIGHTS RESERVED.
NO PART OF THIS PUBLICATION MAY BE REPRODUCED OR
TRANSMITTED IN ANY FORM OR BY ANY MEANS, ELECTRONIC
OR MECHANICAL, INCLUDING PHOTOCOPY, RECORDING, OR ANY
INFORMATION STORAGE AND RETRIEVAL SYSTEM, WITHOUT
PERMISSION IN WRITING FROM THE PUBLISHER.

ACADEMIC PRESS, INC.
111 Fifth Avenue, New York, New York 10003

United Kingdom Edition published by
ACADEMIC PRESS, INC. (LONDON) LTD.
24/28 Oval Road, London NW1

LIBRARY OF CONGRESS CATALOG CARD NUMBER: 74-187228

PRINTED IN THE UNITED STATES OF AMERICA

CONTENTS

Evolution of the Path Number of a Graph: Covering and Packing in Graphs, II

FRANK HARARY AND ALLEN J. SCHWENK

The Production of Graphs by Computer

B. R. HEAP

A Graph-Theoretic Programming Language

C. A. KING

Algebraic Isomorphism Invariants for Graphs of Automata
JOHN F. MEYER

The Coding of Various Kinds of Unlabeled Trees
RONALD C. READ

A Graph-Theoretic Study of the Numerical Solution of Sparse Positive Definite Systems of Linear Equations
DONALD J. ROSE

Intelligent Graphs: Networks of Finite Automata Capable of Solving Graph Problems
P. ROSENSTIEHL, J. R. FIKSEL, AND A. HOLLIGER

An Algorithm for a General Constrained Set Covering Problem

B. Roy

Tripartite Path Numbers

R. G. Stanton, L. O. James, and D. D. Cowan

Non-Hamiltonian Planar Maps

W. T. Tutte

A Top-Down Algorithm for Constructing Nearly Optimal Lexicographic Trees

W. A. Walker and C. C. Gotlieb

LIST OF CONTRIBUTORS

Numbers in parentheses indicate the pages on which the authors' contributions begin.

CLAUDE BERGE (1), Faculté des Sciences, Paris, France

N. G. DE BRUIJN (15), Technological University, Eindhoven, The Netherlands

D. D. COWAN (285), Applied Analysis and Computer Science Department, University of Waterloo, Waterloo, Ontario

J. R. FIKSEL (219), Institut de Programmation, Paris, France

SOLOMON W. GOLOMB (23), University of Southern California, University Park, Los Angeles, California

C. C. GOTLIEB (303), Department of Computer Science, University of Toronto, Toronto, Ontario, Canada

FRANK HARARY (39), Research Center for Group Dynamics, Institute for Social Research, The University of Michigan, Ann Arbor, Michigan

B. R. HEAP (47), Division of Numerical Analysis and Computing, National Laboratories, Teddington, Middlesex, England

A. HOLLIGER (219), École Polytechnique, Paris, France

JOEL D. ISAACSON (109), Department of Mathematical Studies, Southern Illinois University at Edwardsville, Edwardsville, Illinois

L. O. JAMES (285), Department of Computer Science, University of Manitoba, Winnipeg, Manitoba, Canada

C. A. KING, (63), University of the West Indies, Kingston, Jamaica

D. E. KNUTH (15), Stanford University, Stanford, California

W. KUICH* (77), IBM Laboratory, Vienna, Austria

W. F. LUNNON† (87, 101), Atlas Computer Laboratory, Chilton, Didcot, Berkshire, England

GEORGE MARBLE (109), Department of Applied Mathematics and Computer Science, Washington University, St. Louis, Missouri

DAVID W. MATULA (109), Department of Applied Mathematics and Computer Science, Washington University, St. Louis, Missouri

JOHN F. MEYER (123), Department of Electrical and Computer Engineering, The University of Michigan, Ann Arbor, Michigan

RONALD C. READ (153), Department of Combinatorics and Optimization, University of Waterloo, Waterloo, Ontario, Canada

S. O. RICE (15), Bell Telephone Laboratories, Inc., Murray Hill, New Jersey

DONALD J. ROSE‡ (183), Department of Mathematics, University of Denver, Denver, Colorado

P. ROSENSTIEHL (219), École Pratique des Hautes Études, Paris, France

B. ROY (267), Groupe METRA, Université Paris-Dauphine, Paris, France

ALLEN J. SCHWENK (39), Research Center for Group Dynamics, Institute for Social Research, The University of Michigan, Ann Arbor, Michigan

R. G. STANTON (285), Department of Computer Science, University of Manitoba, Winnipeg, Manitoba, Canada

W. T. TUTTE (295), Faculty of Combinatorics and Optimization, University of Waterloo, Waterloo, Ontario, Canada

W. A. WALKER§ (303), Department of Computer Science, University of Toronto, Toronto, Ontario, Canada

---

* Present address: Technische Hochschule Wien, Vienna, Austria.

† Present address: Department of Computing in Mathematics, University College, Cardiff, Wales.

‡ Present address: Aiken Computation Laboratory, Harvard University, Cambridge, Massachusetts.

§ Present address: Ontario Hydro, Toronto, Ontario, Canada.

PREFACE

It has become more and more clear in recent years that the two disciplines of graph theory and computer science have much in common, and that each is capable of assisting significantly in the development of the other. Thus, graph theorists are increasingly finding that many of their problems can be solved, or their research furthered by the use of computing techniques; while computer scientists are realizing that the language of graph theory is a convenient one in which to express many of the concepts with which they have to deal, and that standard results in graph theory are often very relevant to the problems that concern them. Despite this, the number of publications in which this interdependence of the two subjects is explicitly recognized is still quite small.

The purpose of this book is largely to draw attention to some of the problems and applications which straddle these two disciplines. It is a collection of invited papers in which computing techniques are applied to graph-theoretical problems, or in which problems in computing are treated by graph-theoretical methods. Thus, mathematicians from several different fields of research will discover within these pages material that is of interest to them. The "pure" graph theorist, the computer scientist who is interested in the study of algorithms, and the researcher into the theory of automata will all find papers relating to these theoretical topics, while mathematicians engaged in operations research, and other practical applications of graph theory and computing, will discover that they have not been forgotten.

Some of the papers in this book were presented at an International Conference held in January 1969 at the University of the West Indies in Kingston,

Jamaica; some others represent later development of work that was discussed at that conference. Most of them, however, were invited for inclusion in this volume with the aim of giving the reader a representative sample of topics in this joint field of research, whose nature is indicated by the title—which was also the name of the conference—Graph Theory and Computing.

ALTERNATING CHAIN METHODS: A SURVEY

Claude Berge

Faculté des Sciences
Paris, France

1. Historical Background

Consider a graph, and partition the set of all its edges into two classes, the *heavy* edges and the *light* edges. An *alternating chain* is a chain whose edges are alternately light and heavy. This concept was introduced in 1891 by Petersen to prove that, in some cubic graphs, any linear factor can be modified in order to use a given edge of the graph.

By 1957, when many people were trying to solve new classes of linear programming problems in integers, I considered two new optimization problems of this type:

(1) *Maximum Matching Problem:*   given a family $\mathscr{E}$ of subsets of a given set X, what is the largest number of members of $\mathscr{E}$ which are pair-wise disjoint;

1

(2) *Minimum Covering Problem:*   what is the smallest number of members
of $\mathscr{E}$ whose union covers X?

When X is the set of vertices of a graph, and $\mathscr{E}$ is the set of its edges, a simple
way to solve these problems is offered by the concept of an alternating chain.
This was discovered nearly simultaneously by us [4], for the maximum
matching problem, and by Norman and Rabin [15], for the minimum covering
problem. The extension of these methods to the most general matching
problems was found independently by Edmonds [9] and Ray-Chaudhuri [16].

The proofs given in the above papers were often unnecessarily complicated.
In this didactical paper, we shall only give some simpler proofs of these results,
and discuss the computational procedures suggested by them, as well as
possible extensions of the method.

2. The Maximum Matching Problem

We shall consider here a simple graph $G = (X, E)$, that is, undirected,
without loops or multiple edges; the set of its vertices is denoted by X, and
the set of its edges by E. A *matching* $E_0 \subset E$ is a set of edges such that no two
meet at the same vertex. The problem is to find a maximum matching.

A vertex x is *saturated* in E_0 if there exists an edge of E_0 incident to x. An
alternating chain is a simple chain whose edges are alternately in E_0 and in
$E - E_0$.

LEMMA 1.   If E_0 and E_1 are two matchings of G, a connected component
of the partial graph $\left(X, (E_0 - E_1) \cup (E_1 - E_0) \right)$ is of one of the three following
types:

Type 1:   isolated vertex;
Type 2:   even elementary cycle, whose edges are alternately in E_0 and in E_1;
Type 3:   elementary chain, whose edges are alternately in E_0 and in E_1,
and whose extremities are unsaturated for one of the two matchings, E_0 or E_1.

The proof is a straightforward verification.

THEOREM 1.   A matching E_0 is maximum, if and only if there exists no
alternating chain connecting two distinct unsaturated vertices.

Proof: (1)   If E_0 is a matching with such an alternating chain, we can
interchange the heavy lines and the light lines along this chain. We obtain a
new matching E_1, and $|E_1| = |E_0| + 1$. Therefore, E_0 is not a maximum
matching.

(2) If E_0 is a matching without such an alternating chain, consider a maximum matching E_1. By (1), E_1 is a matching without such an alternating chain, and by Lemma 1 we have

$$|E_0 - E_1| = |E_1 - E_0|,$$

since the connected components of $(X, (E_0 - E_1) \cup (E_1 - E_0))$ of Type 3 are even chains. This shows that $|E_0| = |E_1|$, and that E_0 is a maximum matching.

According to this theorem, the maximum matching problem can be solved by searching for all alternating chains from each unsaturated vertex. However Edmonds [10] showed that it was not necessary to develop all alternating chains. He described a procedure, more economical, that involved shrinking of parts of the graph. His algorithm was experimented with by Witzgall. Although helpful to one's intuitive understanding, shrinking is difficult to implement on an electronic computer. For that reason, Witzgall and Zahn [18] presented a modification of Edmond's maximum matching algorithm, which displayed a tree-like arrangement of alternating chains using all the vertices reachable by alternating chains issuing from a given unsaturated vertex.

The existence of such a tree is of great interest for the theory, but its construction, as described by Witzgall and Zahn [18], is not simple. Therefore, it is often advisable to return to a suitable *fanning-out algorithm*. For a better understanding of the procedure, consider, instead of G, a labeled graph H, obtained from G as follows. If the n vertices of G are $a, b, \ldots$, draw n points denoted by $a, b, \ldots$. If in G we have $[x, y] \in E - E_0$ and $[y, z] \in E_0$, draw a path of length 2 going from the point x to the point z, and passing through an additional point that we mark with the symbol $\bar{y}$. If a is an unsaturated vertex of G, and if B denotes the set of vertices of G adjacent to an unsaturated vertex different from a, we join by an arc the points of H in B to an additional point z (see Fig. 1). At the end, we mark the n points $a, b, \ldots$, by the symbols $\bar{a}, \bar{b}, \ldots$.

An alternating chain in G joining a to another unsaturated vertex is, in H, a path from a to z whose vertices are all marked differently, and conversely. The problem is to find such a path, which we call *admissible*.

Denote by $s(x)$ the symbol attached to a vertex x of H. We shall construct a sequence $\mu_1, \mu_2, \ldots$, of admissible paths, by a labeling procedure (inspired by Tremaux; see [7]), with the following rules:

Rule 1: denote by x_1 the initial vertex with $s(x_1) = \bar{a}$; set $\mu_0 = [x_1]$, a path of length 0.

Rule 2: if $\mu_i = [x_1]$, and if there exists an unlabeled arc (x_1, x_2), set $\mu_{i+1} = [x_1, x_2]$. Label arc (x_1, x_2) with a $+$. If such a vertex x_2 does not exist, stop the procedure.

Rule 3: if $\mu_i = [x_1, x_2, \ldots, x_k]$, $k > 1$, we shall consider two cases:

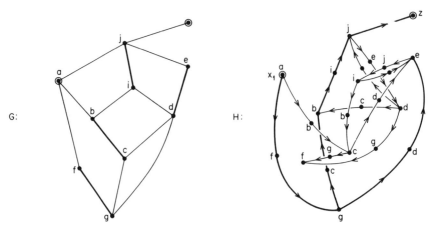

Fig. 1. In graph G, the heavy lines represent the edges of E_0, a matching which is to be improved. In H, the heavy lines represent an admissible path going from a to z.

Case 3.1: if $s(x_k) \neq s(x_1), s(x_2), \ldots, s(x_{k-1})$, and if there exists a vertex x_{k+1} such that (x_k, x_{k+1}) is an unlabeled arc, set $\mu_{i+1} = [x_1, x_2, \ldots, x_k, x_{k+1}]$. Also, label with a $+$ the new arc (x_k, x_{k+1}).

Case 3.2: Otherwise, set $\mu_{i+1} = [x_1, x_2, \ldots, x_{k-1}]$; also, remove the labels of all the labeled arcs (x_k, y), but not of the arc (x_{k-1}, x_k). We stop the procedure as soon as μ_i is a path leading to z.

If vertex z has not been reached, then every possible admissible path has been encountered exactly twice. The proof is identical to the Tremaux proof. In the classical Tremaux procedure, the number of steps is bounded by twice the number of arcs, but this is not true for the above algorithm. Though no practical tests were done to compare it to Edmond's algorithm, some experiments conducted in 1965, while we were at the International Computation Center in Rome, show only that the above procedure leads rapidly to a solution for a medium-size problem.

3. The Maximum c-Matching Problem

The maximum matching problem can be generalized. Consider a multigraph G, with multiple edges but no loops, whose vertices are $x_1, x_2, \ldots, x_n$, and an n-tuple $c = (c_1, c_2, \ldots, c_n)$, where c_i is a nonnegative integer less than or equal to $d_G(x_i)$, the degree of vertex x_i. A set $E_0 \subset E$ is said to be a *c-matching* if the set $E_0(x_i)$ of all the edges incident to x_i verifies

$$|E_0(x_i)| \leqslant c_i, \qquad i = 1, 2, \ldots, n.$$

The problem we shall consider now is to find a c-matching of maximum cardinality. I proved [5] the following theorem:

If the edges of a c-matching are denoted by heavy lines, and if we add a new vertex x_0, linked to x_i by $|E_0(x_i)|$ light edges and $c_i - |E_0(x_i)|$ heavy edges, E_0 is maximum, if and only if no alternating chain leaves x_0 by a heavy edge and comes back to x_0 by a heavy edge.

An alternating chain is not permitted to use the same edge more than once, but may visit the same vertex several times. For that reason the procedures described by Dantzig *et al.* [8] or Witzgall and Zahn [18] are not valid, as was noticed by Witzgall and Zahn [18].

However, we shall now show that, by constructing a new graph $\bar{G}$, this new problem can be reduced to the problem considered in Section 2. A simple graph $\bar{G}$ is obtained from G. For every vertex x_i of G, construct two disjoint sets of points

$$A_i = \{a_i^e \mid e \in E(x_i)\} \quad \text{and} \quad B_i = \{b_i^k \mid 1 \leqslant k \leqslant d_G(x_i) - c_i\}.$$

The set of vertices of $\bar{G}$ will be $\bar{X} = (\bigcup A_i) \cup (\bigcup B_i)$. For every i, link every element $a_i^e \in A_i$ to all the elements of B_i. For every edge $e = [x_i, x_j]$ of G, link the elements $a_i^e \in A_i$ and $a_j^e \in A_j$ by an edge denoted $\bar{e}$ (see Fig. 2).

THEOREM 2. A maximum matching $\bar{E}_0$ of $\bar{G}$ which saturates all the vertices of $\bigcup_i B_i$, induces in G a maximum c-matching E_0 of G, and vice versa.

Proof: (1) Let $\bar{E}_0$ be a maximum matching of $\bar{G}$. We can assume that it saturates every element of $\bigcup B_i$. If not, we interchange the heavy and light lines along alternating chains of length 2. $\bar{E}_0$ induces in G a set $E_0 \subset E$ with

$$|E_0(x_i)| \leqslant d_G(x_i) - |B_i| = d_G(x_i) - [d_G(x_i) - c_i] = c_i.$$

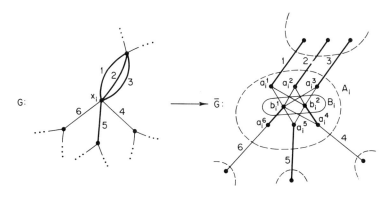

Fig. 2

E_0 is thus a c-matching of G, and we have

$$|E_0| = |\bar{E}_0| - \sum_{i=1}^{n} |B_i|.$$

(2) Let E_1 be a maximum c-matching of G. It gives in $\bar{G}$ a matching $\bar{E}_1$ which saturates all the elements of $\bigcup B_i$, and we have

$$|\bar{E}_1| = |E_1| + \sum_{i=1}^{n} |B_i|.$$

As we have $|E_0| \leqslant |E_1|$, we obtain

$$|\bar{E}_1| = |E_1| + \sum_{i=1}^{n} |B_i| \geqslant |E_0| + \sum_{i=1}^{n} |B_i| = |\bar{E}_0|.$$

Since $\bar{E}_0$ is a maximum matching of $\bar{G}$, this implies $|\bar{E}_1| = |\bar{E}_0|$, and therefore $|E_1| = |E_0|$. This proves that E_0 is a maximum c-matching of G.

This theorem shows immediately how to find a maximum c-matching by using alternating chain methods.

There are other optimization problems that one can solve by these methods. Consider a multigraph $G = (X, E)$ and a n-tuple $d = (d_1, d_2, ..., d_n)$. A set $F_0 \subset E$ is said to be a d-cover if we have

$$|F_0(x_i)| \geqslant d_i, \qquad i = 1, 2, ..., n.$$

If $d_i = 1$ for all i, F_0 is a *cover* of the vertices of G. Norman and Rabin [15] found that the minimum cover problem reduces to the maximum matching problem. We noticed [5] that the problem of the minimum c cover reduces also to the maximum matching problem. More precisely, we have

THEOREM 3. Consider a multigraph $G = (X, E)$ without loops, and consider integers c_i, d_i, with $c_i + d_i = d_G(x_i)$ for $i = 1, 2, ..., n$. A set $F_0 \subset E$ is a minimum d-cover, if and only if $E - F_0$ is a maximum c-matching.

Proof: First, note that if we set $E_0 = E - F_0$, then $|F_0(x_i)| \geqslant d_i$ is equivalent to

$$|E_0(x_i)| \leqslant d_G(x_i) - d_i = c_i.$$

If F_0 is a minimum d-cover, then $E_0 = E - F_0$ is a c-matching. If E_1 is a maximum c-matching, then $F_1 = E - E_1$ is a d-cover. Hence, $|E_1| \geqslant |E_0|$ and $|F_0| \leqslant |F_1|$. On the other hand, we have

$$|F_1| = |E - E_1| = |E| - |E_1| \leqslant |E| - |E_0| = |F_0| \leqslant |F_1|.$$

Therefore, we have $|F_1| = |F_0|$ and $|E_0| = |E_1|$. This shows that F_1 is a minimum d-cover, and that E_0 is a maximum c-matching.

A generalization of Theorem 1 is due to Balinski [2]. Suppose we are given a graph in which every edge is assigned a weight $w(e) > 0$, and we want a c-matching E_0 for which $w(E_0) = \sum_{e \in E_0} w(e)$ is maximum. If $c_i = 1$ for all i, we have the problem of the maximum weighted matching. Define an *augmenting chain* μ relative to E_0, as an alternating cycle, or alternating chain, such that no edge of $E_0 - \mu$ is incident to a vertex of μ, and such that

$$\sum_{\substack{e \in \mu \\ e \notin E_0}} w(e) > \sum_{\substack{e \in \mu \\ e \in E_0}} w(e).$$

THEOREM 4. A matching E_0 of G is a maximum weighted matching, if and only if there is no augmenting chain.

Proof: (1) If E_0 is a matching with an augmenting chain, it is not a maximum weighted matching. This is obvious, as in Theorem 1.

(2) If E_0 is a matching without an augmenting chain, and if E_1 is a maximum weighted matching, consider the partial graph $G' = (X, (E_0 - E_1) \cup (E_1 - E_0))$. If $w(E_0) < w(E_1)$, there exists in G' a connected component H, with

$$\sum_{\substack{e \in H \\ e \in E_0 - E_1}} w(e) < \sum_{\substack{e \in H \\ e \in E_1 - E_0}} w(e).$$

By Lemma 1, H is in G an augmenting chain relative to E_0, which yields the contradiction.

COROLLARY. If G is a weighted multigraph, a c-matching E_0 is of maximum weight, if and only if the graph $\bar{G}$ obtained as in Theorem 2, with $w(\bar{e}) = w(e)$ for every $e \in E$, has no augmenting chain relative to $\bar{E}_0$.

The proof is the same as that for Theorem 4.

Note that when the weight of each edge is equal to 1, Theorem 4 reduces to Theorem 1.

Another variation, discussed by Glover [11], is to find amongst all the maximum matchings, a matching of minimum weight. Define a *reducing chain*, relative to a maximum matching E_0, as an alternating cycle, or alternating even chain issuing from an unsaturated vertex μ, such that

$$\sum_{\substack{e \in \mu \\ e \notin E_0}} w(e) < \sum_{\substack{e \in \mu \\ e \in E_0}} w(e).$$

THEOREM 5. A maximum matching E_0 is of minimum weight, if and only if there is no reducing chain.

Proof: (1) If E_0 is a maximum matching with a reducing chain, it is not of minimum weight. This is obvious.

(2) If E_0 is a maximum matching without a reducing chain, and if E_1 is a

maximum matching of minimum weight, then each connected component of $G' = (X, (E_0 - E_1) \cup (E_1 - E_0))$ is an alternating cycle, or an even alternating chain with one extremity unsaturated in E_0 and the other extremity unsaturated in E_1, by Lemma 1. If $w(E_0) > w(E_1)$, one of these connected components is a reducing chain, relative to E_0, which contradicts the assumption made about E_0.

This theorem permits us to show how this new problem can be reduced to the well-known problem of the shortest elementary path in a graph with a *length* $l(e)$, positive or negative, assigned to each edge e.

Let $G = (X, E)$ be a graph, and let $E_0 \subset E$ be a maximum matching. Assume that all the unsaturated vertices are contracted into one single vertex a. Construct as in Section 2 an oriented graph H associated with G. If in G we have $[x, y] = e_1 \in E - E_0$ and $[y, z] = e_2 \in E_0$, we shall assign the lengths $l(\bar{x}, \bar{y}) = w(e_1) - w(e_2)$ and $l(\bar{y}, \bar{z}) = 0$ to the two corresponding arcs $(\bar{x}, \bar{y})$ and $(\bar{y}, \bar{z})$ of H.

In G, E_0 is a maximum matching of minimum weight, if and only if in H there is no circuit, and no path issuing from $\bar{a}$ with a negative length and all its vertices marked differently.[†]

4. The Maximum Stable Set Problem

In a graph G, a set B of vertices is said to be *stable* (or *independent*) if no two vertices in B are adjacent. We now seek a maximum stable set.

This problem is equivalent to the general matching problem with a family of sets, if we consider the *representing graph* of this family of sets. Several methods of solution are known for this problem, [see 13, 14, 16], and one can always use the tool provided by the theory of alternating chains. The following result, often used in the theory of hypergraphs, is in fact a special case of Ray-Chaudhuri [16] and a consequence of Edmonds [9]. Denote by $\Gamma_G(x)$ the set of all the vertices adjacent to x. Define an *alternating sequence*

[†] There exist in the literature several algorithms to determine for every x the shortest elementary path from $\bar{a}$ to $\bar{x}$, or to detect at least one negative circuit. They easily can be adapted to our matching problem. For the general case, see Dantzig *et al.* [8] and Roy [17].

It should be noticed that, if we want a fast procedure, we must not claim to obtain all the shortest elementary paths of H, since this would solve, as it is well-known, the traveling salesman problem. The claim is only to find a negative circuit in a graph H if one exists, or if none exists, then to find all the shortest paths from the given vertex a.

The inductive method developed by Dantzig *et al.* [8], is especially easy to adapt to the matching problems. It gives, for the number of additions to perform, an upper bound equal to $m + n(n+1)h/2$, where m is the number of arcs of H, n the number of vertices, and h the maximum out degree.

relative to a stable set B as a sequence $(a_1, b_1, a_2, b_2, a_3, \ldots)$ of distinct vertices of G, belonging alternately to $A = X - B$ and to B, with

(1) $a_1 \in A$;
(2) b_i is chosen in $B - \{b_1, b_2, \ldots, b_{i-1}\}$ so that

$$\Gamma_G(b_i) \cap \{a_1, a_2, \ldots, a_i\} \neq \varnothing;$$

(3) a_{i+1} is chosen in $A - \{a_1, a_2, \ldots, a_i\}$ so that

$$\Gamma_G(a_{i+1}) \cap \{b_1, b_2, \ldots, b_i\} \neq \varnothing,$$

$$\Gamma_G(a_{i+1}) \cap \{a_1, a_2, \ldots, a_i\} = \varnothing.$$

To extend the alternating chain theorem we need two lemmas.

LEMMA 2. If G is a tree, and if (A, B) is a bicoloring of its vertices with $|B| \geq |A|$, then G has at least one pendant vertex in B.

Proof: A bicoloring (A, B) is a partition of X into two stable sets. Suppose that the set of all pendant vertices of G is a set $A_1 \subset A$. We shall show that this leads to a contradiction. In the tree G_{X-A_1}, the set of all pendant vertices is $B_1 \subset B$; in $G_{X-A_1-B_1}$, the set of pendant vertices is $A_2 \subset A$; etc. We have $|A_1| \geq |B_1|$, since every pendent vertex of G_{X-A_1}, can be mapped into one of its neighbors in A_1 by an injection.

Thus, we have for some $q \geq 1$

$$|A_1| \geq |B_1| \geq |A_2| \geq |B_2| \geq \cdots \geq |B_q| \geq |A_{q+1}|,$$

$$B_q \neq \varnothing, \quad \text{and} \quad B_{q+1} = \varnothing.$$

If $A_{q+1} \neq \varnothing$, we have

$$|A| > \sum_{i=1}^{q} |A_i| \geq \sum_{i=1}^{q} |B_i| = |B|,$$

which contradicts the assumption that $|A| \leq |B|$.

If $A_{q+1} = \varnothing$, the set B_q reduces to a single vertex, which is not pendant in $G_{A_q \cup B_q}$. Therefore $|A_q| > |B_q|$, and

$$|A| = \sum_{i=1}^{q} |A_i| > \sum_{i=1}^{q} |B_i| = |B|.$$

In both cases, we find a contradiction.

LEMMA 3. If G is a tree of order n, and if (A, B) is a bicoloring of its vertices with $|A| = |B|$ or $|A| = |B| + 1$, then there exists an alternating sequence $(a_1, b_1, a_2, b_2, \ldots)$ using once and only once every vertex of G.

Proof: (1) For $n = 1$ or $n = 2$, this is obvious.
(2) If the statement is true for $n = 2k$, let us show that it is true for a tree

G of order $n = 2k + 1$. As $|A| = |B| + 1 > |B|$, there exists by Lemma 2 a pendant vertex $a_{k+1} \in A$. For $G_{X - \{a_{k+1}\}}$ there exists an alternating sequence $(a_1, b_1, ..., b_k)$ using all its vertices, by the induction hypothesis, and $(a_1, b_1, ..., b_k, a_{k+1})$ is the desired alternating sequence of G.

(3) If the statement is true for $n = 2k + 1$, then it is true for $n = 2k + 2$. This is obvious from part (2) of the proof.

THEOREM 6. A stable set B is maximum, if and only if there is no maximal alternating sequence of odd length.

Proof: (1) If such an alternating sequence σ existed, then B would not be a maximum stable set, since $B' = (B - \sigma) \cup (\sigma - B)$ is a stable set with a greater cardinality.

(2) Let A be a maximum stable set, and let B be a stable set with $|B| < |A|$. Let us show the existence of a maximal alternating sequence of odd length relative to B. Set $B_0 = B - A$ and $A_0 = A - B$. Thus, $|B_0| < |A_0|$. In the subgraph $G_{A_0 \cup B_0}$, let $A_1 \cup B_1, A_2 \cup B_2, ..., A_k \cup B_k$ be the different components, where $A_i \subset A_0$ and $B_i \subset B_0$ for $i = 1, 2, ..., k$. We have

$$\sum_{i=1}^{k} |B_i| = |B_0| < |A_0| = \sum_{i=1}^{k} |A_i|.$$

Thus, there exists an index, let us say $i = 1$, with $|B_1| < |A_1|$.

(3) If $|B_1| + 1 = |A_1|$, a spanning tree of the subgraph $G_{A_1 \cup B_1}$, which is connected, admits (A_1, B_1) as a bicoloring. By Lemma 3, its vertices constitute a maximal odd alternating sequence σ, relative to B_1. This sequence σ is also for G an alternating sequence relative to B. Also, σ is a maximal alternating sequence. If $b \in B - \sigma$, we have either $b \in B - A = B_0$, and b is adjacent to no $a_i \in \sigma$, or $b \in B \cap A$, and b is adjacent to no $a_i \in \sigma$, since A is a stable set.

If $|B_1| + 1 < |A_1|$, one can remove from the spanning tree of $G_{A_1 \cup B_1}$ as many pendant vertices in A_1 as necessary, to obtain $|B_1| + 1 = |A_1|$. This is always possible by Lemma 2, and the theorem follows.

There is an interesting application of Theorem 6. Denote by $\alpha(G)$ the stability number of $G = (X, E)$, that is, the maximum cardinality of a stable set. G is said to be *α-critical* if for every edge $e \in E$, the partial graph $G - e$, obtained from G by deleting edge e, verifies $\alpha(G - e) = \alpha(G) + 1$. The structure of α-critical graphs has been extensively studied in the literature. One of the main results was obtained by Beineke *et al.* [3], who proved that in an α-critical graph, through any two adjacent edges, there passes an odd cycle. Andrásfai [1] proved that through a nonisolated edge of G there passes an odd cycle without a chord. I proved [6] a generalization of these two results by showing: in an α-critical graph, through two adjacent edges there passes an odd elementary cycle without a chord. We shall give now a simple proof of this result by using Theorem 6.

THEOREM 7.   In an α-critical graph G, with $\alpha(G) = k$, for any two adjacent edges $[a, b]$ and $[b, x]$, there exists an odd elementary cycle without a chord that uses these two edges.

Proof: (1)   If we remove edge $[b, x]$, we create a stable set S_{bx} of cardinality $k + 1$ with $b, x \in S_{bx}$. Let $B = S_{bx} - \{b\}$. Then B is a maximum stable set. We have $a, b \notin B$ and $x \in B$. Also, b is linked with B by only one edge $[b, x]$.

(2)   In the partial graph $G - [a, b]$, the stable set B is not maximum. By Theorem 6, there exists in $G - [a, b]$ a maximal alternating sequence $\sigma = (a_1, b_1, a_2, b_2, ..., a_q)$ with $a_i \in X - B$ and $b_i \in B$ for all i. In $G - [a, b]$, the set $T = (B - \sigma) \cup (\sigma - B)$ is a maximum stable set. Therefore, $a, b \in T$. Hence, $a, b \in \sigma - B$. The subgraph of $G - [a, b]$ induced by σ is connected, and has $(\sigma \cap B, \sigma - B)$ as a bicoloring. Let μ be the shortest chain connecting a and b. As $a, b \in \sigma - B$, this chain μ, together with the edge $[a, b]$, is in G an odd elementary cycle without a chord. This cycle uses $[a, b]$ and $[b, x]$, since b is linked to $\sigma \cap B$ only by this edge.

From a computational viewpoint, the search for all alternating sequences is more difficult to implement than the search for alternating chains, though the principle is similar. To simplify the problem, one can use a subprocedure, involving the algorithm described in Section 2 [4], that we shall discuss now. Suppose that we have obtained a maximum matching E_0, and that there exists at least one unsaturated vertex.[†]

(1)   If a vertex x is unsaturated, or if x can be reached only by an alternating chain issuing from an unsaturated vertex and terminating in a heavy edge incident to x, we shall say that x is a *heavy vertex*.

(2)   If x can be reached only by an alternating chain issuing from an unsaturated vertex and terminating in a light edge incident to x, we shall say that x is a *light vertex*.

(3)   If x can be reached by both an alternating chain terminating in a light edge and an alternating chain terminating in a heavy edge, we shall say that x is a *mixed vertex*.

(4)   If x cannot be reached by an alternating chain issuing from an unsaturated vertex, we shall say that x is an *inaccessible vertex*.

The set H of all the heavy vertices, the set L of all the light vertices, the set M of all the mixed vertices, and the set I of all the inaccessible vertices, verify $H \cup L \cup M \cup I = X$.

LEMMA 4.   If $H \cup L = X$, then H is a maximum stable set of G. Also the maximum stable set is unique.

---

† If there exists no unsaturated vertex, we shall add an additional vertex x_0 that we link by a light edge to a given vertex x_1 of G. The procedure will give a stable set of G, to which x_1 does not belong, and which is maximum under these assumptions.

Proof: (1) Define a *covering* C as a set of vertices such that every edge of G has at least one extremity in C. If E_0 is a matching and C is a covering, then we always have $|E_0| \leqslant |C|$. Therefore, if a given matching E_0 and a given covering C satisfy $|E_0| = |C|$, then C is necessarily a minimum covering, and $X - C$ is a maximum stable set.

(2) If $H \cup L = X$, the set L is a covering, since two heavy vertices cannot be adjacent, and every light vertex is the terminal of one heavy edge leading into H. Thus, $|L| = |E_0|$, and $H = X - L$ is a maximum stable set.

(3) Let us show the uniqueness of the maximum stable set that is, if $h \in H$, then every maximum stable set contains h. That is, if $h \in H$, then $H - \{h\}$ is a maximum stable set of the subgraph $G_{X - \{h\}}$. As the subgraph $G_{X - \{h\}}$ admits L as a covering, we have only to show that L is a minimum covering. If h is an unsaturated vertex of G, then E_0 is a maximum matching of $G_{X - \{h\}}$ with $|E_0| = |L|$. Therefore, L is a minimum covering of $G_{X - h}$.

If h is a saturated vertex of G, it is incident to a heavy edge $e = [h, x]$, and there exists in $G_{X - \{h\}}$ an alternating chain between the unsaturated vertex x and another unsaturated vertex. By Theorem 1, $E_0 - e$ is not a maximum matching, and there exists in $G_{X - \{h\}}$ a maximum matching E_1 with $|E_1| = |E_0|$. As $|L| = |E_0| = |E_1|$, this shows L is a minimum covering of $G_{X - \{h\}}$.

LEMMA 5. If I_1 is a connected component of the subgraph G_1 induced by the inaccessible vertices, then every vertex adjacent to I_1 is a light vertex, attached to I_1 by a light edge.

The proof is obvious.

LEMMA 6. If M_1 is a connected component of the subgraph G_M induced by the mixed vertices, then every vertex adjacent to M_1 is a light vertex. Also, there is one and only one heavy edge going out from M_1.

This is the Petersen–Gallai Lemma. For a proof see Berge [7, Theorem 9, Chapter 8].

THEOREM 8. Let $M_1, ..., M_p$ be the p connected components of G_M, let $I_1, I_2, ..., I_q$ be the q connected components of G_I, and let $S(A)$ denote a maximum stable set of G_A for $A \subset X$. The set $S = H \cup [\bigcup_{i=1}^{p} S(M_i)] \cup [\bigcup_{j=1}^{q} S(I_j)]$ is a stable set of G, and if $p \leqslant 1$, it is a maximum stable set.

Proof: It is obvious that S is a stable set by Lemmas 4, 5, and 6.

(1) If $p = 0$, remove all the light edges going out from the sets I_k. We obtain a graph G, having S as a maximum stable set, by Lemmas 4 and 5. We have

$$|S| = \alpha(G') \geqslant \alpha(G) \geqslant |S|.$$

Therefore, $|S| = \alpha(G)$, and S is a maximum stable set of G.

(2) If $p = 1$, we obtain a graph G'' by contracting the unique component M_1 into a single vertex z_1 and by removing the sets I_k. It is obvious the E_0 induces again a maximum matching, by Theorem 1, and G'' has only heavy and light vertices. As $H'' = H \cup \{z_1\}$ is the unique maximum stable set of G'', by Lemma 4, H is a maximum stable set of G''_{X-z_1}, and, equivalently, of the graph G' obtained from G by removing M_1 and the sets I_k. Again we see as before that S is a maximum stable set of G.

For constructing the classes H, L, M, and I the procedure described in Section 2 can be applied without modifications. Thus, Theorem 8 permits us to simplify the maximum stable set problem if the number of connected components of G_M is 0 or 1. A simple backtracing procedure can permit us to attack the other cases.

References

1. Andrásfai, B., On critical graphs, *in* "Théorie des graphes, Rome ICC" (P. Rosenstiehl, ed.), p. 9. Dunod, Paris, 1967.
2. Balinski, M., On maximum matching, minimum covering and their connections, *J. Combinatorial Theory* (in press).
3. Beineke, W., Harary, F., and Plummer, M. D., On the critical lines of a graph, *Pacific J. Math.* **21**, 205–212 (1967).
4. Berge, C., Two theorems in graph theory, *Proc. Nat. Acad. Sci. U.S.A.* **43**, 842 (1957).
5. Berge, C., Sur le couplage maximum d'un graphe *C. R. Acad Sci. Paris* **247**, 258–259 (1958).
6. Berge, C., Sur une propriété des graphes k-stables critiques, *in* "Combinat. Structures" (R. K. Guy, H. Hanani, N. Sauer, and J. Schonheim, eds.), p. 7–11. Gordon & Breach, New York, 1970.
7. Berge, C., "Graphes et hypergraphes." Dunod, Paris, 1970.
8. Dantzig, G. B., Blattner, W. O., and Rao, M. R., All the shortest routes from a fixed origin in a graph, *in* "Théorie des graphes, Rome ICC," pp. 85–92. Dunod, Paris, 1967.
9. Edmonds, J., Covers and packings in a family of sets, *Bull. Amer. Math. Soc.* **68**, 494–499 (1962).
10. Edmonds, J., Paths, trees and flowers, *Canad. J. Math.* **17**, 449–467 (1965).
11. Glover, F., "Shortest Alternating Paths in a Graph with Negative Edges but with No Negative Alternating Cycles." University of Texas, 1969 (unpublished).
12. Johnson, E., "Programming in Networks and Graphs." Op. Res. Center, Berkeley, 1965.
13. Lawler, E. L., Covering problems, *SIAM J. Appl. Math.* **14**, No. 5, (1966).
14. Maghout, K., Sur la détermination des nombres de stabilité et du nombre chromatique d'un graphe, *C. R. Acad. Sci. Paris* **248**, 3522–3523 (1959).
15. Norman, R. Z., and Rabin, M. O., An algorithm for a minimum cover of a graph, *Proc. Amer. Math. Soc.* **10**, 315–319 (1959).
16. Ray-Chaudhuri, D. K. An algorithm for a minimum cover of an abstract complex, *Canad. J. Math.* **15**, 11–24 (1963).
17. Roy, B., "Algèbra moderne et théorie des graphes," Vol. 2, Chapter 6, Section A, pp. 2–19. Dunod, Paris, 1970.
18. Witzgall C., and Zahn, C. T., Modification of Edmonds' maximum matching algorithm, *J. Res. Nat. Bur. Standards Sect. B* **69**, 91–98 (1965).

THE AVERAGE HEIGHT OF PLANTED PLANE TREES

N. G. de Bruijn

Technological University
Eindhoven, The Netherlands

D. E. Knuth†

Stanford University
Stanford, California

S. O. Rice

Bell Telephone Laboratories, Inc.
Murray Hill, New Jersey

A *planted plane tree*, sometimes called an ordered tree, is a rooted tree which has been embedded in the plane so that the relative order of subtrees at each branch is part of its structure. In this paper we shall say simply *tree* instead of *planted plane tree*, following the custom of computer scientists.

The *height* of a tree is the number of nodes on a maximal simple path starting at the root. For example, there are exactly five trees with five nodes and height 4, namely

† This research was supported in part by the National Science Foundation, under grant number GJ-992, and the Office of Naval Research under grant number N-00014-67-A-0112-0057 NR 044-402. Reproduction in whole or in part is permitted for any purpose of the United States Government.

15

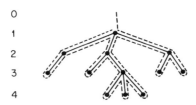

Fig. 1. A tree as a random walk.

The height of a tree is of interest in computing because it represents the maximum size of a stack used in algorithms that traverse the tree [3, pp. 317–318]. Our goal in this paper is to study the average height of a tree with n nodes, assuming that all n-node trees are equally likely. The corresponding problem for oriented, that is, rooted, unordered, trees has been solved by Rényi and Szekeres [6]. Our principal results are stated in Eqs. (32) and (34).

Trees appear in many disguises, and in particular there is a natural correspondence between trees of height less than or equal to h and discrete random walks in a straight line, with absorbing barriers at 0 and $h+1$. If we wander around a tree with n nodes, as shown by the dotted lines in Fig. 1, the vertical component of successive positions describes a path of length $2n-1$ from 1 to 0. For example, the path in Fig. 1 is

$$1, 2, 3, 2, 1, 2, 3, 2, 3, 4, 3, 4, 3, 4, 3, 2, 1, 2, 3, 2, 3, 2, 1, 0.$$

This is one way a gambler can lose \$1 before winning \$5. This construction, suggested by Harris [2] in 1952, is clearly reversible.

The height of trees plays a similar role in the classical ballot problem. How many ways are there to arrange n ballots for candidate A and n for candidate B in such a way that the number of votes for A never lags behind the number for B, as the ballots are counted, but A is never more than h votes ahead? The answer is the number of trees with $n+1$ nodes and height less than or equal to $h+1$, again by the construction indicated in Fig. 1. The ballot sequence corresponding to that tree is AABBAABAABABABBBAABABB.

We shall begin our study of the asymptotic properties of height by reviewing some known results. Let A_{nn} be the number of trees with n nodes and height less than or equal to h, and let

(1) $$A_h(z) = \sum A_{nh} z^n$$

be the corresponding generating function. We obtain all trees with height less than or equal to $h+1$ by taking a root node and attaching zero or more subtrees each of which has height less than or equal to h. Therefore,

(2) $$A_{h+1}(z) = z(1 + A_h(z) + A_h(z)^2 + A_h(z)^3 + \cdots)$$
$$= z/(1 - A_h(z)), \qquad h \geqslant 0.$$

Clearly $A_0(z) = 0$. This relation yields a simple recurrence for the numbers A_{nh},

(3) $\quad A_{n,h+1} = A_{n-1,h+1} A_{1,h} + A_{n-2,h+1} A_{2,h} + \cdots + A_{1,h+1} A_{n-1,h}$,

$$n \geqslant 2, \quad h \geqslant 0,$$

from which the first few values are easily calculated, as shown in Table I.

TABLE I

	$n = 1$	2	3	4	5	6	7	8
$h = 1$	1	0	0	0	0	0	0	0
2	1	1	1	1	1	1	1	1
3	1	1	2	4	8	16	32	64
4	1	1	2	5	13	34	89	233
5	1	1	2	5	14	41	122	365
6	1	1	2	5	14	42	131	417

Since no tree with n nodes can have a height greater than n, we have

(4) $$A_{nh} = A_{nn} = \binom{2n-2}{n-1} \frac{1}{n}, \quad h \geqslant n,$$

which is the well-known formula for the total number of trees with n nodes [3, p. 389].

Iteration of (2) yields a continued fraction representation of $A_h(z)$. For example,

(5) $$A_4(z) = \cfrac{z}{1 - \cfrac{z}{1 - \cfrac{z}{1-z}}}.$$

This suggests expressing the generating function as a quotient of polynomials

(6) $$A_h(z) = z p_h(z)/p_{h+1}(z),$$

where

(7) $\quad p_0(z) = 0, \quad p_1(z) = 1, \quad p_{h+1}(z) = p_h(z) - z p_{h-1}(z)$.

The solution to this recurrence is

(8) $$p_h(z) = (1-4z)^{-\frac{1}{2}} \left(\left(\frac{1+(1-4z)^{\frac{1}{2}}}{2} \right)^h - \left(\frac{1-(1-4z)^{\frac{1}{2}}}{2} \right)^h \right),$$

and the form of this solution suggests setting $z = 1/(4\cos^2\theta)$. We obtain

(9) $p_h(4\cos^2\theta)^{-1} = \sin h\theta/(\sin\theta(2\cos\theta)^{h-1})$,

              $A_h(4\cos^2\theta)^{-1} = \sin h\theta/(2\cos\theta\sin(h+1)\theta)$.

Incidentally it is easy to verify that $p_h(-1)$ is the Fibonacci number F_h, and that

(10) $p_h(z) = \sum_{0 \leqslant k < h} \binom{h-1-k}{k}(-z)^k, \qquad h \geqslant 1.$

This leads to another recurrence for the A_{nh}.

Since $p_h(z)^2 - p_{h+1}(z)p_{h-1}(z) = z^{h-1}$, there is a simple generating function for the number of trees with n nodes and height exactly h,

(11) $A_h(z) - A_{h-1}(z) = z^h/p_{h+1}(z)p_h(z).$

This formula was recently derived by Kreweras [4, p. 37].

Since p_h is a polynomial of degree $\lfloor(h-1)/2\rfloor$, the roots of $p_h(z) = 0$ are $(4\cos^2(j\pi/h))^{-1}$, for $1 \leqslant j < h/2$. We obtain a partial fraction expansion of the generating function

(12) $A_h(z) = \sum_{1 \leqslant j \leqslant h/2} \dfrac{\tan^2\theta_{jh}}{(h+1)(1-(4\cos^2\theta_{jh})z)} + a_h + b_h z,$

where

              $\theta_{jh} = j\pi/(h+1),$

and

(13) $a_{2m} = -m,$ $b_{2m} = 0,$

      $a_{2m+1} = -m(2m+1)/6(m+1),$ $b_{2m+1} = (m+1)^{-1}, \qquad m \geqslant 1.$

This leads immediately to the explicit formula

(14) $A_{nh} = (h+1)^{-1}\sum_{1 \leqslant j \leqslant h/2} 4^n\sin^2(j\pi(h+1))\cos^{2n-2}(j\pi(h+1)), \qquad n \geqslant 2.$

It is rather remarkable that this formula gives a constant value for fixed n and all $h \geqslant n$. It is perhaps even more remarkable that Lagrange derived a formula in 1775 which essentially includes this as a special case, see Lagrange [5, p. 247]. Feller [1, p. 322] observes that the formula has been rediscovered many times, although it appears in many texts on probability in connection with the equivalent gambler's ruin problem. As a special case of (14) we have the asymptotic formula

(15) $A_{nh} \sim (4^n/(h+1))\tan^2(\pi/(h+1))\cos^{2n}(\pi/(h+1)), \qquad \text{fixed} \quad h, \quad n \to \infty.$

Another interesting expression for A_{nh} can be derived by applying complex variable theory. We have

$$(16) \qquad A_{nh} = (2\pi i)^{-1} \int^{(0+)} \frac{dz}{z^{n+1}} A_h(z)$$

$$= (2\pi i)^{-1} \int^{(0+)} \frac{dz}{z^n} (1+u) \frac{1-u^h}{1-u^{h+1}},$$

where

$$(17) \qquad u = \frac{1-(1-4z)^{\frac{1}{2}}}{1+(1-4z)^{\frac{1}{2}}},$$

by (6) and (8). Since

$$(18) \qquad z = u/(1+u)^2,$$

we have $u \approx z$ when $|z| \ll 1$. Hence, we may change variables in (16) to obtain

$$(19) \qquad A_{nh} = (2\pi i)^{-1} \int^{(0+)} \frac{du}{u^n} (1-u)(1+u)^{2n-2} \frac{1-u^h}{1-u^{h+1}}.$$

In other words A_{nh} is the coefficient of u^{n-1} in $(1-u)(1+u)^{2n-2}(1-u^h)/(1-u^{h+1})$. Some simplification now occurs when we consider the number of trees with height *greater* than h,

$$(20) \qquad B_{nh} = A_{nn} - A_{nh}$$

$$= (2\pi i)^{-1} \int^{(0+)} \frac{du}{u^{n+1}} (1-u)^2 (1+u)^{2n-2} \frac{u^{h+1}}{1-u^{h+1}}.$$

It follows that

$$(21) \qquad B_{n+1,h-1} = \sum_{k \geqslant 1} \left(\binom{2n}{n+1-kh} - 2\binom{2n}{n-kh} + \binom{2n}{n-1-kh} \right).$$

The *average height* of a tree with n nodes is S_n/A_{nn}, where S_n is the finite sum

$$(22) \qquad S_n = \sum_{h \geqslant 1} h(A_{nh} - A_{n,h-1})$$

$$= \sum_{h \geqslant 1} h(B_{n,h-1} - B_{nh})$$

$$= \sum_{h \geqslant 0} B_{nh}$$

$$= (2\pi i)^{-1} \int^{(0+)} \frac{du}{u^{n+1}} (1-u)^2 (1+u)^{2n-2} \sum_{h \geqslant 1} \frac{u^h}{1-u^h}$$

$$= (2\pi i^{-1}) \int^{(0+)} \frac{du}{u^{n+1}} (1-u)^2 (1+u)^{2n-2} \sum_{k \geqslant 1} d(k) u^k.$$

As usual, $d(k)$ denotes the number of positive divisors of k. Therefore,

$$(23) \qquad S_{n+1} = \sum_{k \geq 1} d(k) \left(\binom{2n}{n+1-k} - 2 \binom{2n}{n-k} + \binom{2n}{n-1-k} \right).$$

We shall now proceed to obtain an asymptotic series for the sum

$$(24) \qquad f_a(n) = \sum_{k \geq 1} \left(\binom{2n}{n+a-k} \middle/ \binom{2n}{n} \right) d(k), \qquad \text{fixed } a, \quad n \to \infty,$$

and this will lead to an asymptotic series for S_n.

Let $x = (k-a)/n$. By Stirling's approximation we have

$$(25) \qquad \binom{2n}{n+a-k} \middle/ \binom{2n}{n} = \exp\left(-2n \left(\frac{x^2}{1 \cdot 2} + \frac{x^4}{3 \cdot 4} + \cdots \right) + \left(\frac{x^2}{2} + \frac{x^4}{4} + \cdots \right) \right.$$
$$\left. - \frac{1}{6n}(x^2 + x^4 + \cdots) + O(x^2 n^{-3}) \right),$$

when $-\frac{1}{2} < x < \frac{1}{2}$, and

$$\binom{2n}{n+a-k} \middle/ \binom{2n}{n} = O(\exp(-n^{2\varepsilon}))$$

when $k \geq n^{\frac{1}{2}+\varepsilon}+a$, for all fixed $\varepsilon > 0$. Therefore the sum of all terms for $k \geq n^{\frac{1}{2}+\varepsilon}+a$ in (24) is negligible, being $O(n^{-m})$ for all $m > 0$, and we may take $x = O(n^{-\frac{1}{2}+\varepsilon})$ in (25).

We now turn to the asymptotic behavior of the function

$$(26) \qquad g_b(n) = \sum_{k \geq 1} k^b d(k) \exp(-k^2/n), \qquad \text{fixed } b, \quad n \to \infty.$$

Again the terms for $k \geq n^{\frac{1}{2}+\varepsilon}$ are negligible, so we can use (25) to express f in terms of g:

$$(27) \quad f_a(n) = g_0(n) + \frac{2a}{n} g_1(n) - \frac{a^2}{n} g_0(n) + \frac{4a^2+1}{2n^2} g_2(n) - \frac{1}{6n^3} g_4(n)$$

$$- \frac{2a^3+a}{n^2} g_1(n) + \frac{4a^3+5a}{3n^3} g_3(n) - \frac{a}{3n^4} g_5(n) + O(n^{-2+\varepsilon} g_0(n)).$$

In principle such an expansion could be carried out as far as we like. Hence, the problem of obtaining an asymptotic expansion for $f_a(n)$ reduces to the analogous problem for $g_b(n)$.

The behavior of $g_b(n)$ can be derived by starting with the well-known formula

$$(28) \qquad e^{-x} = (2\pi i)^{-1} \int_{c-i\infty}^{c+i\infty} \Gamma(z) x^{-z}\, dz, \qquad c > 0, \quad x > 1,$$

obtained, for example, by Fourier inversion of $\Gamma(c+2\pi i t)$. Then since $\zeta(z)^2 = \sum_{k \geqslant 1} d(k)/k^z$, we find

$$(29) \qquad g_b(n) = \sum_{k \geqslant 1} (2\pi i)^{-1} \int_{c-i\infty}^{c+i\infty} n^z \Gamma(z) k^{b-2z}\, d(k)\, dz$$

$$= (2\pi i)^{-1} \int_{c-i\infty}^{c+i\infty} n^z \Gamma(z) \zeta(2z-b)^2\, dz,$$

where now $c > \frac{1}{2}(b+1)$. Let q be a fixed positive number. When $\mathrm{Re}(s) \geqslant -q$, $\zeta(s) = 0(|s|^{q+\frac{1}{2}})$ as $s \to \infty$. Since $n^z \Gamma(z)$ gets small on vertical lines we can shift the line of integration to the left as far as we please if we only take the residues into account. There is a double pole at $z = \frac{1}{2}(b+1)$, and possibly some simple poles at $z = 0, -1, -2, \dots$. Let $w = z - \frac{1}{2}(b+1)$, we have

$$n^z \Gamma(z) \zeta(2z-b)^2 = n^{(b+1)/2} \Gamma(\tfrac{1}{2}(b+1))(1 + w\ln n + O(w^2))$$

$$\times (1 + w\psi(\tfrac{1}{2}(b+1)) + O(w^2))((2w)^{-2} + \gamma/w + O(1)),$$

where $\psi(z) = \Gamma'(z)/\Gamma(z)$, hence the residue at the double pole is

$$(30) \qquad n^{\frac{1}{2}(b+1)} \Gamma(\tfrac{1}{2}(b+1))(\tfrac{1}{4}\ln n + \tfrac{1}{4}\psi(\tfrac{1}{2}(b+1)) + \gamma).$$

The residue at $z = -k$ is

$$(31) \qquad n^{-k}(-1)^k \zeta(-2k-b)^2/k! = n^{-k}(-1)^k B_{2k+b+1}^2/(2k+b+1)^2\, k!$$

which is almost always zero when b is even. The sum of (30) and (31) for all $k \geqslant 0$ gives an asymptotic series for $g_b(n)$. Hence, we have, for all $m > 0$,

$$g_0(n) = \tfrac{1}{4}(\pi n)^{\frac{1}{2}}\ln n + (\tfrac{3}{4}\gamma - \tfrac{1}{2}\ln 2)(\pi n)^{\frac{1}{2}} + \tfrac{1}{4} + O(n^{-m});$$

$$(32) \qquad g_1(n) = \tfrac{1}{4}n\ln n + \tfrac{3}{4}\gamma n + (\tfrac{1}{144}) - (\tfrac{1}{14400})n^{-1} + O(n^{-2});$$

$$g_2(n) = (n/8)(\pi n)^{\frac{1}{2}}\ln n + (\tfrac{1}{4} + \tfrac{3}{8}\gamma - \tfrac{1}{4}\ln 2)n(\pi n)^{\frac{1}{2}} + O(n^{-m});$$

etc. These formulas have been verified by computer calculation. For example, when $n = 10$, $g_0(n) = 3.96042$ and $\tfrac{1}{4}(\pi n)^{\frac{1}{2}}\ln n + (\tfrac{3}{4}\gamma - \tfrac{1}{2}\ln 2)(\pi n)^{\frac{1}{2}} + \tfrac{1}{4} = 3.96041$.

Returning to our original problem about trees, we have

$$(33) \quad S_{n+1}/(n+1) A_{n+1,n+1} = f_1(n) - 2f_0(n) + f_{-1}(n)$$

$$= (-2/n)g_0(n) + (4/n^2)g_2(n) + O(n^{-3/2}\log n)$$

by (4), (23), (24), and (27), and this equals $(\pi n)^{-1/2} - \tfrac{1}{2}n^{-1} + O(n^{-3/2}\log n)$. We have proved the following result.

THEOREM. The average height of a planted plane tree with n nodes, considering all such trees to be equally likely, is

(34) $$(\pi n)^{1/2} - \tfrac{1}{2} + O(n^{-1/2}\log n).$$

The same method can be used to obtain as many further terms of the expansion as desired. The factor $\log n$ in the error term turns out to be unnecessary.

References[†]

1. Feller, W., "An Introduction to Probability Theory and its Applications," Vol. 1, 2nd ed. Wiley, New York, 1957.
2. Harris, T. E., First passage and recurrence distributions, *Trans. Amer. Math. Soc.* **73**, 471–486 (1952).
3. Knuth, D. E., "The Art of Computer Programming," Vol. 1. Addison-Wesley, Reading, Massachusetts, 1968.
4. Kreweras, G., Sur les eventails de segments, *Cahiers du Bureau Universitaire de Recherche Operationelle* **15**, 1–41 (1970).
5. Lagrange, J. L., Recherches sur les suites récurrentes, *in* "Oeuvres de Lagrange," Vol. 4, pp. 149–251. Paris, 1869.
6. Rényi, A., and Szekeres, G., On the height of trees, *Austral. J. Math.* **7**, 497–507 (1967).
7. Riordan, J., The Enumeration of Trees By Height and Diameter, *IBM J. Res. Develop.* **4**, 473–478 (1960).
8. Riordan, J., Ballots and trees, *J. Combinatorial Theory* **6**, 408–411 (1969).

† We wish to thank Prof. John Riordan for pointing out references [2] and [4].

HOW TO NUMBER A GRAPH†

Solomon W. Golomb

University of Southern California
University Park
Los Angeles, California

> *It is a Tree of Life to them that grasp it,*
> *...and all its Paths are Peace.*
>
> Proverbs III

1.   A Statement of the Problem

Let Γ be a *graph*, with n nodes and e edges. By the term *graph*, we under-stand a connected, undirected graph without loops or double connections.

† This research was supported in part by the United States Air Force under Grant AFOSR-68-1555.

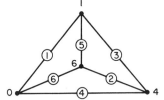

Fig. 1.   The graph K_4, successfully numbered.

We wish to associate n distinct nonnegative integers to the n nodes of Γ in such a way that the e edges receive e distinct positive integers by the assignment of $|a_i - a_j|$ to a given edge, where a_i and a_j are the numbers assigned to its end points. Moreover, we wish to minimize the value of the largest integer assigned to any node of Γ. We will call this minimized value $G(\Gamma)$. The problem of *numbering a graph* is to assign integers to the nodes so as to achieve $G(\Gamma)$.

In Fig. 1 we see the complete graph on four nodes, K_4, with the nodes numbered $\{0, 1, 4, 6\}$, and the edges numbered $\{1, 2, 3, 4, 5, 6\}$. Since the edge numbers must be distinct positive integers, and K_4 has six edges, this numbering must be optimum and $G(K_4) = 6$. Thus, we also have the general lower bound $G(\Gamma) \geqslant e$ for all graphs Γ. The principal questions which arise in the theory of numbering the nodes of graphs revolve around the relationship between $G(\Gamma)$ and e, for example, identifying classes of graphs for which $G(\Gamma) = e$ and other classes for which $G(\Gamma) > e$ and looking for bounds on $G(\Gamma) - e$. A graph for which $G(\Gamma) = e$ will be called a *graceful graph*, and the numbering which achieves $G(\Gamma) = e$, a *graceful numbering*.

2. A Context for the Problem

Think of the graph Γ as a communication network with n terminals and e interconnections between terminals. We wish to assign a distinct identifying number to each terminal, in such a way that each interconnection is then uniquely identified by the absolute value of the difference between the numbers assigned to its two end terminals. For economy, the largest number assigned to any node is to be minimized. This is clearly the same problem as in the preceding section.

Suppose we require that the minimum number assigned to any node be at least a_0. The resulting problem is trivially isomorphic to the problem in Section 1, since all the node numbers may be increased by a constant amount a_0, with no effect whatever on the edge numbers $|a_i - a_j|$. The chief advantage of setting $a_0 = 0$ is that the relation $G(\Gamma) = e$ for graceful graphs then refers to both the number of edges, the highest edge number, and the highest node number.

3. A History of Subproblems

Several years ago it was conjectured[†] that every tree can have its n nodes numbered from 1 to n in such a way that each of the $n-1$ edges gets a distinct number from 1 to $n-1$ as the absolute difference of the numbers at its end points. In our terminology, this conjecture asserts that every tree is a graceful graph. This conjecture is still unproved in 1971, although it has been proved for special types of trees called *caterpillars*, and for other assorted flora and fauna.

In 1968, I generalized the tree problem to that of characterizing those graphs Γ for which $G(\Gamma) = e$. I presented some results at the January, 1969, conference on graph theory and computing at the University of the West Indies in Kingston, Jamaica. Those results, previously unpublished, constitute a significant portion of the present article.

There is a classical combinatorial problem involving the notching of a metal bar of length k at integer points in such a way that all the distances between two notches, or between a notch and an end point, are distinct. If there are $n-2$ notches and two end points, then there are $\binom{n}{2}$ lengths which must be distinct. This problem is isomorphic to numbering the nodes of the complete graph K_n, and the smallest k for which the notch problem has a solution is equal to $\Gamma(K_n)$.

4. Necessary Conditions for Graceful Graphs

THEOREM 1. Let Γ be a graph with n nodes and e edges. A necessary condition for Γ to be graceful is that it be possible to partition the nodes into two sets $\mathscr{E}$ and $\mathcal{O}$, such that the number of edges connecting nodes in $\mathscr{E}$ with nodes in $\mathcal{O}$ is exactly $[(e+1)/2]$.

Proof: If Γ is graceful, the n nodes can be partitioned into two sets having respectively *even* ($\mathscr{E}$) and *odd* ($\mathcal{O}$) node numbers. The e edges end up numbered from 1 to e, and $[(e+1)/2]$ of these edge numbers are odd. However, an odd-numbered edge must have one even end point and one odd end point.

DEFINITION. A successful partitioning of the nodes of Γ into sets $\mathscr{E}$ and $\mathcal{O}$ with $[(e+1)/2]$ interconnecting edges is called a *binary labeling* of Γ.

Example: The complete graph K_n on n nodes has $\binom{n}{2} = n(n-1)/2$ edges. If we assign m nodes to class $\mathscr{E}$ there will be $n-m$ nodes in class $\mathcal{O}$, and $m(n-m)$ even–odd interconnections. For K_n, we have $[(e+1)/2] = [(n^2-n+2)/4]$,

† An unpublished but widely circulated conjecture, attributed to Gerhard Ringel.

and K_n has a binary labeling if and only if there is a choice of m for which $m(n-m) = [(n^2 - n + 2)/4]$. Such a choice exists for $n = 2, 3, 4, 6, 9, 11, 16,...$, but fails to exist for $n = 5, 7, 8, 10, 12, 13, 14, 15, 17,....$ While this enables us to prove that many graphs K_n are not graceful, a stronger result is given in Theorem 4.

THEOREM 2. Suppose that integers, not necessarily distinct, are assigned to the nodes of a graph Γ, and that each edge of Γ is given an *edge number* equal to the absolute difference of the node numbers at its end points. Then the sum of the edge numbers around any *circuit* of Γ is *even*.

Proof: Let the consecutive node numbers around a circuit be $a_1, a_2,..., a_r$. Then the consecutive edge numbers are

$$|a_1 - a_2|, \quad |a_2 - a_3|,..., \quad |a_{r-1} - a_r|, \quad |a_r - a_1|,$$

and their sum satisfies

$$\sum_{i=1}^{r} |a_i - a_{i+1}| \equiv \sum_{i=1}^{r} (a_i - a_{i+1}) \equiv 0 \bmod 2$$

as asserted.

THEOREM 3. Let Γ be an Eulerian graph, that is, with an even number of edges at each node, with e edges. A necessary condition for Γ to be graceful is that $[(e + 1)/2]$ be even. That is, if $e \equiv 1 \bmod 4$, or $e \equiv 2 \bmod 4$, then Γ cannot be graceful. In fact, Γ cannot be binary labeled.

Proof: An Eulerian graph may be regarded as a union of edge-disjoint circuits, or in fact as one big circuit involving each edge once. By Theorem 2, the sum of the edge numbers around each circuit must be even, and hence the sum of all the edge numbers must be even. For a graceful graph, there will be $[(e + 1)/2]$ odd edges. Thus, if Γ is Eulerian and $[(e + 1)/2]$ is an odd number, then Γ cannot be graceful. In fact, in this case no labeling of the nodes of Γ as even and odd can lead to $[(e + 1)/2]$ odd edges. Whence, Γ cannot be binary labeled.

Examples: (1) There are three graphs having 5 nodes which cannot be binary labeled, by Theorem 3. It happens that these are the only nongraceful graphs with fewer than 6 nodes. In Fig. 2, we see these three graphs, each numbered so as to verify $G(\Gamma) = e + 1$.

(2) There are also non-Eulerian graphs which cannot be binary labeled, such as K_8, K_{10}, K_{12}, and K_{14} (see the example following Theorem 1).

(3) There are Eulerian graphs which cannot be binary labeled even though $[(e + 1)/2]$ is even, for example K_{17}. The graph K_n is Eulerian if and only if $n > 1$ is odd. In the case of K_{17}, there are $e = \binom{17}{2} = 136$ edges, and $[(e + 1)/2]$

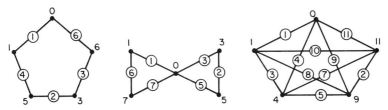

Fig. 2. The three nongraceful graphs with 5 nodes, numbered to illustrate $G(\Gamma) = e+1$.

$= 68$ is even. Since there is no choice of m, $0 \leqslant m \leqslant 17$, for which $m(17-m)$ $= 68$, we see that K_{17} cannot be binary labeled.

(4)   There are graphs which are neither Eulerian nor complete which cannot be binary labeled. For example, if one edge is removed from K_{10}, the resulting graph H is neither Eulerian nor complete. There are $\binom{10}{2} - 1 = 44$ edges in H, and $[(e+1)/2] = 22$. If we label m nodes in H as odd and the other $10-m$ as even, the number of odd edges will be either $m(10-m)$ or $m(10-m)-1$, depending on whether the deleted edge of K_{10} would be even or odd. But there is no choice of m, $0 \leqslant m \leqslant 10$, for which either $m(10-m) = 22$ or $m(10-m) = 23$.

THEOREM 4.   If $n > 4$, the complete graph K_n cannot be graceful.

Proof: For $n > 4$, the graph K_n has $e = \binom{n}{2} \geqslant 10$ edges. If K_n were graceful, we could assign a subset of the numbers $\{0, 1, 2, ..., e\}$ to the nodes in such a way that the edges receive each of the numbers $\{1, 2, ..., e\}$. We shall show that the assumption that this is possible leads to a contradiction.

In order for K_n to have an edge numbered e, both 0 and e must be node numbers. For there to be an edge numbered $e-1$, either 1 or $e-1$ must also be a node number. In any graceful graph Γ with e edges, the replacement of every node number a_i by $e-a_i$ leaves all edge numbers unchanged, and is the equivalent *inverse node numbering*. Hence, we may pick the node number 1 for K_n, instead of $e-1$, with no loss of generality.

Next, to obtain an edge numbered $e-2$, we must adjoin the node number $e-2$. If we adjoined $e-1$ to get $e-2$ as the difference of $e-1$ and 1, we would have two edges numbered 1, namely, between nodes 0 and 1, and between nodes $e-1$ and e. If we adjoined 2 to get $e-2$ as the difference of e and 2, we would again have two edges numbered 1, namely, between nodes 0 and 1, and between nodes 1 and 2.

With nodes numbered 0, 1, $e-2$, and e, we have edges numbered 1, 2, $e-3$, $e-2$, $e-1$, and e. To get an edge numbered $e-4$, we must adjoin the node number 4. All other choices are quickly ruled out, as above.

With nodes numbered 0, 1, 4, $e-2$, and e, we have edge numbers 1, 2, 3, 4, $e-6$, $e-4$, $e-3$, $e-2$, $e-1$, and e. Note that for K_4, with $e = 6$, this gives us the numbering of K_4 shown in Fig. 1. There is now no way to obtain an edge numbered $e-5$, because each of the ways to obtain $e-5$ as a difference of two numbers contains at least one impossible node number. The reader may quickly verify that the following circled numbers are not possible choices as node numbers:

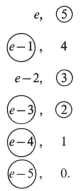

$$e, \quad ⑤$$
$$Ⓔ{-}1, \quad 4$$
$$e-2, \quad ③$$
$$Ⓔ{-}3, \quad ②$$
$$Ⓔ{-}4, \quad 1$$
$$Ⓔ{-}5, \quad 0.$$

This contradicts the assumption that K_n is graceful for all cases in which $e-5>4$, which corresponds to $n \geqslant 5$.

Remarks: (1) If a metal bar of length 6 is notched at the points 1 and 4, then each of the lengths $1, 2, 3, 4, 5, 6$ can be obtained in one and only one way as the distance between two notches, including end points as notches. It is a classical combinatorial result, equivalent to Theorem 4, that no bar of length $\binom{n}{2}$ for $n > 4$ can be similarly notched.

(2) If we require all distances between notches to be distinct, and ask for the shortest bar with n notches, still counting end points as notches, we have the problem of determining $G(K_n)$. In Fig. 3, we see the optimum

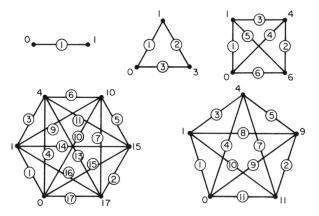

Fig. 3. Numberings of the graphs K_n which achieve $G(K_n)$ for $2 \leqslant n \leqslant 6$.

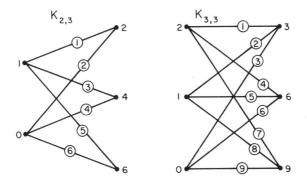

Fig. 4. Numbering of complete bipartite graphs.

numberings for K_2, K_3, K_4, K_5, and K_6. Note that $G(K_n) = \binom{n}{2}$ for $n = 2, 3, 4$, while $G(K_5) = \binom{5}{2} + 1$ and $G(K_6) = \binom{6}{2} + 2$. It follows from the proof of Theorem 4 that there is no node numbering of K_n, $n > 4$, for which all edge numbers are distinct and for which the edge numbers g, $g-1$, $g-2$, $g-3$, $g-4$, $g-5$ all occur, where g is the largest node number used. In the numberings shown for both K_5 and K_6 in Fig. 3, the edge number $g-5$ fails to occur.

(3)   Theorem 4 can be extended to show that graphs which are nearly complete cannot be graceful. That is, if a graph on n nodes has more than $\binom{n}{2} - \alpha(n)$ edges, it does not have a graceful numbering, where $\alpha(n)$ appears to grow rapidly in n. It would be interesting to determine the function $\alpha(n)$ precisely.

5.   Classes of Graceful Graphs

The *complete bipartite graph* $K_{a,b}$ is the graph with $n = a + b$ nodes and $e = ab$ edges, obtained by connecting each of a nodes with each of b nodes in all possible ways. For this class of graphs we have the following result.

THEOREM 5.   For all positive integers a and b, the complete bipartite graph $K_{a,b}$ is graceful.

Proof: It suffices to exhibit a numbering. Consider the two sets of nodes, A and B, containing a and b elements, respectively. Assign the nodes in set A the numbers $0, 1, 2, \ldots, a-1$, and assign the nodes in set B the numbers $a, 2a, 3a, \ldots, ba$. In this way, every integer from 1 to ab has a unique representation as a difference between a number in B and a number in A. Examples of this numbering are shown in Fig. 4.

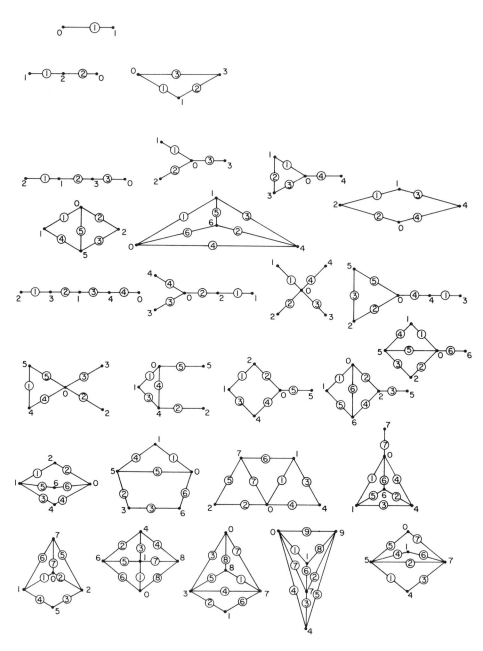

Fig. 5. The graceful graphs with $n \leqslant 5$ modes.

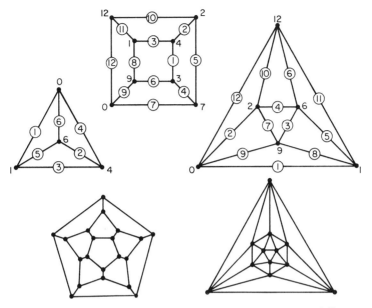

Fig. 6. Graphs of the five Platonic solids. Are they all graceful?

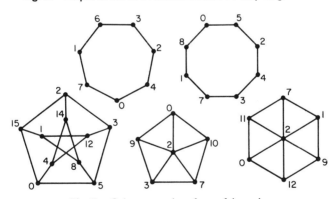

Fig. 7. Other examples of graceful graphs.

Note: Since $K_{3,3}$ and K_5 are the two irreducible examples of nonplanar graphs, and since $K_{3,3}$ is graceful while K_5 is not, we may conclude that planarity is unnecessary and insufficient for gracefulness.

As previously mentioned, all graphs with $n \leqslant 5$ nodes are graceful except for the three graphs shown in Fig. 2. This is verified by the numberings given in Fig. 5. Among other graphs which have been shown to be graceful are the graphs of three of the five Platonic solids (see Fig. 6), the 10-node Petersen graph, and a great many miscellaneous examples (see Fig. 7).

6. Some General Questions

Although numerous examples of infinite families of graceful graphs are known (see Theorem 5 and Fig. 8a), a general necessary and sufficient condition for gracefulness has not been found. In particular, the 7-node graph in Fig. 8b does not have a graceful numbering, although it is not covered by any of the theorems mentioned thus far.

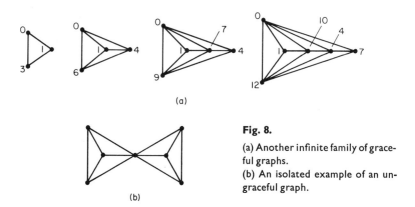

(a)

Fig. 8.
(a) Another infinite family of graceful graphs.
(b) An isolated example of an ungraceful graph.

(b)

A particularly interesting unsolved problem is to determine, asymptotically, as $n \to \infty$, the percentage of graphs on n nodes which are graceful. At the present time, it has not been shown that this limit exists, nor, if it does, that any value on $[0, 1]$ is excluded. It is likewise possible to consider the percentage of graphs with e edges which are graceful. It is reasonable to conjecture that the limit of this percentage, as $e \to \infty$, is the same as the corresponding limit taken nodewise.

If graceful graphs could be characterized, the question of whether all trees are graceful graphs would be settled. In the absence of such a result, it is interesting to note the following theorem (first proved by A. Lempel).

THEOREM 6. Let T be a tree with n nodes and $e = n - 1$ edges. Then there exists a binary labeling of T for which $[n/2]$ of the nodes are odd (set $\mathcal{O}$) and $[(n+1)/2]$ of the nodes are even (set $\mathcal{E}$).

Proof: We may observe from Fig. 5 that this is at least true for all trees with 5 or fewer nodes. To complete the proof by induction, suppose that T_0 is a tree which does not satisfy the theorem, and for which the number of nodes $n_0 > 5$ is a minimum. Such a tree has at least one of the following two features: either a pair of terminal nodes joined to the same preterminal node (Case A), or a terminal node joined to a preterminal node which connects to only one other node (case B). This is proved by taking a maximum-length path through the tree, from one terminal point X to another terminal point Y, and looking at the preterminal point Z to which Y connects. By maximality of the path from X to Y, the node Z connects only (1) to a node leading back toward X, (2) directly to Y, and (3) possibly directly to other terminal nodes. These two cases are illustrated in Fig. 9. At least one of these two cases occurs in every tree with $n \geqslant 3$ nodes.

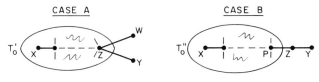

Fig. 9

In Case A, we consider the tree T_0' with $n_0 - 2$ nodes, from which Y and W and the edges connecting them to Z have been dropped. By the inductive hypothesis, T_0' can be labeled satisfying all conditions of the theorem. We then adjoin Y and W with opposite parities to one another, to complete a valid labeling of T_0.

In Case B, we consider the tree T_0'' with $n_0 - 2$ nodes, from which Y and Z and the edges connecting Y to Z and Z to some interior point P, have been dropped. We label T_0'' by the inductive hypothesis, and suppose that P has been assigned a parity p, even or odd. We then assign parity p to Z, and the complementary parity $\bar{p}$ to Y, to complete a satisfactory binary labeling of T_0.

7. Euclidean Models and Complete Graphs

DEFINITION. By the *Euclidean model* of a numbered graph, we mean the result of placing the numbered nodes on the corresponding positions along the real axis and connecting them as in the original graph.

Examples of such Euclidean models are shown in Fig. 10. The Euclidean model is the same basic idea as the notched metal bar previously mentioned. The Euclidean model of a graceful graph Γ with e edges consists of e line segments, of respective lengths 1 through e, and joined at end points to be isomorphic to Γ. This viewpoint may facilitate the computation of a significant upper bound to the number of graceful graphs.

The Euclidean model is frequently a convenient tool in visualizing or simplifying problems involving numbering of graphs. We shall consider

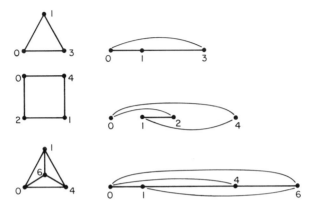

Fig. 10. Some Euclidean models of graphs.

specifically the case of assigning numbers to the nodes of the complete graph K_n so as to achieve $G(K_n)$. We note that the numbering assigned to a complete graph is completely specified by the consecutive distances along the real axis in the Euclidean model. Thus, we see in Fig. 10 that the numbering for K_3 is specified by the sequence of lengths $1, 2$; and the numbering for K_4 by $1, 3, 2$. The best sequences known for K_n, $2 \leqslant n \leqslant 10$, are shown in Fig. 11.

n	$\binom{n}{2}$	GOLOMB'S TRIANGLE	$G(K_n)$?
2	1	1	1
3	3	1 2	3
4	6	1 3 2	6
5	10	1 3 5 2	11
6	15	1 3 6 5 2	17
7	21	1 3 6 8 5 2	25
8	28	1 3 6 11 8 5 2	36
9	36	1 3 6 12 11 8 5 2	48
10	45	1 3 6 12 16 11 8 5 2	64

Fig. 11. Best known numberings for the complete graphs K_n.

The semiempirical numerical triangle which arises in this context has many remarkable and mysterious properties. The test for validity of a row is to consider all possible sums of consecutive terms, and verify that the $\binom{n}{2}$ numbers which result are all distinct. The verifications for $n = 4$, $n = 6$, and $n = 9$ are shown in Fig. 12. The bottom number is the sum of the entire row, and is thus the conjectured value for $G(K_n)$. Every partial sum is the number assigned to some edge of the complete graph. Representing K_n as a regular n-gon with all diagonals drawn in, the perimeter edges receive the consecutive numbers of the top row of Fig. 12, which is simply the row of Fig. 11, with the remaining outside edge receiving the number identified as $G(K_n)$?, the sum of the entries in this row.

```
I  3  2      I  3  6  5  2      I   3   6  12  II   8   5   2
   4  5         4  9  II  7         4   9  18  23  I9  I3   7
      6        10 14 13            10  21  29  31  24  I5
              I5 I6               22  32  37  36  26
              I7                  33  40  42  38
                                 4I  45  44
                                 46  47
                                 48
```

Fig. 12. Verification of the numberings for $n = 4$, 6, and 9.

The following problems all remain unsolved:

(1)   specify the precise rules for the formation of Golomb's triangle;
(2)   prove or disprove that the corresponding numbering of K_n is optimal;
(3)   determine $G(K_n)$, either precisely or between suitable bounds.

It is easy to produce examples of the unfortunate phenomenon that a subgraph of a graceful graph need not be graceful. For example, the ungraceful regular pentagon is a subgraph of several of the graceful graphs in Fig. 5. More convenient is Theorem 7.

THEOREM 7.   If Γ is any graph, and if H is a subgraph of Γ, then $G(H) \leqslant G(\Gamma)$.

Proof: We need never do worse than assign the same node numbers to H that were used in Γ.

From this we have the immediate corollary, Theorem 8.

THEOREM 8.   If Γ is any graph with n nodes, then $G(\Gamma) \leqslant G(K_n)$. This result adds further importance to the study of $G(K_n)$, which is thus the least upper bound on G for all graphs on n nodes.

8.   Numbered Graphs and Difference Sets

The numberings previously exhibited (see Fig. 5) for K_3 and K_4 can be interpreted as the constructions for the finite projective planes with $v = 7$ and $v = 13$, respectively. Thus, for K_3, the node numbers $0, 1, 3$ have as signed differences the numbers $-3, -2, -1, 1, 2, 3$, while for K_4, the node numbers $0, 1, 4, 6$ have as (signed) differences the numbers $-6, -5, -4, -3, -2, -1,$ $1, 2, 3, 4, 5, 6$.

The complete graph K_n has n nodes and $e = \binom{n}{2}$ edges. If we set $v = 2e + 1 = n^2 - n + 1$, $k = n$, and $\lambda = 1$, then the corresponding difference set with parameters (v, k, λ) may be visualized as an assignment of n distinct integers to the nodes of K_n in such a way that the set of all $n(n-1)$ signed differences represent all the nonzero residue classes modulo v.

This relationship suggests several directions for further investigation. For example, the class of graphs which are *graceful modulo m*, in particular when $m = 2e + 1$, seems worthy of study, and is clearly larger than the class of graceful graphs previously treated, namely, the modulo 0 case. Another possibility is that new kinds of finite geometries may be suggested by graph-numbering problems, in analogy with the connection between finite planes and the numbering of complete graphs.

9. Summary of Unsolved Problems

1. Characterize the class of *graceful graphs*. In particular, are all trees graceful?

2. Determine $G(K_n)$ for all n.

3. Determine the function $\alpha(n)$ of Section 4.

4. Determine the asymptotic percentage of graphs on n nodes, and/or of graphs on e edges, which are graceful.

5. Investigate the numerical triangle (Fig. 11).

6. $G(\Gamma)$ is defined as the highest node number which must be used in numbering the graph Γ. Prove or disprove, that for every graph Γ, there is an edge numbered $G(\Gamma)$.

10. Postscript

The following unpublished asymptotic results have recently been obtained by P. Erdös:

1. $G(K_n) \sim n^2$.

2. $\alpha(n) \sim cn^2$, where probably $c = \frac{1}{4}$.

3. 0% of all graphs are graceful.

Moreover, the numerical triangle (Fig. 11) has been superseded for all rows beyond $n = 7$. The revised triangle exhibits no discernable regularity.

Acknowledgments

Helpful suggestions were received from several colleagues and from members of the various audiences to whom I have presented portions of this material. I wish specifically to thank F. Harary, A. Lempel, L. Welch, and especially R. C. Read for suggestions, help, and encouragement.

EVOLUTION OF THE
PATH NUMBER OF A GRAPH:
COVERING AND PACKING IN GRAPHS, II[†]

Frank Harary
Allen J. Schwenk

Research Center for Group Dynamics
Institute for Social Research
The University of Michigan
Ann Arbor, Michigan

1. History

The development of the path number of a graph or digraph was a direct result of our attending the FILE 68 conference in November 1968, in Helsingør. There we met David Hsiao and held extensive conversations which

† This research was supported in part by a grant to the Research Center for Group Dynamics of the University of Michigan from the NIH Biomedical Sciences Division.

39

led to our formal system [3] for information retrieval from files, in terms of directed graphs (see Harary *et al.* [4]). In this context, the points of a digraph D stand for records in a file structure and there is an arc (directed line) from u to v whenever the record u points to the address of record v. It is then natural to ask for the smallest number of record addresses needed to trace through the entire file structure, that is, all the arcs of D. In terms of (ordinary, undirected, Michigan) graphs G, *the path number* of G, $\pi(G)$, is the smallest number of line-disjoint paths which cover all of G. Not surprisingly, we shall follow the graph theoretic notation and terminology of the book [1].

When Ralph Stanton heard the talk in Kingston, Jamaica, on which this article is based, he enjoyed the concept of the path number of a graph sufficiently that in collaboration with Cowan and James, he calculated the path number for certain classes of graphs including trees, cubic graphs, complete graphs, and complete bipartite graphs. These results, delivered at a conference in Louisiana [5], are summarized below, and alternate shorter proofs will be offered.

The concepts of packing and covering were explored in a lecture [2] given in New York City, as a generalization of path number, arboricity, and several other graphical invariants. This approach suggested the definition of the "linear arboricity" of a graph, which has an interpretation in file structures.

2. Results on the Path Number

We now present the principal results of Stanton, Cowan, and James. The proof of Theorem 1 is both shorter and simpler than theirs in [5]. Let p_0 be the number of points of odd degree in a graph G.

THEOREM 1.   The path number of a tree T is given by $\pi(T) = p_0/2$.

Proof.   Since every point of odd degree must be the end point of at least one path, we immediately see that $\pi(T) \geqslant p_0/2$. We obtain the opposite inequality by induction on p_0. Form a forest T' by deleting the lines of the path joining any two end points of T. If no lines remain, we are finished. Otherwise, T' has $p_0 - 2$ odd points. Thus, by induction, $\pi(T') = (p_0 - 2)/2$ and consequently $\pi(T) = p_0/2$.

Let $m = m(G) = q - p + 1$ be the cycle rank of a connected graph G, and let $f = f(G)$ be the number of blocks which are not bridges.

THEOREM 2.   If G is a connected graph with no end points and G is not a cycle, then $p_0/2 \leqslant \pi \leqslant m + f - 1$.

Sketch of Proof. The lower bound was already obtained in the proof of Theorem 1. The upper bound is proved by induction on f and hence is established when $f(G) = 1$. Let T be a spanning tree of G. Then the number of lines in $G - T$ is $m(G)$. Taking these as paths of length one in G, it can be demonstrated that sufficiently many of these can be extended to cover the lines of T, completing the proof for $f(G) = 1$.

For the inductive step, the idea of the proof is to remove an endblock B and the path in G joining a point of B with a point of degree at least 3 (such a path is called a "tendril" by Stanton *et al.* [5]).

THEOREM 3.   If G is cubic, then $\pi(G) = p/2$.

Sketch of Proof. Since every point of G has degree 3, we see by Theorem 2 that $\pi(G) \geqslant p/2$. The equality is attained when $p = 4$, because then $G = K_4$, which is the union of two spanning paths. The result now is obtained for all p by induction, the basic approach being the deletion of a line followed by the suppression of the resulting two points of degree 2. The altered graph G' is cubic and has $p - 2$ points, so by induction, $\pi(G') = (p-2)/2$. Now using the deleted line as an additional path, we have $\pi(G) = p/2$. This simple-minded argument, however, must be presented more carefully to be rigorous, since the deletion–suppression operation can create loops and multiple lines.

The brace symbol $\{x\}$ for a real number x is related to the maximum integer function $[x]$ by $\{x\} = -[-x]$. The proof of the next theorem is also neater than that of Stanton *et al.* [5].

THEOREM 4.   The path number of the complete graph is given by $\pi(K_p) = \{p/2\}$.

Proof. Since each path can cover at most $p-1$ lines, we must have $\pi(K_p) \geqslant \binom{p}{2}/(p-1) = p/2$. Since the path number is an integer, we strengthen this to read $\pi(K_p) \geqslant \{p/2\}$. For even p, Beineke [see 1, p. 91] constructed $p/2$ spanning paths covering K_p. For odd $p = 2n+1$, K_p is known [1, p. 89] to be the sum of n spanning cycles of the form $C_i = wv_iv_{i+1} \cdots v_{i-n+1}v_{i+n}w$, where the subscripts are taken modulo $2n$. From these cycles, we can construct a family of paths $P_i = C_i - v_iv_{i+1}$ for $1 \leqslant i \leqslant n$ and take $P_0 = v_1v_2v_3 \cdots v_{n+1}$ to obtain $n+1$ paths covering K_{2n+1}, which completes the proof. The construction works because it is possible to omit one line from each cycle in such a way that the omitted lines induce the last path P_0.

According to Stanton *et al.* [5], the proof of results giving the path number of any complete bipartite graph $K_{m,n}$ goes through twelve lemmas. Our main new contribution here is to state the result in recognizable closed form, rather than merely say, "The pattern is apparent."

THEOREM 5.   Let $m \geqslant n$ and consider the product mn. Then

$$\pi(K_{m,n}) = \begin{cases} \dfrac{m+n}{2} & \text{for} \quad mn \text{ odd} \\[2ex] \left\{\dfrac{mn}{2n-\delta(m,n)}\right\} & \text{for} \quad mn \text{ even,} \end{cases}$$

where $\delta(m,n)$ is the conventional Kronecker delta.

Sketch of Proof.   In the case that mn is odd, $(m+n)/2 = p_0/2 \leqslant \pi(K_{m,n})$. In the even case, $2n - \delta(m,n)$ is the length of a longest path in $K_{m,n}$. Hence, since there are mn lines to be covered, we may conclude that

$$\pi(K_{m,n}) \geqslant \left\{\frac{mn}{2n-\delta(m,n)}\right\}.$$

In both cases, equality is attained by a brutal construction, which we shall omit.

3.   The Unrestricted Path Number

An alternative path-covering invariant of a graph can be defined as the minimum number of paths, unrestricted in that they are not necessarily line disjoint, needed to cover the lines of G. We proceed to relate this modified path number $\pi^*(G)$ to the original path number.

THEOREM 6.   The path numbers π and π^* satisfy the inequalities, $\pi^*(G) \leqslant \pi(G) \leqslant 2\pi^*(G) - 1 + m(G)$, and the bounds are the best possible.

Proof.   The first inequality is trivial because any line-disjoint covering is also one of the candidates for a minimal nondisjoint covering.

We prove the second inequality by constructing a line-disjoint covering, the size of which does not exceed that given by the formula. We first describe the idea of the proof. Start with two paths in a nondisjoint path covering of G, and alter it to obtain a line-disjoint covering. We can take the first of these two paths as it is, but the second may overlap the first from time to time. Hence, we need to remove from the second path all the lines in their intersection, thus forming additional line-disjoint paths. This procedure is formalized in the following argument.

Let $P_1, P_2, \ldots, P_n$ with $n = \pi^*(G)$ be a covering of G by paths. Define G_r to be the graph the point set of which is $V(G)$ and the line set of which is $X(G_r) = \bigcup_{i=1}^{r} X(P_i)$. We let $Q_{11} = P_1$ be the first path in our line-disjoint cover, and let $Q_{21}, Q_{22}, \ldots, Q_{2k_2}$ be the k_2 subpaths of P_2 not already in G_1. In general,

$Q_{r1}, Q_{r2}, ..., Q_{rk_r}$ are the k_r subpaths of P_r not already in G_{r-1}. Evidently, the union of these Q_{ij} covers G with line-disjoint paths, so we need only count the number of Q_{ij}'s to get the desired upper bound. In fact, for each r we find $\pi(G_r) \leqslant |\{Q_{ij} : i \leqslant r\}| \leqslant 2r - 1 + m(G_r)$. The result is obtained inductively, for in passing from G_{r-1} to G_r we add k_r elements to the set of Q_{ij}'s. But $m(G_r)$ exceeds $m(G_{r-1})$ by at least $k_r - 2$, and $2r$ exceeds $2(r-1)$ by 2, so the right-hand side has also increased by k_r. For $r = n$, this yields $\pi(G) \leqslant |\{Q_{ij} : i \leqslant n\}| \leqslant 2n - 1 + m(G)$. But recalling that $n = \pi^*(G)$, we see that $\pi(G) \leqslant 2\pi^*(G) - 1 + m(G)$, completing the proof.

The complete graphs K_p and the stars $K_{1,n}$ are simple examples of graphs attaining the lower bound $\pi^*(G) = \pi(G)$. The upper bound is attained for the following two infinite families of graphs.

The first family is given by G_i as illustrated in Fig. 1 for $i = 4$. Here G_i has $4i + 6$ points and $5i + 5$ lines. Evidently $\pi(G_i) = 3 + i$ while $\pi^*(G_i) = 2$ for all i. Since $m(G_i) = i$, we note that $\pi(G_i) = 2\pi^*(G_i) - 1 + m(G_i)$ as claimed.

The second family consists of the standard 2-branching trees T_i illustrated in Fig. 2. It is easy to see that $\pi(T_i) = 2^i - 1$ while $\pi^*(T_i) = 2^{i-1}$. Thus, $\pi(T_i) = 2\pi^*(T_i) - 1$, as desired.

Let e be the number of end points of a tree. We may now determine π^* for trees.

THEOREM 7. The nondisjoint path number of a tree is given by $\pi^*(T) = \{e/2\}$.

Proof. Since each path in a minimal collection can cover at most two end lines, we immediately have $\pi^*(T) \geqslant \{e/2\}$. We shall now show that equality is attained. Let $v_1, v_2, ..., v_e$ be the end points of T, and let $P_{i,j}$ denote the unique path joining v_i and v_j. Consider a collection $\mathscr{P}$ consisting of the

Fig. 1. A representative of an infinite family of graphs attaining the upper bound in Theorem 6.

Fig. 2. Trees that apply to Theorem 6.

paths $P_{k,[e/2]+k}$ for $k = 1, 2, ..., \{e/2\}$. Suppose T has a line x that is not covered by $\mathscr{P}$. Since all the end lines of T are covered, x is not an end line, and so $T - x = T_1 \cup T_2$, where neither component T_i is trivial. Consequently, there exist distinct endpoints v_i, v_j, v_r, v_s such that $P_{i,j} \subset T_1$ and $P_{r,s} \subset T_2$ and both these paths belong to $\mathscr{P}$. Remove these two paths from $\mathscr{P}$ and replace them by $P_{i,r}$ and $P_{j,s}$. The altered cover $\mathscr{P}'$ has the same number of paths as $\mathscr{P}$ and covers every line previously covered by $\mathscr{P}$, but in addition, line x is covered. Thus, if we repeat this procedure, in at most $p - 1$ iterations we will obtain a collection of $\{e/2\}$ paths covering T.

COROLLARY. For a tree T, $\pi^*(T) = \pi(T)$ if and only if T has at most one odd point which is not an end point.

Proof. In the light of Theorems 1 and 7, this equivalence is obvious since $\{e/2\} = p_0/2$ if and only if T has at most one odd point that is not an end point.

4. Unsolved Problems

1. Expressions for $\pi(G)$ and $\pi^*(G)$ are not known for graphs other than trees. It is known that $\pi(G) = \pi^*(G)$ for those trees specified in the corollary, for all complete graphs K_p, and for those complete bigraphs $K_{m,n}$ for which the product mn is even. However, it appears to be difficult to characterize the graphs for which $\pi(G) = \pi^*(G)$.

2. The *arboricity* of a graph G is the minimum number of line-disjoint spanning subforests which cover G (see [1, p. 90]). Equivalently, it is the minimum number of spanning subtrees (not necessarily line-disjoint) which cover G. We have seen that for complete graphs K_p, the arboricity and the path number are equal. For which graphs does this hold?

3. As defined in [2] the *linear arboricity* of a graph G is the smallest number of line-disjoint linear subforests (in which every component is a path) needed to cover G. Denoting as usual the arboricity by capital upsilon Υ, we write $\overline{\Upsilon}$ for the linear arboricity of G. It is easy to see that for any graph, $\Upsilon \leqslant \overline{\Upsilon} \leqslant \pi^* \leqslant \pi$. Thus, the preceding question asks when all four of these invariants are equal. The first question asked when $\pi = \pi^*$. Hence, it remains to ask for which graphs is

(a) $\overline{\Upsilon} = \pi^*$?   (b) $\Upsilon = \pi^*$?   (c) $\overline{\Upsilon} = \pi$?   (d) $\Upsilon = \overline{\Upsilon}$?

The smallest tree for which these four values are distinct has 8 points and is obtained from the 5-point path P_5 by adding an end line at each of the three nonend points. For this tree T, the values are 1, 2, 3, and 4.

4. Still another covering invariant is the tree number of a graph obtained when the subgraphs are subtrees. We denote by $\tau(G)$ the minimum number of line-disjoint subtrees that cover G. Similarly, let τ^* be the corresponding number of subtrees, not necessarily line-disjoint, needed to cover G. Obviously $\Upsilon = \tau^* \leqslant \tau \leqslant \pi$, so that one can ask for the class of graphs satisfying additional equalities.

5. We have discussed five different covering invariants of a graph:

(a) π = path number, disjoint;
(b) π^* = path number, unrestricted;
(c) Υ = arboricity;
(d) $\overline{\Upsilon}$ = linear arboricity;
(e) τ = tree number, disjoint.

To the best of our knowledge, there do not as yet exist effective and convenient computer algorithms for determining the values of these five invariants for a given graph.

References

1. Harary, F., "Graph Theory." Addison-Wesley, Reading, Massachusetts, 1969.
2. Harary, F., Covering and packing in graphs, I. *Ann. N. Y. Acad. Sci.* **175**, 198–205 (1970).
3. Harary, F., and Hsiao, D., A formal system for information retrieval from files. *Comm. ACM* **13**, 67–73 (1970).
4. Harary, F., Norman, R., and Cartwright, D., "Structural Models: An Introduction to the Theory of Directed Graphs." Wiley, New York, 1965.
5. Stanton, R., Cowan, D. D., and James, L. O., Some results on path numbers, *Proc. Louisiana Conf. Combinatorics, Graph Theory and Computing, Baton Rouge, 1970,* 112–135.

THE PRODUCTION OF GRAPHS BY COMPUTER

B. R. Heap

Division of Numerical Analysis and Computing
National Physical Laboratory
Teddington, Middlesex
England

1. Introduction

Over the last few years a number of computer programs have been written at the National Physical Laboratory for the production of lists of graphs and for the enumeration of the number of different ways these graphs can be placed on crystal lattices (the *lattice constants* of the graphs). In this article we shall discuss a number of problems associated with the first of these two topics, namely the production of lists of graphs by computer. The actual

47

enumeration of the lattice constants is performed by means of sophisticated programs produced by J. L. Martin (unpublished), these being extensions of his previous work on the exact enumeration of walks on crystal lattices [12]. This research work formed part of a now completed joint project of the National Physical Laboratory and King's College, University of London, for the systematic enumeration of lattice constants.

Lattice constants are of considerable use in the calculation of exact series expansions for the physical properties of interacting systems and sophisticated extrapolation techniques have been developed for the prediction of the critical properties of such expansions. In many cases, the coefficients in the series expansions are equivalent to summations with appropriate weights over a restricted class of undirected graphs known as stars [17]. These graphs also occur widely in other physical problems, notably the Mayer theory of condensation [18]. It was mainly with a view to the production of all star graphs having specified numbers of points and edges that the work described in this article was carried out, though the possession of a full set of seven-point graphs stored in a computer allowed Cameron [2] to investigate the existence and nonexistence of complete subgraphs in the graphs.

Where graphs have only a small number of points, less than seven, they can readily be drawn and listed by hand. In fact, it has even been possible to produce drawings of the full set of seven-point graphs (D. W. Crowe and F. Harary, unpublished), though errors in this list of drawings only came to light after a full set of the graphs had been produced by computer. When the graphs are larger and more complex, the assistance of a computer is essential in order to avoid errors. We shall discuss the general problem of the representation and identification of graphs in a computer and describe how these difficulties were overcome in our programs. We shall also discuss three specific problems, namely the production of all simple graphs having specified numbers of points and edges and in particular the production of all eight-point simple graphs, the production of sets of star topologies (see Section 2) and the production of all simple stars having a specified number of points and a specified topology. Much of this work has been described in earlier publications, [1, 4, 8, 9, 10, 11, 13, 17], some of which are not readily available. This contribution is meant to be a summary of these publications and readers requiring more detail should refer to the original articles.

Graph theory has many roots and branches and as yet no uniform and standard terminology has been agreed. In a recent book, Harary [7] has attempted to produce a standard set of definitions and these will probably be familiar to many readers. However, workers in other fields are accustomed to different definitions and terminology. As the work described here was carried out in conjunction with theoretical physicists, we feel justified in using, instead of Harary's terminology, the terminology of Essam and Fisher [5], who have also recently attempted to produce a standard set of definitions, albeit from a physicist's viewpoint. For those readers who are unfamiliar with these definitions, and for the sake of clarity, we shall briefly state the main graph-theoretical definitions in the next section.

2. Definitions and Terminology

A *graph* G consists of a set P of *points*, together with a set E of undirected *edges* or *lines* joining certain pairs of points. However, an edge may not join a point to itself (a *loop*). In a *multigraph*, a pair of points may be joined by more than one edge (*multiple* or *parallel edges*), but in a *simple graph* a pair of points may be joined by at most one edge. For brevity, in this paper we shall use the term graph to refer to either multigraphs or simple graphs or both, only distinguishing between them where it is not clear from the context. A *labeled graph* is a graph the n points of which are labeled with the integers $1, 2, ..., n$, in some way.

A graph is said to be *disconnected* if it is possible to divide the set of points P into two subsets, P_1 and P_2, such that there are no edges joining any point in P_1 to any point in P_2; otherwise the graph is said to be *connected*. A point belonging to a connected graph G the removal of which from G, together with the edges emanating from it, produces a disconnected graph is referred to as an *articulation point* or *cut point*. A connected graph containing an articulation point is *separable*. If a connected graph does not contain an articulation point, then it is *nonseparable*, or *multiply connected*. If it contains at least two points, then it is referred to as a *star*. Two graphs are said to be *isomorphic* if there is a one-to-one correspondence between their point sets that induces a one-to-one correspondence between their edge sets. Each graph is then said to be an *isomorph* of the other.

The number of edges emanating from a point is known as the *degree* of that point. If a graph G contains a point A of degree two, which is joined to the two distinct points B and C, then a new graph can be formed, which has the same point set as G, apart from A, and the same edge set except that edges AB and AC are removed and an edge BC added. This process is known as the *suppression* of the point A. The reverse process of replacing an edge BC by a new point A and new edges AB and AC is known as the *insertion* of the point A on the edge BC. Two graphs are *homeomorphic*, and one is said to be a *homeomorph* of the other, if they can be made isomorphic by the insertion and suppression of points of degree two. Such graphs are said to have the same *basic topology*. A graph of a given basic topology which is homeomorphic to no graph with fewer points is said to *represent faithfully* the *topological type* of all its homeomorphs. Such a graph is referred to as a *basic topological type* and, in this article, we shall usually abbreviate this as *topology*. In previous papers such a graph was referred to as *homeomorphically irreducible*. The *cyclomatic number c* of a connected graph containing n points and m lines is defined by

$$c = m - n + 1.$$

Since the process of inserting a point of degree two increases both the number of points and the number of edges in the graph by unity, it is clear that all homeomorphs have the same cyclomatic number.

This completes the general graph definitions needed here. A number of other definitions concerned with the representation and identification of a graph in a computer are given in Section 4 when this topic is discussed.

3. Problems

We are now in a position to state in the graph-theoretical terms of Section 2 the problems that we shall discuss. They are

PROBLEM 1.   The production by computer, in some representation or other, of one copy of each simple graph containing a specified number of points and edges, and in particular, one copy of each simple graph having eight points.

PROBLEM 2.   The production by computer, in some representation or other, of one copy of each star topology containing a specified number of points and edges.

PROBLEM 3.   The production by computer, in some representation or other, of each simple star having a specified topology and a specified number of points and edges.

A vital need for all these problems is a method for the representation of both simple graphs and multigraphs in a computer and the identification of a graph in a list of possibly thousands of graphs. This will be considered in the next section. Problem 1 will be dealt with in Section 5, Problem 2 in Section 6, and Problem 3 in Section 7.

4.   Representation and Identification of Graphs in a Computer

The representation and identification of a graph in a computer is possibly the most important problem associated with the automatic production of graphs. The methods and techniques that we describe in this section were evolved to deal with specific problems. Since this research work was completed, Corneil and Gotlieb [3] have described a computer algorithm for testing whether or not two graphs are isomorphic. However, their method does not appear to have been applied to the case that we are mainly interested in, namely the identification of a graph in a list of many graphs.

There is no difficulty in the representation of a graph in a computer if its points can be labeled with the integers $1, 2, ..., n$, since the adjacency matrix of the graph can be used to represent it. The *adjacency matrix* is a $n \times n$ matrix the (i, j)th element of which is the number of edges leading from point i to point j. For the graphs we are concerned with, the adjacency matrix is symmetric, since the edges are undirected, and its diagonal terms are zero, since loops are forbidden. Thus, it is only necessary to actually store the upper triangular part of the matrix, which is a total of $n(n-1)/2$ elements. This does not necessarily mean that we must set aside this number of computer words to store a single graph. Since the elements of the matrix are small integers, several of them may be packed into one computer word and the storage requirements are reduced by a factor which depends on the maximum possible element and also on the length of the computer word. Of course, the routines dealing with individual elements of the matrix are then rather more complex, but those dealing with the matrix as a whole are often very much simpler.

When n is large and the number of edges in the graph is not too large, other methods of storing the matrix can possibly use less storage space. Thus, a straight list of the nonzero elements of the matrix, each entry in the list consisting of i, j and the number of edges joining i to j can be used. This only uses $3k$ words, where k is the number of nonzero elements. This can also be substantially reduced by packing as described above. For very large graphs even more economic storage methods can be evolved, but these are outside the scope of this article.

For the graphs we are concerned with here, a labeling was produced (see below), and the adjacency matrix stored in one of two ways. As it was known from the outset, in the case of multigraphs, that the number of edges joining any two points would be small, certainly less than 15, and that the number of points in the graphs would be at most ten, four bits were assigned to each element of the adjacency matrix. As the computer word length was 48 bits, each row of the matrix occupied one word. The full matrix was stored, although as stated above, this was not strictly necessary. However, in this way, the program routines could be made somewhat simpler. An additional word was used to store the degrees of the points, packed as above. Another word was used to hold information about the graph, that is, the number of points, number of edges, and an identification number. Finally, a third additional word was used for addressing purposes, as will be explained later. Thus, $n + 3$ words were needed to store a multigraph having n points.

For simple graphs, the same storage method could have been used, but because of the necessity to hold up to 1646 eight-point graphs in the main store of the computer at the same time, it was decided to use a more condensed storage method. As there could be at most one edge joining any two points in a simple graph, a single bit only was needed to say whether the edge was present or not. By restricting the matrix to its upper triangular part only 28 bits were required for storing an eight-point graph, and this was easily accommodated in one computer word. An identification word and an address word were still needed as in the case of multigraphs so that three words were needed for the storage of each simple graph.

However, the graphs with which we were dealing were unlabeled. In order to represent them by means of adjacency matrices, it was necessary to produce some method of labeling their points. Since we required the representation of a graph to be unique, this labeling had to be unique, apart from symmetry. Perhaps the simplest way of possibly distinguishing one point in a graph from another is by means of their degrees, since points must be different if they have different degrees. In addition, the degrees are readily available. Thus, the set of points of a graph was divided into subsets, all points belonging to each subset having the same degree. The subsets then were ordered according to the magnitude of the degrees so that if there were n_1 points with degree d_1, n_2 with degree d_2, ..., n_k with degree d_k, then

$$d_1 > d_2 > d_3 > \cdots > d_k,$$

and the subsets contained $n_1, n_2, n_3, ..., n_k$ points, respectively. In the case of multigraphs these subsets were divided further. Consider a point with degree d_i for some i. The edges emanating from this point, consist, in general, of a collection of single edges, double edges, triple edges, and other multiple edges. We refer to this collection of edges as the *degree set* of the point. Clearly two points can be distinguished if their degrees have different degree sets, even though their degrees are equal. If the point has d_{i1} single edges, d_{i2} double edges, and so forth, so that

$$d_i = \sum_{r=1}^{r=t} r \, d_{ir},$$

where $t = d_i - 1$, then we specify the degree set by means of a t-tuple *degree specification*

$$(d_{it}, ..., d_{i3}, d_{i2}, d_{i1}).$$

Just as the degrees of the points were ordered in descending order of magnitude, we order the degree specifications, also in descending order of magnitude. To do this, we define an ordering of the t-tuples by saying that a t-tuple

$A = (A_1, A_2, A_3, ..., A_t)$ is larger than t-tuple A', if there exists an r, $1 \leqslant r \leqslant t$, such that $A_s = A_s'$, $1 \leqslant s \leqslant r - 1$, and $A_r > A_r'$. Thus, all points having a certain degree can be assigned to subsets according to their degree specifications. Two points are in the same subset if and only if they have the same degree specification. Unfortunately this method of further subdividing the degrees requires more storage since for each different degree set, a degree specification must be stored. This can conveniently be done in one word but a total of n words are required for the storage of the specifications of each subset.

In this way the points of both simple graphs and multigraphs were divided into ordered subsets $\{S_i\}$, each containing m_i points, $i = 1, 2, ..., p$. In the case of simple graphs S_i contained all points having degree d_i, $m_i = n_i$, and $p = k$. For multigraphs, the designation was not as simple, and S_i contained all points having a specified degree and a specified degree set. Labels $1, 2, ..., m_1$ were now assigned to the m_1 points in S_1, labels $m_1 + 1, m_1 + 2, ..., m_1 + m_2$ to the points in S_2, etc., and the adjacency matrix formed for each possible method of assignment. There were $m_1! m_2! m_3! \cdots m_p!$ possible assignments and adjacency matrices. If a_{ij} is the (i, j)th element of the adjacency matrix for some labeling, then we define the *canonical labeling* of the graph as that labeling which makes the n^2-tuple

$$(a_{11}, a_{12}, a_{13}, ..., a_{1n}, a_{21}, ..., a_{nn})$$

a maximum. This was the labeling that was applied to the graph. The appropriate adjacency matrix is referred to as the *canonical matrix*.

It is clear from the above that in order to produce the canonical matrix, it might be necessary to consider a large number of possible labelings of the points of a graph. It is important that the minimum possible disturbance be made to the adjacency matrix in going from one labeling to another. This was achieved by a system of interchanging two labels according to a scheme which eventually produces all possible labelings. A simpler form of this scheme is described in Heap [8].

When identifying a graph in a list of possibly some hundreds of graphs, the canonical matrix was first produced and only compared with those of all graphs in the list which had the same degree subsets (simple graphs), or the same degree and degree specification subsets (multigraphs). This was done simply by storing separately an index of the different degree subsets and degree specification subsets, and with each one the address of the last graph in the list having these degrees and degree specifications, as well as the number of such graphs in the list. Then with each graph was stored the address of the previous graph having the same set of degrees and degree specifications. Thus, it was extremely easy to scan through only those graphs which could conceivably be the one we were identifying.

5. Production of Simple Graphs

Simple graphs having a specified number of points n and edges m can be constructed easily from the full sets of those having either n points and $m+1$ edges or those having n points and $m-1$ edges. In the former method, a graph having n points and $m+1$ edges is examined, a single edge removed, and the resultant graph, having n points and m edges, is identified and stored as described in the previous section. The process is repeated for each of the $m+1$ possible ways of removing a single edge from the original graph and is then repeated for each graph having n points and $m+1$ edges. It is clear that each graph with n points and m edges must be produced at least once by this process. Starting from the complete graph having each of its $n(n-1)/2$ distinct pairs of points joined by edges, the full set of n-point graphs can be progressively built up. In the latter method, each graph with $m-1$ edges is examined in turn, and new graphs are formed by adding an edge in all possible ways consistent with its remaining a simple graph, a total of $n(n-1)/2 - m + 1$ possible ways. The identification is carried out as before. This time we start with the graph containing n points and no edges and the full set of n-point graphs again is built up progressively.

An alternative method of constructing the graphs with n points and m edges is to produce them from the full set of graphs having $n-1$ points. In order to achieve this, we must consider all possible ways of adding one point and joining it to j of the original points of each graph having $n-1$ points and $m-j$ edges, $j = p(1)q$, where

$$p = \max(0, m - (n-1)(n-2)/2)$$

and

$$q = \min(n-1, m).$$

The efficiency of this method compared to that of the previous methods can be estimated by calculating the number of graphs that have to be constructed in order to produce the full set of eight-point graphs. It turns out that a total of 133,632 are formed using this last method as opposed to 172,845 for either of the other methods. Thus, this method is slightly more efficient. However, an advantage of the first method that we described is that, for each graph, a complete list of its partial graphs can be produced. A *partial graph* of a graph G is a graph produced from G by the removal of edges alone. This was important in the original problem for which the graphs were required [17], and is also useful in evaluating some of the properties of the graphs. For example, it is useful in deciding whether a graph is planar or not. From Kuratowski's theorem (see Harary [7] or Ore [14]), we know that a graph is nonplanar, if and only if it contains as a partial graph, any graph which is homeomorphic to the complete graph on five points, (K_5), or the complete bipartite graph on six points ($K_{3,3}$). In $K_{3,3}$ the six points are divided into two subsets of three points each, and each point in one subset is joined to all the points in the other subset, but not to any of the other points in its own subset. It is a simple matter to produce by hand those graphs having eight points that are homeomorphic to K_5 and $K_{3,3}$. By examining only the sets of immediate partial graphs, that is, those with only one edge removed, of all graphs, it is possible to build up a full list of nonplanar graphs. Details of the numbers of planar and nonplanar graphs having six, seven, and eight points do not appear to have been previously published and they are given in Tables I, II, and III.

A knowledge of the partial graphs of a graph also allows us to decide whether it is connected or not, and whether it is separable or not, because a disconnected graph having eight points must be a partial graph of at least one of four disconnected graphs, which are easily constructed by hand. If K_n is the complete graph having n points, these graphs can be described by $K_1 + K_7$, $K_2 + K_6$, $K_3 + K_5$, and $K_4 + K_4$. Similarly, a separable graph having eight points must be a partial graph of at least one of three separable graphs. The numbers of connected and nonseparable (star) eight-point graphs are given in Table IV.

As was stated in Section 1, a set of drawings of all seven-point graphs has been made by Crowe and Harary. The total number of eight-point graphs, 12,346, prohibits any possibility of a similar scheme for their systematic representation as a set of drawings. However, it has been found possible to prepare a set of punched cards on which the full set of eight-point graphs are represented by their canonical matrices. This is achieved by using only one row of a punched card to store the matrix, in which a hole represents an edge. In this way 12 graphs can be stored on a single card and only 1044 cards are required to store the full set of 12,346 eight-point graphs. The graphs also are stored on magnetic tape for use in the computer.

6. Production of Star Topologies

In this section we restrict attention to stars and discuss Problem 2, which is the production of all star topologies having a specified number of points and edges. We note, from Section 2, that a useful means of classifying such a graph is by means of its cyclomatic number, denoted by c. For $c = 2$, there is only one such topology, a *theta graph*, consisting of two points joined by three edges. For $c = 3$, there are four topologies, usually described as *alpha, beta, gamma*, and *delta graphs*. For $c = 4$, there are 17 such graphs. All these graphs are illustrated in the paper by Essam and Sykes [6]. Diagrams can also be found in the full catalog [10].

TABLE I

The Numbers of Planar $P(m)$ and Nonplanar $N(m)$ Simple Graphs Having
Six Points and m Edges

m	$P(m)$	$N(m)$	m	$P(m)$	$N(m)$	m	$P(m)$	$N(m)$
0	1	0	6	21	0	11	5	4
1	1	0	7	24	0	12	2	3
2	2	0	8	24	0	13	0	2
3	5	0	9	20	1	14	0	1
4	9	0	10	13	2	15	0	1
5	15	0						

TABLE II

The Numbers of Planar $P(m)$ and Nonplanar $N(m)$ Simple Graphs Having
Seven Points and m Edges

m	$P(m)$	$N(m)$	m	$P(m)$	$N(m)$	m	$P(m)$	$N(m)$
0	1	0	8	97	0	15	5	36
1	1	0	9	130	1	16	0	21
2	2	0	10	144	4	17	0	10
3	5	0	11	135	13	18	0	5
4	10	0	12	98	33	19	0	2
5	21	0	13	51	46	20	0	1
6	41	0	14	16	49	21	0	1
7	65	0						

TABLE III

The Numbers of Planar $P(m)$ and Nonplanar $N(m)$ Simple Graphs Having
Eight Points and m Edges

m	$P(m)$	$N(m)$	m	$P(m)$	$N(m)$	m	$P(m)$	$N(m)$
0	1	0	10	658	5	20	0	221
1	1	0	11	956	24	21	0	115
2	2	0	12	1217	95	22	0	56
3	5	0	13	1264	293	23	0	24
4	11	0	14	1038	608	24	0	11
5	24	0	15	619	938	25	0	5
6	56	0	16	255	1057	26	0	2
7	115	0	17	56	924	27	0	1
8	221	0	18	10	653	28	0	1
9	401	1	19	0	402			

TABLE IV

The Numbers of Connected Graphs $C(m)$ and Nonseparable (Star) Graphs $N(m)$ Having Eight Points and m Edges

m	$C(m)$	$N(m)$	m	$C(m)$	$N(m)$	m	$C(m)$	$N(m)$
0	0	0	10	486	40	20	220	215
1	0	0	11	814	161	21	114	112
2	0	0	12	1169	429	22	56	55
3	0	0	13	1454	780	23	24	24
4	0	0	14	1579	1076	24	11	11
5	0	0	15	1515	1197	25	5	5
6	0	0	16	1290	1114	26	2	2
7	23	0	17	970	885	27	1	1
8	89	1	18	658	622	28	1	1
9	236	6	19	400	386			

These topologies were obtained using paper-and-pencil methods and it is not difficult to do this. However when we turn to the production of the star topologies with $c = 5$, such methods are prone to error, and more systematic methods are required. It is perhaps of interest that Nagle [13] made an attempt to produce these graphs by hand and quickly found a total of 116. However, our computer method had previously discovered that there were 118. For topologies with $c > 5$, it is totally impracticable to produce the graphs other than by computer.

There are three ways in which star topologies with n points and m edges, and thus having $c = m - n + 1$, can be produced from previously obtained topologies having $c = m - n$. These are

1. by joining any two distinct points of a topology having n points and $m - 1$ edges;
2. by inserting a point of degree two on any edge of a topology having $n - 1$ points and $m - 2$ edges, and then joining this point to any other point;
3. by inserting two points of degree two on any edge or edges of a topology having $n - 2$ points and $m - 3$ edges, and then joining these two points.

A little thought shows that these are the only possible ways of constructing star topologies, and a topology cannot exist that cannot be constructed in this fashion. However it is necessary to consider all possible ways of performing the operations in order to make absolutely certain that none are missed. The star topologies having $c = 5$ were constructed using these methods. The storage and identification are as described in Section 4.

The same general method was used to produce most of the star topologies having $c = 6$. However some of these have as many as ten points and all of these points have degree three. If all the degree specifications of the points are the same, then they are computationally indistinguishable, as far as the method of representation described in Section 4 is concerned. Therefore, for graphs having more than eight points, special methods were derived for producing the canonical matrices. We shall illustrate these by a consideration of the production of all star topologies having ten points, each with degree three, and each being joined to three distinct points. To produce the canonical matrix for such a graph using the methods of Section 4 would have required the consideration of 10! labelings and thus 10! = 3,628,800 adjacency matrices. As this would have taken about 100 min for a single graph, the method was out of the question.

However, in the canonical matrix, we note that point 1 must necessarily be joined to points 2, 3, and 4 for such a graph. Thus, in finding the canonical labeling we run through all possible ways of:

1. assigning label 1 to a point (10 possible ways);
2. assigning labels 2, 3, and 4 to the three points to which point 1 is joined (3! = 6 possible ways);
3. assigning labels $5, 6, \ldots, 10$ to the remaining points (6! = 720 possible ways).

Thus, only $10 \times 6 \times 720 = 43,200$ different labelings need to be considered. This takes about 1 min and is a considerable reduction. A penalty is that a special computer program is required.

Similar reductions in time, at the expense of specially written programs, allowed all topologies having $c = 6$ with nine and ten points to be readily obtained. A full description of these methods is too specialized for this article, but can be found in the full catalog of the 1198 topologies with $c = 6$ [11]. In addition, drawings of all star topologies with $c < 6$ are available [9, 10]. The numbers of star topologies having $c \leqslant 6$, classified according to their numbers of points and edges, are given in Table V.

7. Production of Stars Having a Given Topology

In the last section, we discussed the production of star topologies. These may be either simple graphs or multigraphs. We now wish to take one of these topologies and produce all simple star graphs having this topology, and a specified number of points, by the simple expedient of the insertion of points of degree two on the edges of the topology.

In fact, a formal method for this enumeration process can be derived using the well-known methods of Pólya [15]. Detailed descriptions of Pólya's

TABLE V

The Numbers $T_c(m)$ of Star Topologies Having Cyclomatic Number $c \leqslant 6$, n Points, and $m = c + n - 1$ Edges

c	n	m	$T_c(m)$	c	n	m	$T_c(m)$
2	2	3	1	5	6	10	38
3	2	4	1	5	7	11	23
3	3	5	1	5	8	12	16
3	4	6	2	6	2	7	1
4	2	5	1	6	3	8	4
4	3	6	2	6	4	9	26
4	4	7	5	6	5	10	84
4	5	8	4	6	6	11	216
4	6	9	5	6	7	12	314
5	2	6	1	6	8	13	325
5	3	7	3	6	9	14	162
5	4	8	13	6	10	15	66
5	5	9	24				

enumeration techniques can be found in Harary [7], Riordan [16] and Uhlenbeck and Ford [18]. Let G be a star topology having m edges and let $T_G(t_1, t_2, t_3, ..., t_m)$ denote the *cycle index* of the group of permutations of the edges of G. Now stars homeomorphic to G are produced by the insertion of points of degree two on the edges of G. As any number of points may be inserted on any edge, the *enumerator* for the process is

$$x + x^2 + x^3 + \cdots = x/(1 - x).$$

If $U_G(k)$ is the number of stars having k edges, homeomorphic to G, we then have

$$T_G(Y_1, Y_2, Y_3, ..., Y_m) = \sum_{k=m} U_G(k) x^k,$$

where

$$Y_r = x^r + x^{2r} + x^{3r} + \cdots = x^r/(1 - x^r).$$

These equations represent the formal technique for the calculation of $U_G(k)$. However, since a knowledge of the cycle index of G is required, the method is not in general practical. Note also that $U_G(k)$ enumerates both simple graphs and multigraphs, if G is a multigraph. Some means of deleting the multigraphs is necessary, if they are not to be included in the enumeration.

An alternative enumeration method devised by Domb and described in Domb and Heap [4] and Heap [10] is of more practical interest. For this we introduce the concept of a *colored graph*, by which we mean a graph whose edges carry identifying "colors." Two colored graphs are said to be the same if they are both isomorphic and if corresponding edges carry the same color.

Now let $m_1, m_2, m_3, \ldots, m_r$, $m_1 \geqslant m_2 \geqslant m_3 \geqslant \cdots \geqslant m_r$, be some partition of m into r parts. We define a *symmetry factor* $W_G(m_1, m_2, m_3, \ldots, m_r)$ of G as the number of different ways of coloring the edges of G to produce differently colored graphs, with the proviso that m_1 of the edges are colored with color 1, m_2 of the edges are colored with color 2, ..., and m_r of the edges are colored with color r. Let $P(k; m_1, m_2, m_3, \ldots, m_r)$ be the total number of partitions of k into m parts such that m_1 are equal to some integer k_1, m_2 are equal to some different integer k_2, etc. We now make a correspondence between a colored edge of G and the number of points inserted on that edge in order to produce a graph homeomorphic to G. It follows that

$$U_G(k) = \sum_{(m)} P(k; m_1, m_2, m_3, \ldots, m_r) \, W_G(m_1, m_2, \ldots, m_r),$$

where the summation extends over all partitions of m. Since the symmetry factors are fairly easy to obtain by hand or can be derived from the cycle index, and tables of $P(k; m_1, m_2, m_3, \ldots, m_r)$ are available [1], it is possible to enumerate $U_G(k)$ when the topology is simple. However, multigraphs may still be included in the enumeration and these have to be deleted. For further details of how this is achieved the reader is referred to Domb and Heap [4] or Heap [10]. Expressions for the numbers of simple stars homeomorphic to the 22 star topologies having cyclomatic numbers less than five, together with extensive tables, are given in Heap [10]. In addition, Domb's method can also be used for the enumeration of star topologies themselves. Full details, formulas, and tables are also given in Domb and Heap [4] and Heap [10].

The methods of actually producing simple stars with a given topology are akin to Domb's method of enumeration. Let a given topology G have n points and m edges. Assume that we wish to produce all stars having k edges and the topology of G. Normally, k is not much larger than m. This is achieved by considering all partitions of $k - m$ into m parts, not all parts being necessarily nonzero, and all possible ways of inserting points on the edges of G consistent with the partitions. For example, if a given topology has six edges and $k = 9$, then we consider all possible ways of inserting three points on one of the edges, all ways of inserting two points on one edge and one point on another edge, and all ways of inserting one point on each of three different edges. For most topologies this can be carried out by hand provided $k - m$ is not too large and the topology is fairly simple.

However, where k is fairly large and the topology is complex, usually when the topology is fairly symmetric, the computer must be used. As an example, consider the case where G contains no multiple edges, and so each nonzero element in the adjacency matrix of G is equal to unity. We now insert points on the edges of G according to the appropriate partition as described above. However, instead of constructing a new adjacency matrix of size $n + k - m$ from the adjacency matrix of G of size n, we construct a new $n \times n$ matrix in which the (i,j)th element is equal to 1 plus the number of points inserted on the edge joining i to j. One way of describing this matrix is as a *chain matrix*, for now the (i,j)th element is equal to the length of the shortest chain from i to j, with only the points of degree three or more being labeled. This matrix is an alternative method of representing the graph and can be used exactly like the adjacency matrix. If there are no points of degree two, it is identical with the adjacency matrix. A canonical matrix can be defined and formed just as before, and graphs can be stored and identified as described in Section 4. The great advantage of this representation is that we are representing a graph with $n + k - m$ points by an $n \times n$ matrix. By running through all possible ways of inserting the points consistent with the partition and then running through all partitions, the full set of stars can be found.

If G has multiple edges, the above procedure cannot be used directly, though it is sometimes possible to use a slightly amended method. Recently, J. L. Martin (private communication) has produced a technique which deals with both simple graphs and multigraphs, but in many cases it is rather time consuming, especially where the topology has a high degree of symmetry. In these cases, it is necessary to resort to special programs which take into account the symmetry of the topology. As an example, consider the production of simple beta graphs. The basic topology of a beta graph is a graph with four points, which are labeled 1, 2, 3, and 4, and six edges, $(1,2)$, $(1,2)$, $(3,4)$, $(3,4)$, $(1,3)$, and $(2,4)$. If we denote by a, b, c, d, e, and f the numbers of points inserted on these edges respectively to produce a simple beta graph having n points, then the graph can be specified uniquely by values of a, b, c, d, e, and f satisfying the relations

$$a + b + c + d + e + f = n - 4, \qquad a \geqslant 0;$$

$$b > a, \qquad a = 0, \quad c \geqslant a,$$

$$b \geqslant a, \qquad a \neq 0, \quad c \geqslant a;$$

$$d \geqslant b, \qquad c = a, \quad f \geqslant e \geqslant 0,$$

$$d \geqslant c, \qquad c \neq a, \quad f \geqslant e \geqslant 0.$$

The graphs can now be specified, as the reader may verify, by

$$a = 0 \; (1) \; [(n-4)/4],$$

$$c = a \; (1) \begin{cases} [(n-5)/2], & a = 0 \\ [(n-4-2a)/2], & a \neq 0 \end{cases},$$

$$b = \begin{cases} 1, & a = 0 \\ a, & a \neq 0 \end{cases} (1) \begin{cases} [(n-a-c-4)/2], & a = c \\ n - a - 2c - 4, & a \neq c \end{cases}$$

$$d = \begin{cases} b, & a = c \\ c, & a \neq c \end{cases} (1) n - a - b - c - 4,$$

$$e = 0(1) \left[(n-a-b-c-d-4)/2 \right].$$

$$f = n - a - b - c - d - e - 4.$$

where $[x]$ denotes the largest integer less than or equal to x. A computer program to run through all these possibilities is straightforward to produce.

References

1. Arrowsmith, J. M., and Heap, B. R., Partition symmetries, *Nat. Phys. Lab., Div. Numerical Appl. Math. Rep.* Ma 64, (1966).
2. Cameron, J. B. Initial sieves for complete subgraphs, unpublished manuscript.
3. Corneil, D. G., and Gotlieb, C. C., An efficient algorithm for graph isomorphism, *J. Assoc. Comput. Mach.* **17**, 51–64 (1970).
4. Domb, C., and Heap, B. R., The classification and enumeration of multiply connected graphs, *Proc. Phys. Soc. London* **90**, 985–1001 (1967).
5. Essam, J. W., and Fisher, M. E., Some basic definitions in graph theory, *Rev. Modern Phys. Supp.* **42**, 271–288 (1970).
6. Essam, J. W., and Sykes, M. F., Percolation processes I. Low density Expansion for the mean number of clusters in a random mixture, *J. Mathematical Phys.* **7**, 1573–1581 (1966).
7. Harary, F., "Graph Theory." Addison-Wesley, Reading, Massachusetts, 1969.
8. Heap, B. R., Permutations by interchanges, *Comput. J.* **6**, 293–294 (1963).
9. Heap, B. R., The enumeration of homeomorphically irreducible star graphs, *J. Mathematical Phys.* **7**, 1582–1587 (1966).
10. Heap, B. R., The Production and Use of Homeomorphically Irreducible Star Graphs, *Nat. Phys. Lab., Div. Numerical Appl. Math. Rep.* Ma 57 (1967).
11. Heap, B. R., A Catalogue of homeomorphically irreducible star graphs with cyclomatic number six, *Nat. Phys. Lab., Div. Numerical Appl. Math. Rep.* Ma 82 (1969).
12. Martin, J. L., The exact enumeration of self-avoiding walks on a lattice, *Proc. Cambridge Philos. Soc.* **58**, 92–101 (1962).
13. Nagle, J. F., On ordering and identifying undirected linear graphs, *J. Mathematical Phys.* **7**, 1588–1592 (1966).
14. Ore, O., "Theory of Graphs." Amer. Math. Soc., Providence, Rhode Island, 1962.
15. Pólya, G., Kombinatorische Anzahlbestimmungen für Grüppen, Graphen und chemische Verbindungen, *Acta. Math.* **68**, 145–254 (1937).
16. Riordan, J., "An Introduction to Combinatorial Analysis." Wiley, New York, 1958.
17. Sykes, M. F., Essam, J. W., Heap, B. R., and Hiley, B. J., Lattice constant systems and graph theory, *J. Mathematical Phys.* **7**, 1557–1572 (1966).
18. Uhlenbeck, G. E., and Ford, G. W., The theory of linear graphs with application to the virial development of the properties of gases, *in* "Studies in Statistical Mechanics" (J. de Boer and G. E. Uhlenbeck, eds.), Vol. 1. North-Holland Publ., Amsterdam, 1962.

A GRAPH-THEORETIC PROGRAMMING LANGUAGE

C. A. King

University of the West Indies
Kingston, Jamaica

1. Introduction

The past few years have witnessed the development of a large number of problem-oriented programming languages. In the field of mathematics, the languages developed include ALPAK [2] and FORMAC [12] for algebraic manipulation, a language for polynomial arithmetic [6], and the language NAPSS [11] for numerical analysis, among others.

Many authors have presented algorithms for solving various problems in graph theory, but it is only in very recent times that attempts have been made to provide a programming facility which will enable the graph theorist to state his problem in a programming language that is natural to the subject. Wolfberg [13] has developed an interactive system for graphs which emphasizes graphic display and Cresspi-Gregatti and Mortuga [3] have outlined an extension of ALGOL for handling graphs.

In the present paper, we introduce and discuss a programming language for graph theory, GTPL, which is an extension of FORTRAN for handling graphs. A compiler has been written for GTPL, and implemented on an IBM 1620 computer with 20k storage and one disk drive.

2. Design Considerations

The types of algorithms required to solve graph-theoretical problems on a computer resemble more the algorithms used in list processing and similar non-numerical applications of computers. However, a certain amount of numerical computation can be expected in most graph-theoretical problems. In designing a language to meet both these objectives, we have taken an existing language, FORTRAN, and added a number of extra statements, definitions, and the grammatical structure necessary to handle graph-theoretical manipulations on a computer.

The Graph-Theoretic Programming Language (GTPL) we now describe is thus a dialect of FORTRAN, and for simplicity Basic FORTRAN has been used. In the following discussion, Basic FORTRAN is intended whenever reference is made to FORTRAN. We first make a few general comments on the FORTRAN characteristics of GTPL and then describe the graph-theoretical statements. By way of examples, two programs, which have been compiled and executed using the GTPL system, are included.

3. FORTRAN Characteristics of GTPL

We mentioned in the last section that GTPL uses FORTRAN as its host language. There are however, a few departures from FORTRAN and these are set out below.

ASSIGNMENT STATEMENT. The arithmetic assignment statement of GTPL is of the form

$$v = l,$$

where v is the symbolic name of a variable and l is a signed or unsigned expression, consisting of at most one binary operator. Further, a statement may contain parentheses only when referencing subscripted variables. In all other respects, the assignment statement of GTPL is exactly like that of FORTRAN. Although this version of the assignment statement may appear somewhat restrictive, it has proved quite adequate for the purpose for which the language has been designed.

IF STATEMENT. In addition to the IF statement of FORTRAN, there is a logical IF statement which is used when testing for certain graph-theoretic properties. The form of the logical IF statement and the properties which may be tested are given in Section 4.

CHARACTER SET. A symbolic name that identifies a variable has a type established by the first character of the name, and the only departure from FORTRAN in this respect lies in the use of the letters G and H to identify graphs. Also, in addition to the usual characters of FORTRAN, two special characters @ and $ have been introduced. The character @ always precedes the label of a graph, for example,

$$@GA,$$

which may generally be interpreted as *of the graph GA.*

An interesting feature of GTPL is the facility to handle *collections* of graphs. The label which identifies a collection of graphs is prefixed by the character $, and one may refer to the entire collection of graphs or to a particular member of the collection. The character $ also has another use that will be described in due course.

INPUT/OUTPUT STATEMENTS. The READ, PRINT, and TYPE statements of GTPL are not formatted. The READ statement has the form

$$READ, LIST$$

where LIST consists of any combination of simple or subscripted variables or of labels of graphs. Data fields are separated by one or more blank columns. The form of the PRINT and TYPE statements is similar to that of the READ statement.

Alphameric information may be printed by bracketing the information to be printed within a pair of @ signs as follows:

$$PRINT, @MESSAGE@$$

The control statements DO, GO TO, CONTINUE, STOP, PAUSE, and END are exactly as in FORTRAN.

4. The Graph-Theoretical Statements of GTPL

The most important difference between GTPL and FORTRAN lies in the incorporation in GTPL of a number of graph-theoretical statements. These statements will now be described, and for the purpose of discussion they have been grouped into eight categories.

GROUP I. Statements for enumerating some characteristic of a given graph are

NCOMPS @ G	(number of components of the graph G);
NEDGES @ G	(number of edges of the graph G);
NNODES @ G	(number of nodes of the graph G);
NNODES (N) @ G	(number of nodes of the graph G of valency N);
VALENCE (I) @ G	(valency of node I of the graph G);
MINVAL @ G	(minimum valency of nodes of the graph G);
MAXVAL @ G	(maximum valency of nodes of the graph G);
NCUTND @ G	(number of cutnodes of the graph G);
NBLOCKS @ G	(number of blocks of the graph G).

Sample Statement:

Form: K = NCOMPS @ G
Meaning: Find the number of components of the graph G and store the result as an integer constant at K.

GROUP II. Statements for deriving a graph from a given graph are

COMPLEM @ G	(find the complement of the graph G);
CENTER @ G	(find the center or bicenter of the graph G, a tree);
SPANTREE @ G	(find a graph which is a spanning tree of G);
SEQUENCE @ G	(resequence the nodes of the graph G);
LABEL @ G	(construct the canonical labeling[†] for G and relabel the nodes of G accordingly).

Sample Statement:

Form: GXYZ = COMPLEM @ G
Meaning: Find the complement of the graph G and store it as $GXYZ$.

GROUP III. Statements for deriving certain collections of graphs from a given graph are

COMPNTS @ G	(find the components of the graph G);
BLOCKS @ G	(find the blocks of the graph G).

† Canonical labeling is defined in Section 5.3.

Sample Statement:

Form: $GX = COMPNTS @ G

Meaning: Find the components of the graph G, and store the result as a collection of graphs. This collection is referenced by the label $*GX*, where the initial character $ indicates a collection of graphs subsumed under a common name.

GROUP IV. Below are statements for performing an operation on a given graph so as to derive a modified graph. The derived graph replaces the original graph in store.

INSERT (I, J) @ G, n   (insert edge I, J);
DELETE (I, J) @ G, n   (delete edge I, J);
DELETE (I) @ G, n   (delete node I and incident edges);
INNODE (I, J) @ G, n   (add a node valency 2 in edge I, J).

The statement number *n* is a default transfer of control.

Sample Statement:

Form: DELETE (I, J) @ G, n
Meaning: Delete the edge (I, J) of the graph G. If there is no such edge, transfer control to the statement labeled *n*.

GROUP V. Statements for defining certain special graphs are

$K(N)†   (define the complete graph on N nodes);
$KBAR(N)   (define the empty graph on N nodes);
$K(M, N)   (define the complete bipartite graph on M, N nodes).

Sample Statement:

Form: GABC = $K(M, N)
Meaning: Define the complete bipartite graph on (M, N) nodes, and store at *GABC*.

GROUP VI. Below are the logical IF statements. These statements allow us to test whether or not a graph exhibits some particular property and to transfer control accordingly. The truth conditions are in parentheses.

IF (G, CONNEX) m, n   (graph G is connected);
IF (G, REGULAR) m, n   (graph G is regular);
IF (G, REGULAR (K)) m, n   (graph G is K regular);
IF (G, TREE) m, n   (graph G is a tree);
IF (G, FOREST) m, n   (graph G is a forest);

† This is the other use for the $ sign.

IF (G, PLANAR) m, n   (graph G is a planar);
IF (G, POLYGON) m, n   (graph G is a polygon);
IF (G, COMPLETE) m, n   (graph G is complete);
IF (G, EMPTY) m, n   (graph G is empty).

Sample Statement:

Form: IF (G, PLANAR) m, n
Meaning: Test whether the graph G is planar. If it is, transfer control to the statement labeled m. If it is not, transfer control to the statement labeled n. In addition, there is the logical statement

$$\text{IF (G, ISOMOR) m, n,}$$

which tests whether or not the graph G is isomorphic to the graph currently in working store.

Note that all graphs handled by GTPL are, in a sense, labeled graphs, since a label, an integer, has to be associated with each node in order that the graph can be input to the computer (see the remarks below on input format). ISOMOR tests for isomorphism between *labeled* graphs, that is, for each pair i, j of nodes it checks that either an edge (i, j) exists in both graphs, or that it exists in neither graph. We shall see in Section 5.3 how this routine can be used to test for isomorphism between unlabeled graphs.

The following group of logical IF statements differs from those in Group VI in that these statements refer to a subgraph of a given graph.

GROUP VII.   The truth conditions are in parentheses.

IF (G, HASCUTND) m, n   (graph G has a cut node);
IF (G, HASEDGE (I, J)) m, n   (graph G has an edge joining node I to node J).

GROUP VIII.   Statements for the transfer of data on graphs are

GET, G   (call graph G from backing store);
PUT, G   (put graph G into backing store);
READ, G   (read a graph G and store it in backing store);
PRINT, G   (print the graph G).

Commentary on Statements in Group VIII: All graphs are read in node-pair format as follows: $nneee, xxyy, xxyy, \ldots, xxyy$, where nn is the number of nodes, eee is the number of edges and $xxyy$ is an edge incident with node xx and node yy. This format allows for graphs on up to 99 nodes and 999 edges. Graphs on up to 49 nodes can be accommodated by the program currently written for GTPL.

Graphs normally are kept in the backing store of the computer and brought into working store as required. The GTPL compiler keeps track of the name of the graph currently in working store. When a new graph is required it is brought in from the backing store. If, however, the next graph name which is encountered in a statement is the same as that of the graph in working store, this is not done. This has an important consequence.

The statements mentioned in Group IV modify a graph and leave the modified graph in working store. If, after executing one of these statements, the next graph referred to is currently in working store, then the operations are performed on the modified graph. If, however, the operations are to be performed on the original graph, then the statement

GET, G

which calls the graph G from backing store, should be used.

The statement

PUT, G

causes the graph currently in working store to be stored in backing store. If a graph was previously stored in the area reserved for the graph G, then it is overwritten.

5. Notes on Graph Theory Algorithms

We next describe two programs which illustrate some of the most important features of GTPL. We first give a brief description of the algorithms used, and in Section 6 we present listings of the actual programs together with some explanatory notes.

For the functions NNODES, COMPLEM, and MAXVAL the description given in Section 4 is adequate. We therefore restrict our discussion to

1. COMPNTS and NCOMPS;
2. PLANAR;
3. LABEL and ISOMOR.

5.1. COMPNTS and NCOMPS

Our algorithm for finding the components of a graph G is as given by Read [10], and derives from a procedure for constructing the spanning tree of a graph. The method assigns the same label to each node of a connected component of the graph G. Hence, if G is not connected, each component of G is determined by an equivalence class of labels of G. The number of distinct labels of G gives the number of components of G.

5.2. PLANAR

The algorithm used to test whether or not a given graph G is planar is that of Fisher and Wing [4] as modified by Read [9]. Briefly, the algorithm is as follows:

1. Choose any circuit of the graph G; call this circuit K.
2. Of those edges not belonging to K, there may be transversals, that is, edges with both end nodes on K; we insert a node in each transversal and obtain graph G'. Clearly, G' is planar, if and only if G is planar.
3. Remove from G' all the edges of K together with all other edges of G' that are incident with nodes of K; call these edges *link edges*. Then what is left will be a number of connected components F_i, which we shall call fragments. Each F_i is connected to K by a set L_i of link edges. Denote by H_i the subgraph of G' consisting of K, F_i and L_i.
4. In G', shrink each F_i to a point, thus obtaining a reduced graph G_r.
5. Two fragments F_i and F_j are said to be *incompatible* if they cannot both be placed inside, or both outside, of K without causing some link edges to intersect. If all fragments cannot be placed in such a way that all pairs of incompatible fragments are differently placed, then G_r, and hence G, is nonplanar. If the fragments F_i can be so placed, one then proceeds to apply the above test for planarity to each of the subgraphs H_i, and eventually a verdict is obtained.

The above is an oversimplification, and one is referred to Fisher and Wing [4] for a detailed discussion of the algorithm, and to King [7] and Read [9] for a discussion of the modification.

5.3. LABEL *and* ISOMOR

The method which we use to obtain a canonical labeling of a graph was developed by Parris and Read [8], and derives from the construction of a unique code for a given graph. By considering each node, its nearest neighbors, that is, its neighbors of order 1, and its neighbors of orders 2, 3,..., one is able to classify the nodes of a graph G in such a way that each class has exactly one node. A problem arises when the graph has symmetries with respect to certain of its nodes, but this problem is overcome by forcing these nodes to be in different classes. An important consequence of this approach is that the classification thus obtained is independent of the original, arbitrary manner in which the nodes of the graph were labeled.

Since the canonical labeling is independent of the original labeling of the graph, as determined by the nature of its input, it follows that two graphs $G1$ and $G2$ are isomorphic as *unlabeled* graphs, if and only if, after both have been relabeled by the LABEL routine, they are isomorphic as *labeled* graphs. Thus a double use of LABEL, followed by ISOMOR, enables isomorphism between unlabeled graphs to be tested.

The LABEL routine uses an inefficient, and therefore lengthy, algorithm; necessarily so, since no efficient algorithm for this purpose is yet known. It is therefore worthwhile to make a preliminary check for possible obvious nonisomorphism of G_1 and G_2. This whole procedure is displayed more fully in the second of the sample programs which follow.

6. Sample Programs

We now describe two programs that have been compiled and executed, using the GTPL system.

6.1. Sample Program 1

```
      CCCCC      SAMPLE PROGRAM 1

      C          DETERMINE WHICH COMPONENTS OF THE COMPLEMENT

      C          OF A GIVEN GRAPH ARE PLANAR, WHICH NONPLANAR

      C

 1               COLLECTION $GB,20

 2               READ , G

 3               GA = COMPLEM @ G

 4               $GB = COMPNTS @ GA

 5               K = NCOMPS @ GA

 6               I = 0

 7            2 I = I + 1

 8               IF ( $GB, I, PLANAR) 22,24

 9    C          THIS COMPONENT IS PLANAR, PRINT IT

10            22 PRINT, @PLANAR@

11               PRINT, $GB,I

12               GO TO 25

13    C          THIS COMPONENT IS NONPLANAR, PRINT IT

14            24 PRINT,@NON-PLANAR@

15               PRINT, $GB,I

16            25 K = K - 1

17               IF (K) 26, 26, 2

18            26 END
```

The lines are numbered for easy reference in the following discussion. The line numbers are not part of the program.

Line 1: The declarative COLLECTION, not previously mentioned, specifies that the set of graphs GB forms a *collection* of graphs, and may have up to 20 members. In some respects this statement is similar to the DIMENSION statement of FORTRAN. The extent, 20, serves to inform the processor that storage must be reserved, in backing store, for 20 graphs.

Line 2: The graph G is to be read. The processor reserves an area of store for the graph G.

Line 3: The graph G is to be complemented and the complement stored as graph GA.

Line 4: The components of the graph GA are to be found and stored as a collection of graphs, indexed from 1 to N, where N is the number of members in the collection. In the above program, an error results if N exceeds 20, the extent of GB as specified in Line 1.

Line 5: The number of components of GA is to be computed, and this number stored as an integer variable K.

Line 8: This statement specifies that the Ith member of the collection $\$GB$ should be tested for planarity. If the graph tested is planar, then control is to be transferred to statement 22 (Line 10). If the graph is nonplanar, control is to be transferred to statement 24 (Line 14).

Line 10: The PRINT and READ statements of GTPL are not formatted. A blank delimiter is used for the input of numeric information, and the output is according to the standard format F 16.8 or I 5, according as the result is real or integer. In place of the H FORMAT specifications of FORTRAN, a pair of @ signs is used as quotation marks. The statement in Line 10 causes the message PLANAR to be printed. Similarly, the statement in Line 14 causes the message NONPLANAR to be printed.

Line 11: This statement causes the Ith member of the collection GB to be printed. So does the statement of Line 15.

The other lines in the program are normal FORTRAN statements, and require no special explanation.

6.2. Sample Program 2

Line 1: This is the DIMENSION statement of FORTRAN.
Line 2: This statement causes the graph GA to be read.
Line 3: The graph GA is to be stored in backing store.
Line 4: The maximum valency of the graph GA is to be found and stored at MAXA.

```
      CCCCC    SAMPLE PROGRAM 2

      C       THIS IS A TEST FOR ISOMORPHISM BETWEEN TWO GRAPHS

      C       GA AND GB. A PRELIMINARY CHECK IS MADE TO SEE IF

      C       GA AND GB HAVE THE SAME VALENCY SEQUENCE. IF THEY

      C       HAVE, THEN EACH GRAPH IS GIVEN A CANONICAL LABELLING,

      C       USING THE -LABEL- STATEMENT, AND THE RESULTING

      C       LABELLED GRAPHS ARE TESTED FOR ISOMORPHISM USING THE

      C       - IF(ISOMOR) - STATEMENT.

      C

 1            DIMENSION NVA(20), NVB(20)

 2            READ,GA

 3            PUT, GA

 4            MAXA = MAXVAL @ GA

 5            READ, GB

 6            PUT, GB

 7            MAXB = MAXVAL @ GB

 8            DO 12 J = 1, MAXA

 9         12 NVA (J) = NNODES(J) @GA

10            DO 14 J = 1, MAXB

11         14 NVB(J) = NNODES (J) @ GB

12            DO 16 J = MINA,MAXA

13            NDIFF = NVA (J) - NVB (J)

14            IF (NDIFF) 22, 16, 22

15         16 CONTINUE

16    C       VALENCY SEQUENCE FAILS TO DISCRIMINATE BETWEEN GRAPHS

17            GX = LABEL @ GA

18            GC = LABEL @ GB

19            IF( GX, ISOMOR ) 24,22

20         24 TYPE, @ISOMORPHIC@

21            GO TO 99

22         22 TYPE ,@NOT ISOMORPHIC@

23         99 END
```

Lines 5, 6, and 7:   These are similar to Lines 2, 3, and 4.

Line 8:   This is the normal DO statement of FORTRAN.

Line 9:   This statement specifies that for the graph GA, the number of nodes of valency J is to be found and stored as the Jth element of the array NVA. This statement, together with the DO statement in Line 8, causes the valency sequence of the graph GA to be constructed.

Lines 10 and 11:   These are similar to the statements in Lines 8 and 9.

Lines 12 to 16:   These are ordinary FORTRAN statements.

Line 17:   This statement causes the canonical labeling of the nodes of the graph GA to be constructed and the resulting labeled graph to be stored as GX. The statement in Line 18 is similar. In both cases, the labeled graph is also available in a work area of store.

Line 19:   This statement causes the labeled graph GX, now in backing store, to be compared with the labeled graph GC which is still in store. The graphs are isomorphic, if and only if they are equivalent as labeled graphs.

Line 20:   This statement causes the message ISOMORPHIC to be printed. Line 22 is similar.

Lines 21 and 23:   These are normal FORTRAN statements.

7. Concluding Remarks

The operating system for the programming language that we have described, consists of a compiler phase and an execution phase. The compiler phase produces an object deck which must be loaded together with any data, so that the program may be executed. The graph theory routines are kept in backing store and are called into working store only when required. The same is true of graphs, which normally reside in backing store.

An attempt has been made to include a wide variety of graph theory routines in GTPL, but it is expected that use of the language will point to the need for additional routines. For this reason, the system has been designed so as to accommodate additional routines. A full description of the algorithms used and of the design of the compiler and operating system is given by King [7].

References

1. Berge, C., "The Theory of Graphs and its Applications" (A. Doig, transl.). Wiley, New York, 1962.
2. Brown, W. S., A language and system for symbolic algebra on a digital computer, *Proc. Sci. IBM Symp.*, 77–114 (1966).
3. Crespi-Reghizzi, S., and Morpurgo, R., A language for treating graphs, *Comm. ACM* **13**, 319–323, (1970).

4. Fisher, G., and Wing, O., Computer recognition and extraction of planar graphs from the incidence matrix, *IEEE Trans. Circuit Theory* **CT-13**, 2, 154–163 (1966).
5. Goldstein, A. J., An efficient and constructive algorithm for testing whether a graph can be imbedded in a plane, Bell Telephone Labs., unpublished report.
6. Hartt, K., Some analytical procedures for computers and their applications to a class of multidimensional integrals, *J. Assoc. Comput. Mach.* **11**, 416–421 (1964).
7. King, C. A., A graph-theoretic programming language, Doctoral thesis, University of the West Indies, 1970.
8. Parris, R., and Read, R. C., A coding procedure for graphs, *Univ. of the West Indies Comput. Center Sci. Rep.* UWI/CC 10 (1969).
9. Read, R. C., Graph theory algorithms, *in* "Graph Theory and Its Applications" (B. Harris, ed.), pp. 51–78. Academic Press, New York, 1970.
10. Read, R. C., Teaching graph theory to a computer, *in* "Recent Progress in Combinatorics," (W. T. Tutte, ed.), pp. 161–173. Academic Press, New York, 1969.
11. Rice, T., and Rosen, S., NAPSS—A numerical analysis problem solving system, *Proc. 21st Nat. Conf. ACM*, 51–56 (1966).
12. Sammet, J. E., Formula manipulation by computer, *Advan. Computers* **8**, 47–102 (1967).
13. Wolfberg, M. S., An interactive graph theory system, Doctoral thesis, University of Pennsylvania, 1969.

ENTROPY OF TRANSFORMED FINITE-STATE AUTOMATA AND ASSOCIATED LANGUAGES

W. Kuich[†]

IBM Laboratory
Vienna, Austria

1. Introduction

After a short review of the concepts involved in finite-state automata, associated languages, directed multigraphs and nonnegative matrices, an *S* transformation on automata is defined. This transformation replaces a transition between two states of the original automaton by the transitions of an automaton of simple structure. Speaking in terms of language theory, this transformation is equivalent to a language-preserving function called substitution or homomorphism.

[†] Present address: Technische Hochschule Wien, Vienna, Austria.

Defining the entropy of finite-state automata and associated languages, it is then natural to ask for the change in the entropy caused by applying the S transformation. The answer to this question is given in this paper for certain types of automata. It generalizes several results achieved by Izbicki [4, 5].

2. Preliminaries

In this section we briefly review the concepts involved in finite-state automata and associated finite-state languages, following essentially Ginsburg [3].

An alphabet is a finite nonempty set. A word of length $k \geqslant 0$ over an alphabet Σ is a finite sequence $x_1, x_2, ..., x_k$ of elements in Σ. The word of length zero, called the empty word, is denoted by ε. The set of all words, including ε, over an alphabet Σ is denoted by Σ^*. Let U and V be subsets of Σ^*, then the complex product of U and V, written UV, is the set of words $\{w = w_1 w_2 | w_1 \text{ in } U, w_2 \text{ in } V\}$.

A finite-state automaton is specified by a 5-tuple $A = (K, \Sigma, \delta, p_0, F)$, where

(1) K is a finite nonempty set of states;
(2) Σ is an alphabet of input symbols;
(3) δ is a function from a subset of $K \times \Sigma$ into K, the next state function;
(4) p_0 is a distinguished element of K, the start state;
(5) F is a subset of K, the set of final states.

The function δ is extended to a subset of $K \times \Sigma^*$ by defining $\delta(q, \varepsilon) = q$ and $\delta(q, x_1 \cdots x_k) = q_k$, where $q_0 = q$ and $q_i = \delta(q_{i-1}, x_i)$, $1 \leqslant i \leqslant k$.

The behavior of the automaton is deterministic, that is, the next-state function δ defines for each state q in K and each input symbol x at most one next state $\delta(q, x)$. Each automaton A defines a subset $T(A)$ of Σ^*, the set of generated or accepted words, given by

$$T(A) = \{w \in \Sigma^* | \delta(p_0, w) \in F\}.$$

A subset of Σ^*, generated by some finite-state automaton, is called finite-state language. $T(A)$ is called the language generated by A.

Walk [8] defined the informational structure of an automaton A to be the directed multigraph $G(A) = (V(A), E(A))$. The set $V(A)$ of vertices coincides with the set K of states. The set of edges is defined by

$$E(A) = \{e_{p,q}^x = (p, q) | \delta(p, x) = q; x \in \Sigma\}.$$

Loops and multiple edges between two vertices are allowed.

Associated with this informational structure is the adjacency matrix of the graph, defined in the usual manner. Its (i, j) entry is a_{ij} if a_{ij} edges, rooted in vertex p_i are leading to vertex p_j. The elements of this matrix are nonnegative, and hence the powerful theory of nonnegative matrices, developed by Perron and Frobenius, comes into play (see Wielandt [9]).

A matrix B is called irreducible if there exists no permutation matrix P such that

$$P^{-1}BP = \begin{pmatrix} B_{11} & B_{12} \\ \varnothing & B_{22} \end{pmatrix}$$

with square blocks B_{11} and B_{22} and null matrix $\varnothing$. An irreducible matrix B is said to be imprimitive of index h if there exists a permutation matrix P such that

$$P^{-1}BP = \begin{pmatrix} \varnothing & \varnothing & \cdots & \varnothing & B_{1n} \\ B_{21} & \varnothing & \cdots & \varnothing & \varnothing \\ \varnothing & B_{32} & \cdots & \varnothing & \varnothing \\ \vdots & \vdots & & \vdots & \vdots \\ \varnothing & \varnothing & \cdots & B_{n,n-1} & \varnothing \end{pmatrix},$$

with square $\varnothing$'s in the main diagonal. Otherwise it is called primitive.

The theorem of Perron–Frobenius states that a nonnegative irreducible matrix B has a positive eigenvalue λ, which is the simple root of the characteristic equation. The modules of all other eigenvalues of B do not exceed λ. λ is called maximal eigenvalue. In case B is imprimitive of index h, there exist exactly h eigenvalues of modulus λ, which are roots of the equation $x^h - \lambda^h = 0$. For primitive matrices, λ is the only eigenvalue of modulus λ.

For the sequel we need the following definitions:

1. An automaton is called strongly connected if and only if for every pair (p, q) of states there exists a word w such that $\delta(p, w) = q$.

2. An automaton is said to be periodic of period h if and only if its set of states can be partitioned into h sets $V_1, V_2, \ldots, V_h$ such that $\delta(p, x)$ is in V_l only if p is in V_k and $l - k \equiv 1$, modulo h, x in Σ; otherwise the automaton is called aperiodic.

3. An automaton that is strongly connected and aperiodic is called ergodic.[†]

4. An automaton is said to be complete if and only if its next state function δ is defined on the whole set $K \times \Sigma$.

The following statements are equivalent:

1. (a) An automaton is strongly connected;
    (b) its informational structure is strongly connected;
    (c) its adjacency matrix is irreducible.

[†] Note that this definition, according to Shannon [7], differs slightly from that given usually in the theory of Markov chains.

2. (a) An automaton is strongly connected and periodic of period h;
    (b) its informational structure is strongly connected and cyclically
         h-partite;
    (c) its adjacency matrix is imprimitive of index h.
3. (a) An automaton is ergodic;
    (b) its adjacency matrix is primitive.
4. (a) An automaton is complete;
    (b) its informational structure is out regular of degree n;
    (c) its adjacency matrix has constant row sum n, where n is the cardinal
         of the set of states K.

Let A be a strongly connected automaton and $T(A)$ the language generated
by A. Let $u(n)$ be the number of words of length n in $T(A)$. Then the entropy
H of A respectively $T(A)$ is defined to be the quantity

$$H = \lim_{n \to \infty} \sup [\log u(n)/n].$$

This definition covers the aperiodic and periodic case. In the aperiodic case,
the entropy usually is defined to be

$$H = \lim_{n \to \infty} [\log u(n)/n],$$

according to the definition of the channel capacity by Shannon [7]. But if A
is a periodic automaton of period h, then this limit is to be taken only for those
residue classes, modulo h, for which the numbers $u(n)$ are not identically
zero. Hence, in this case, the entropy H is defined by the limit superior.

The entropy is a measure for the amount of information that must be

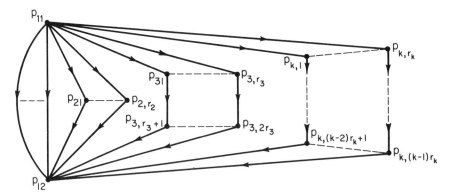

Fig. I.   The informational structure of N_a.

provided on the average in order to specify a particular symbol of any word of the language. This quantity also may be considered as a measure of the uncertainty existing about a symbol in a word before its generation by the automaton.

3. S Transformation of Automata

Given fixed nonnegative integers $r_1, r_2, \ldots, r_k$, $r_k \geqslant 1$ and an abstract symbol a, the automaton $N_a = N(r_1, \ldots, r_k; a)$ is defined to be the 5-tuple $N_a = (K_N, \Sigma_N{}^a, \delta_N, p_{11}, \{p_{12}\})$, where

(1) $K_N = \{p_{11}, p_{12}\} \cup \{p_{ij}| 1 \leqslant i \leqslant k, 1 \leqslant j \leqslant (i-1)r_i\}$;
(2) $\Sigma_N{}^a = \{x_{ij}^a| 1 \leqslant i \leqslant k, 1 \leqslant j \leqslant ir_i\}$;
(3) δ_N is defined by

$$\delta_N(p_{11}, x_{1i_1}^a) = p_{12},$$

$$\delta_N(p_{11}, x_{m,i_m}^a) = p_{m,i_m},$$

$$\delta_N(p_{m,(s-1)r_m+i_m}, x_{m,sr_m+i_m}^a) = p_{m,sr_m+i_m},$$

$$\delta_N(p_{m,(m-2)r_m+i_m}, x_{m,(m-1)r_m+i_m}^a) = p_{12},$$

$$1 \leqslant i_1 \leqslant r_1, \quad 2 \leqslant m \leqslant k, \quad 1 \leqslant i_m \leqslant r_m, \quad 1 \leqslant s \leqslant m-2.$$

N_a is to agree with the definition of an automaton. Hence, all the symbols x_{m,i_m}^a, $1 \leqslant m \leqslant k$, $1 \leqslant i_m \leqslant r_k$, have to be different.

The language $T(N_a) = T(N(r_1, \ldots, r_k; a))$ consists of the $r_1 + r_2 + \cdots + r_k$ words

$$x_{11}^a; \ldots; x_{1r_1}^a; x_{21}^a x_{2,r_2+1}^a; x_{2,r_2}^a x_{2,2r_2}^a; \ldots;$$

$$x_{k,1}^a x_{k,r_k+1}^a \cdots x_{k,(k-1)r_k+1}^a; \ldots; x_{k,r_k}^a x_{k,2r_k}^a \cdots x_{k,kr_k}^a.$$

The informational structure is given by the graph $G(N)$, shown in Fig. 1.

Let $M = (K_M, \Sigma_M, \delta_M, q_M, F_M)$ be an automaton and let

$$\{N(r_1, \ldots, r_k; a)| \ a \in \Sigma_M\}$$

be a family of automata of the type described above, such that $\Sigma_N{}^a \cap \Sigma_N{}^b = \varnothing$ for $a \neq b$ in Σ_M. Then the automaton $M(r_1, \ldots, r_k) = (K, \Sigma, \delta, q_M, F_M)$ is defined by

(1) $K = K_M \cup \{p_{ij}(p, q, a)| 1 \leqslant i \leqslant k, 1 \leqslant j \leqslant (i-1)r_i, \delta_M(p, a) = q\}$;
(2) $\Sigma = \bigcup_{a \in \Sigma_M} \Sigma_N{}^a$;

(3)
$$\delta(p, x^a_{1 i_1}) = q,$$

$$\delta(p, x^a_{m, i_m}) = p_{m, i_m}(p, q, a),$$

$$\delta(p_{m, (s-1) r_m + i_m}(p, q, a), x^a_{m, s r_m + i_m}) = p_{m, s r_m + i_m}(p, q, a),$$

$$\delta(p_{m, (m-2) r_m + i_m}(p, q, a), x^a_{m, (m-1) r_m + i_m}) = q,$$

$$1 \leqslant i_1 \leqslant r_1, \quad 2 \leqslant m \leqslant k, \quad 1 \leqslant i_m \leqslant r_m, \quad 1 \leqslant s \leqslant m - 2,$$

for each transition $\delta_M(p, a) = q$; p, q in K_M, a in Σ_M.

This transformation of the automaton M is called an S transformation of M. Intuitively speaking the automaton $M(r_1, r_2, \ldots, r_k)$ originates from M by replacing each transition $\delta_M(p, a) = q$ of M by the transitions of the automaton $N(r_1, \ldots, r_k; a)$. This produces the following changes in the respective informational structures. Each edge (p, q) in $G(M)$ is replaced by the graph, drawn in Fig. 1, such that p and q coincide with p_{11} and p_{12}, respectively.

To study the effect that an S transformation of an automaton M has on the language generated by M, we have to introduce the concept of substitution (see Ginsburg [3]). For each element a in an alphabet Σ, let Σ_a be an alphabet and $\tau(a)$ a subset of Σ_a^*. Let $\tau(\varepsilon) = \{\varepsilon\}$ and $\tau(x_1 \ldots x_r) = \tau(x_1) \cdots \tau(x_r)$ for each word $x_1 \cdots x_r$ in Σ^*. Then the function τ, mapping Σ^* into the set of subsets of $(\bigcup_{a \in \Sigma} \Sigma_a)^*$ is called a substitution. In case $\tau(a)$ consists of a single word w_a in Σ_a^* for each a in Σ, τ is regarded as a mapping of Σ^* into $(\bigcup_{a \in \Sigma} \Sigma_a)^*$ and is called homomorphism.

The S transformation replaces each transition $\delta_M(p, a) = q$ in M by a transition $\delta(p, w_a) = q$ in $M(r_1, \ldots, r_k)$, where w_a is in $T(N_a)$. Defining a substitution τ_N by $\tau_N(a) = T(N_a)$ for each a in Σ_M, we get the result

$$\tau_N(T(M)) = T(M(r_1, \ldots, r_k)),$$

that is, the language $T(M(r_1, \ldots, r_k))$ originates from the language $T(M)$ by application of the substitution τ_N on $T(M)$.

4. Entropy of S-Transformed Automata

In Section 2 we defined the entropy H of an automaton A to be the quantity

$$H = \lim_{n \to \infty} \sup[\log u(n)/n],$$

where $u(n)$ is the number of words of length n generated by A. This definition of the entropy differs slightly from that one given by Shannon [7], Chomsky and Miller [1], or Walk [8] in order to cover the cases of ergodic and periodic automata.

Walk [8] proved the entropy of an ergodic automaton to equal the logarithm of the maximal eigenvalue of the adjacency matrix associated with its informational structure. By applying his proof method to the definition of entropy given above and including the periodic (hence nonergodic) automata his result remains valid:

The entropy H of a strongly connected automaton equals the logarithm of the maximal eigenvalue λ of the adjacency matrix associated with its informational structure, that is,

$$H = \log \lambda.$$

Given a strongly connected automaton M and nonnegative integers $r_1, r_2, \ldots, r_k$, $r_k \geqslant 1$, we want to evaluate the entropy of the S-transformed automaton $M(r_1, \ldots, r_k)$. We can apply the result achieved before because with M strongly connected and $r_k \geqslant 1$, where $N(r_1, \ldots, r_k; a)$ generates at least one word of length k, $M(r_1, \ldots, r_k)$ remains strongly connected. Hence, we have to find relations between the maximal eigenvalues of the matrices B and C associated with the informational structures of M and $M(r_1, \ldots, r_k)$, respectively.

Let $M = (K_M, \Sigma_M, \delta_M, q_M, F_M)$ and $M(r_1, \ldots, r_k) = (K, \Sigma, \delta, q_M, F_M)$. The set K of states of $M(r_1, \ldots, r_k)$, and hence the set of vertices of its informational structure, is equal to

$$K = K_M \cup \{p_{ij}(p, q, a) \mid 1 \leqslant i \leqslant k, \ 1 \leqslant j \leqslant (i-1)r_i, \ \delta_M(p, a) = q\}$$

To get a clearly arranged form of the adjacency matrix we proceed as follows: We partition K into sets $V_0, V_1, \ldots, V_{k-1}$, which are defined by $V_0 = K_M$ and $V_i = \{s \in K - K_M \mid \delta(s, x) = t, t \in V_{i-1}, x \in \Sigma\}, 1 \leqslant i \leqslant k-1$. Hence

$$V_1 = \{p_{2,1}(p, q, a), \ldots, p_{2,r_2}(p, q, a), p_{3,r_3+1}(p, q, a), \ldots,$$
$$p_{k,(k-1)r_k}(p, q, a) \mid \delta_M(p, a) = q\},$$
$$V_{k-1} = \{p_{k,1}(p, q, a), \ldots, p_{k,r_k}(p, q, a) \mid \delta_M(p, a) = q\}.$$

We partition the adjacency matrix C of the informational structure of $M(r_1, \ldots, r_k)$ into blocks $C_{ij}, 0 \leqslant i, j \leqslant k-1$, such that C_{ij} covers the adjacencies of the vertices of V_i to the vertices of V_j. The only blocks possibly unequal to the null matrix $\emptyset$ are $C_{00}, C_{01}, \ldots, C_{0,k-1}, C_{12}, C_{23}, \ldots, C_{k-2,k-1}$. Hence, C has the form

$$C = \begin{pmatrix} C_{00} & C_{01} & \cdots & C_{0,k-2} & C_{0,k-1} \\ C_{10} & \emptyset & \cdots & \emptyset & \emptyset \\ \emptyset & C_{21} & \cdots & \emptyset & \emptyset \\ \vdots & \vdots & & \vdots & \vdots \\ \emptyset & \emptyset & \cdots & C_{k-1,k-2} & \emptyset \end{pmatrix}.$$

We have now to find a relation between the maximal eigenvalues of the matrices B and C. We have done this in the following two cases.

CASE 1. Consider M strongly connected, $r_1 = \cdots = r_{k-1} = 0$, $r_k = r \geqslant 1$. In this case $C_{00} = C_{01} = \cdots = C_{0k-2} = \varnothing$, which indicates that $M(0,\ldots,0,r)$ is a periodic automaton of period at least k. In case $k = 1$, $C = C_{00} = rB$. To evaluate the maximal eigenvalue of C, we compute the kth power of C,

$$C^k = \begin{pmatrix} C_{0,k-1}C_{k-1,k-2}\cdots C_{10} & \varnothing & \cdots & \varnothing \\ \varnothing & C_{10}C_{0,k-1}\cdots C_{21} & \cdots & \varnothing \\ \vdots & \vdots & & \vdots \\ \varnothing & \varnothing & \cdots & C_{k-1,k-2}C_{k-2,k-3}\cdots C_{0,k-1} \end{pmatrix}.$$

The entries of the left upper block $C_{0,k-1}C_{k-1,k-2}\cdots C_{10}$ are equal to the number of ways of length k from the vertices in V_0 to vertices in V_0. Since two vertices of V_0 are joined exactly by r ways of length k, and no shorter ones, in the informational structure of $M(0,\ldots,0,r)$, if they are adjacent in the informational structure of M,

$$C_{0,k-1}C_{k-1,k-2}\cdots C_{10} = rB.$$

The maximal eigenvalues of all the diagonal blocks of C are equal to the maximal eigenvalue of $C_{0,k-1}C_{k-1,k-2}\cdots C_{10}$, and hence to the maximal eigenvalue of rB. Denoting the maximal eigenvalue of B by λ, that of rB and hence C^k equals $r\lambda$. This yields the maximal eigenvalue of C to be $(r\lambda)^{1/k}$. Hence, the entropy $H_{M(0,\ldots,0,r)}$ of $M(0,\ldots,0,r)$ has the value

$$H_{M(0,\ldots,0,r)} = k^{-1}(\log r + \log \lambda).$$

Denoting the entropy of M by H_M, we get the result

$$H_{M(0,\ldots,0,r)} = k^{-1}(\log r + H_M).$$

This S transformation, transforming M into $M(0,\ldots,0,r)$ generalizes two transformations introduced by Izbicki [4]; $r = 1$ yields the η_k transformation, while $k = 1$ yields the η^r transformation.

In the case $r = 1$, each symbol a of a word in $T(M)$ is replaced by a word w_a of length k. In terms of language theory, this S transformation is a homomorphism $\tau(a) = w_a$, $a \in \Sigma_M$. Hence, a homomorphism, mapping each symbol on a word consisting of k symbols, diminishes the entropy H_M of the original language to the kth part, $k^{-1}H_M$.

CASE 2. Consider M complete and strongly connected, $r_k \geqslant 1$, $k \geqslant 2$. Let the cardinal of Σ_M be n. Since M is complete the informational structure is out regular of degree n. Hence, its adjacency matrix B is generalized stochastic with row sum n, yielding the maximal eigenvalue n. Thus

$$H_M = \log n.$$

In the informational structure of M each edge is replaced by the graph drawn in Fig. 1 to yield the informational structure of $M(r_1,...,r_k)$. Since the informational structure of M is out regular of degree n, in the informational structure of $M(r_1,...,r_k)$ there are exactly nr_{i+1} edges, rooted in any vertex of the set V_0 and leading to vertices in V_i, $0 \leqslant i \leqslant k-1$. That means that all the blocks C_{0i}, $0 \leqslant i \leqslant k-1$, have constant row sums $r_{i+1}n$. From any vertex in V_{i+1} exactly one edge leads to the vertices in V_i, $0 \leqslant i \leqslant k-2$. Consequently $C_{i+1,i}$, $0 \leqslant i \leqslant k-2$, has constant row sum 1.

Hence, C is partitioned into blocks having constant row sums. Kuich and Walk [6] called this type of matrix block stochastic and showed the equality of the maximal eigenvalues of a block-stochastic matrix and the matrix associated with it, having the row sums of the blocks as entries. The matrix of the row sums is

$$\begin{pmatrix} r_1 n & r_2 n & \cdots & r_k n \\ 1 & 0 & \cdots & 0 \\ 0 & 1 & \cdots & 0 \\ \vdots & \vdots & & \vdots \\ 0 & 0 & \cdots & 0 \end{pmatrix},$$

which is the companion matrix of the polynomial

$$x^k - r_1 nx^{k-1} - \cdots - r_k n.$$

Hence, the maximal eigenvalue λ of C is the greatest real root of

$$x^k - r_1 nx^{k-1} - \cdots - r_k n = 0,$$

which yields

$$H_{M(r_1,...,r_k)} = \log \lambda.$$

Again this S transformation is a generalization of a transformation introduced by Izbicki [5]. Let $k = 2$, $r_1 = r$, and $r_2 = s$, then it coincides with the $\xi_{r,s}$ transformation of [5]. The entropy of the transformed automaton is in this case

$$H_{M(r,s)} = \log[rn + (rn^2 + 4sn)^{1/2}/2],$$

a special case of the result achieved above.

References

1. Chomsky, N., and Miller, G. A., Finite state languages, *Information and Control* **1**, 91–112 (1958).
2. Ginsburg, S., "An Introduction to Mathematical Machine Theory." Addison-Wesley, Reading, Massachusetts, 1962.

3. Ginsburg, S., "The Mathematical Theory of Context-Free Languages." McGraw-Hill, New York, 1966.
4. Izbicki, H., Die Entropie η-transformierter gerichteter Graphen, *Sitzungsber. Österr. Akad. Wiss. Math. Naturwiss. Klasse Abt. II* **177**, 227–235 (1969).
5. Izbicki, H., Die Entropie ξ-transformierter gerichteter Graphen, *Sitzungsber. Österr. Akad. Wiss. Math. Naturwiss. Klasse Abt. II* **177**, 215–225 (1969).
6. Kuich, W., and Walk, K., Block-stochastic matrices and associated finite-state languages, *Computing* **1**, 50–61 (1966).
7. Shannon, C. E., A mathematical theory of communication, *Bell System Tech. J.* **27**, 379–423 (1948).
8. Walk, K., Entropy and testability of context-free languages, *in* "Formal Language Description Languages for Computer Programming," (T. B. Steel, Jr., ed.), pp. 105–123. North-Holland Publ. Amsterdam, 1966.
9. Wieland, H., Unzerlegbare negative Matrizen, *Math. Z.* **52**, 642–648 (1950).

COUNTING HEXAGONAL AND
TRIANGULAR POLYOMINOES

W. F. Lunnon†

Atlas Computer Laboratory
Chilton, Didcot
Berkshire, England

1. Introduction

We have discussed the problem of counting p-minos (polyominoes, animals) on the square tessellation [1]. Here we extend those methods to the hexagonal and triangular tessellations. Our account will be self-contained, but less discursive than before.

A *hex*(agonal)/*tri*(angular) *p-mino* is an edge-connected configuration of p

† Present address: Department of Computing in Mathematics, University College, Cardiff, Wales.

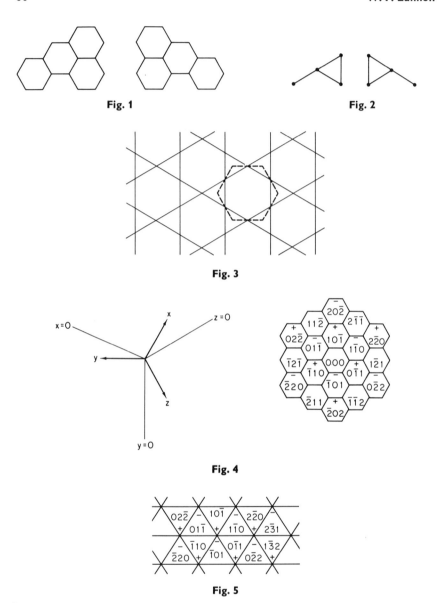

Fig. 1   Fig. 2

Fig. 3

Fig. 4

Fig. 5

cells from the appropriate plane tessellation. A *fixed p-mino* is an equivalence
class under translation; a *free p-mino* is a class under all symmetries of the
lattice. Figure 1 shows a pair of distinct fixed 4-minos belonging to the same
free 4-mino, abbreviated in Fig. 2. We attempt to evaluate the totals $HX(p)$,
$HE(p)$, $TX(p)$, $TE(p)$ of fixed/free hex/tri p-minos: our results are in Section 5.

Working with these tessellations is facilitated by a sanitary system of co-ordinates, which we now develop.

Consider the solid tessellation of cubes with a cube center at each integer point $x = (x, y, z)$, and a plane $x + y + z = 0$ cutting it. This plane intersects each $x + y + z = 0$ cube in a hexagon, and each $x + y + z = \pm 1$ cube in a triangle: the whole tessellation meets the plane in the plane tessellation $6^2 3^2$, of which Fig. 3 is a fragment.

Let us obliterate the triangles and grow the hexagons into the vacancies as shown. The result is the hex tessellation (see Fig. 4). These hexagons correspond one-to-one with the integer triples (x, y, z) such that $x + y + z = 0$.

Now Fig. 4 may be colored in a natural way with 3 colors, indicated by $-$, , $+$, according as $(x - y)(y - z)(z - x) \equiv x - y \equiv -1, 0, +1$ modulo 3. Let us obliterate the color 0 cells and grow the remainder into the vacancies as before. The result is the tri tessellation (see Fig. 5). A cell is colored $-$ or $+$ according to whether it points up or down. Hence tri p-minos are a subset of hex p-minos: ——

STATEMENT 1. If $p > 1$, the tri p-minos correspond to the hex p-minos on just two colors.

The six neighbors of a hex cell (x, y, z) are

(1) $\qquad (x, y+1, z-1), \quad (x-1, y, z+1), \quad (x+1, y-1, z),$

(2) $\qquad (x, y-1, z+1), \quad (x+1, y, z-1), \quad (x-1, y+1, z).$

For a color $+$ triangle, (1) are neighbors; for a color $-$ triangle, (2) are neighbors.

2. Bounding Hexagons

The *bounding hexagon* (bh) of a p-mino, hex or tri, is defined by the lines

$$x = a', \quad x = a, \quad y = b', \quad y = b, \quad z = c', \quad z = c,$$

where a' is the minimum of x over all its cells, a is the maximum, and so forth for y, b and z, c. Figure 6 shows a bh, together with the vertex coordinates and the side lengths s_i: since $x + y + z = 0$, we use * to mean "minus the sum of the other two."

For describing the shape of a bh we use these four independent intrinsic parameters: ——the *diameters*

$$A = a - a', \quad B = b - b', \quad C = c - c',$$

and the *skew*

$$K = a + b + c + a' + b' + c'$$

$$= s_1 - s_4 = \text{difference between any pair of opposite sides.}$$

The sides can be expressed in terms of these four as

(3) $\qquad s_1 = a' + b + c = \tfrac{1}{2}(K - A + B + C),$

$\qquad\qquad s_4 = a + b' + c' = \tfrac{1}{2}(-K - A + B + C), \qquad \text{etc.,}$

(4) $\qquad\qquad s_1 + s_5 + s_3 = \tfrac{1}{2}(K + A + B + C), \qquad \text{etc.}$

We now ask what values of A, B, C, and K are possible. Reflecting and rotating the bh (see Section 3) corresponds to permuting A, B and C and negating K. So we can always arrange, for example, that

(5) $\qquad\qquad 0 \leqslant K \qquad \text{and} \qquad C \leqslant B \leqslant A.$

Furthermore, $|K| \leqslant \min(C, B, A)$: for K is largest compared to C, let us say, in a triangular bh like Fig. 7, and increasing K by 1 inevitably increases C by 1 as well.

Again, since s_1, etc. are nonnegative, by (3)

$$A + K \leqslant B + C, \text{ etc.;}$$

and $s_1 + s_5 + s_3$ is an integer, so by (4) $A + B + C + K$ is even.

To sum up, for a bh of fixed orientation, it is necessary that

(6) $\qquad\qquad 0 \leqslant K \leqslant C \leqslant B \leqslant A \leqslant B + C - K$

with $A + B + C + K$ even. We claim that these conditions are also sufficient.

We now ask what further restrictions a fixed value of p entails. Tri p-minos can occupy only 2 colors, which makes the analysis difficult; so we restrict

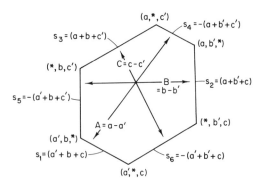

Fig. 6

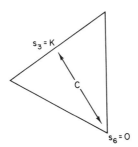

Fig. 7

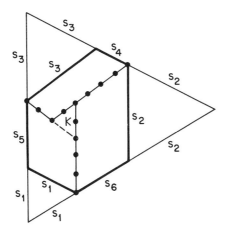

Fig. 8

ourselves to hex. For the upper bound on the bh, consider a *stretched* p-mino like Fig. 8, where $p = 14$, $K = 2$, $A = 9$, $B = 8$, and $C = 7$. We find

$$p - 1 \geqslant \min(s_1, s_4) + \min(s_2, s_5) + \min(s_3, s_6) + 2|K|$$
$$= \max(s_1, s_4) + \max(s_2, s_5) + \max(s_3, s_6) - |K|.$$

Adding,

$$2(p-1) \geqslant \sum s_i + |K| = A + B + C + |K|,$$

or

(7) $\frac{1}{2}(A+B+C+|K|) \leqslant p - 1.$

For the lower bound on the bh, we have $p \leqslant \Delta$, the number of cells inside and on the boundary. To evaluate Δ, extend 3 sides of the bh into a triangle (see Fig. 8). Then

$$\Delta = \binom{s_1+s_5+s_3}{2} - \binom{s_1+1}{2} - \binom{s_2+1}{2} - \binom{s_3+1}{2}$$

Using (3), (4), and some manipulation, eventually

(8) $p \leqslant \Delta = (AB+BC+CA-1) - \frac{1}{4}[(A+B+C-1)^2 + K^2 - 1].$

Note that (7) and (8) are necessary and sufficient for a hex p-mino to exist with bh parameters A, B, C, and K, if they satisfy (6) already.

3. Symmetries

Let G_l be the symmetry group of the hex tessellation, that is, the group of all motions of the plane leaving the tessellation invariant, and let G_t be the normal subgroup of all translations in G_l. G_t is uninteresting because no finite configuration can be invariant under a translation, so we dispatch it in defining a fixed p-mino to be a class under G_t. A free p-mino is a class under all G_l. Under $G = G_l/G_t$ a fixed p-mino may transform into itself or into another fixed p-mino corresponding to the same free p-mino: the subgroup of G leaving it invariant is called its *symmetry*. Of course, conjugate subgroups of G describe the same symmetry in different orientations: so we define a *symmetry type* to be a conjugacy class of subgroups of G, and say that a free p-mino has such-and-such a symmetry type.

In the hex case G turns out to be D_6, the dihedral group of order 12. This is shown in Table I, with descriptions of the operations and suitable coordinate transformations. Referred to the origin, the latter are very simple: for example, a rotation through $\pi/3$ is simply $\mathbf{x} = (x, y, z) \rightarrow (-y, -z, -x) = \mathbf{f}(\mathbf{x})$. We need to refer them to the bh, and so have to add a translation $\mathbf{u}$ determined by requiring a specific corner $\mathbf{x}$ of the bh to transform into another specific corner $\mathbf{y}$, then solving for $\mathbf{u}$ in

$$\mathbf{y} = \mathbf{u} + \mathbf{f}(\mathbf{x}).$$

TABLE I

The Point Group G of the Hex Tessellation

Name	Effect	Transform $(x, y, z) \rightarrow$	Bh symmetry	Bh conditions
1	None	(x, y, z)	I	—
2	$\pi/3$ rotation	$(a+b'-y, b+c'-z, c+a'-x)$	G	$A = B = C, K = 0$
3	$-\pi/3$ rotation	$(c+a'-z, a+b'-x, b+c'-y)$	G	$A = B = C, K = 0$
4	$2\pi/3$ rotation	$(a-c+z, b-a+x, c-b+y)$	DD	$A = B = C$
5	$-2\pi/3$ rotation	$(a-b+y, b-c+z, c-a+x)$	DD	$A = B = C$
6	π rotation	$(a+a'-x, b+b'-y, c+c'-z)$	$R2$	$K = 0$
7	$x = 0$ reflection	$(a+a'-x, b+c'-z, b+c'-y)$	SD	$B = C, K = 0$
8	$y = 0$ reflection	$(c+a'-z, b+b'-y, c+a'-x)$	SD	$A = C, K = 0$
9	$z = 0$ reflection	$(a+b'-y, a+b'-x, c+c'-z)$	SD	$A = B, K = 0$
10	x-axis reflection	$(x, b-c+z, c-b+y)$	D	$B = C$
11	y-axis reflection	$(a-c+z, y, c-a+x)$	D	$A = C$
12	z-axis reflector	$(a-b+y, b-a+x, z)$	D	$A = B$

TABLE II

The Symmetry Types of Hexagonal p-minos

Type	Index in G	Groups	Example
I	12	$\{1\}$	
$R2$	6	$\{1, 6\}$	
S	6	$\{1, 7\} \{1, 8\} \{1, 9\}$	
D	6	$\{1, 10\} \{1, 11\} \{1, 12\}$	
$R3$	4	$\{1, 4, 5\}$	
SD	3	$\{1, 6, 7, 10\} \{1, 6, 8, 11\} \{1, 6, 9, 12\}$	
R	2	$\{1, 2, 3, 4, 5, 6\}$	
SS	2	$\{1, 4, 5, 7, 8, 9\}$	
DD	2	$\{1, 4, 5, 10, 11, 12\}$	
G	1	$\{1 - 12\}$	

Since we have referred a transformation to the bh, this must have sufficient symmetry to remain invariant, whether or not the enclosed p-mino does. So in the final columns of Table I we give the minimum bh symmetry necessary for the transformation to be performed at all, and the corresponding numerical criterion.

Table II lists the various types of symmetry possible for p-minos, which include the types for bh's. We give a name, the set of conjugate subgroups of G, and a hex example. The index of the subgroup in G equals the number of fixed p-minos corresponding to one free p-mino of that type.

By Statement 1 in Section 1 everything above applies to tri p-minos too, except that the examples in Table II are inapplicable where they occupy one $(p = 1)$ or three colors. The apparently simple case of a single cell behaves rather oddly. Its symmetry type as a hex p-mino is G, but as a tri p-mino is only SS. The trouble is that a hex symmetry is only a tri symmetry as well if it maps the third color onto itself. This is only guaranteed provided $p \geqslant 2$, when both $+$ and $-$ must be occupied and so the third color can't map onto $+$ or $-$ under a symmetry.

4. Counting Algorithm

This is similar to that of [1]. Here we shall concentrate on the additional complications.

To start with, suppose we are to find $HX(p)$ by enumerating all fixed hex p-minos. Translations are eliminated by restricting the configurations to the triangular region $x \leqslant 0, y \leqslant 0, z \geqslant p$ (see Fig. 9) with $a = b = 0, 0 \geqslant (a,b) \geqslant -p$,

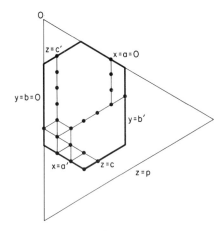

Fig. 9

and $0 \leqslant c' \leqslant c \leqslant p$. We insist on their touching the boundaries $x = 0$ and $y = 0$: that is, for the bh of a complete p-mino,

(9) $\qquad a = b = 0 \qquad$ (when $q = p$, see below).

The choice algorithm is an elementary exercise in backtracking. At level $q - 1$, $1 \leqslant q \leqslant p$, we have constructed some $(q - 1)$-mino, and during level q we attach to it in turn all cells connected to it but unused at previous levels. From now on we shall take all the parameters $a, b, \ldots$ of the bh to apply to this partial configuration as well. Now rather than simply discard completed p-minos which fail to satisfy (9), we impose a further restriction on the choice of a qth cell, known as a *growth criterion*. In this simple instance it is

(10) $\qquad a = 0 \qquad$ and $\qquad b \leqslant p - q;$

that is, cell 1 is chosen on the x axis and subsequent cells are chosen close enough to the y axis to reach it by the time $q = p$.

In practice we only enumerate all free p-minos, computing $HX(p)$ from their symmetry types. First we fix the orientation of the bh, which may in general be in any of 12 positions, by insisting that (5) be satisfied. This is effected by the new growth criteria

(11) $\qquad (B \dot{-} A) - b \leqslant p - q$

and

(12) $\qquad (B \dot{-} A) + (C \dot{-} B) \leqslant p - q$

and

(13) $\qquad (\max(B, A) \dot{-} C) + (0 \dot{-} K) \leqslant p - q.$

Here $x \dot{-} y = \max(0, x - y)$.

These are derived by noticing that, if a bh fails to satisfy (5) and (9) in more than one particular, the addition of a new cell can often only relieve one of its shortcomings at a time (see Fig. 9). If in (11) $b < 0$ and $B > A$, then attaching a new cell on the edge $y = b'$ to decrease b' will also increase B, so cannot decrease $B - A$. If in (12) $B > A$ and $C > B$, a new cell can increase $B - A$ or $C - B$ by 1 but not both. If in (13) B or $C > A$ and $K < 0$, a new cell can increase A by decreasing a', since $a = 0$, but cannot simultaneously increase s_1 (the length of side $x = a'$) or $K = s_1 - s_4$.

If the bh has no symmetry when $q = p$ (the usual case?) we are done. If not, the p-mino inside may have less symmetry than its bh, and we must define a canonical orientation inside its bh. Let the cells of the bh be enumerated in some fixed order, for example, in increasing order of x then y (see Fig. 10).

Let the *weight* of a fixed p-mino be a number, the binary expansion of

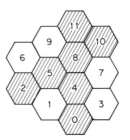

Fig. 10. A 7-mino (shaded) inside its bh.

which has digit *i* equal to 1 if cell *i* is a cell of the *p*-mino, and equal to 0 otherwise. Then a fixed *p*-mino is in canonical form, and counts as a free one, if it
has weight at least as great as all its transforms by symmetries of the bh. In
Fig. 10, bh symmetry is *DD*, 7-mino symmetry is *D*, weight of self is
101011001011, and weights of transforms are 100111101001 and 001111101010,
respectively.

For tri *p*-minos, *p* > 1, the colors of the first two cells chosen are noted and
future cells *q* > 2 chosen only from those two colors.

5. Performance, Results, and Omissions

An algorithm along the lines of Section 4 was written in ALGOL and run
background on the Chilton Atlas I for 6 hr each on tri and hex, reaching
p = 16 and 12, respectively. With machine code and 100-hr runs it should be
possible to reach *p* = 22 and 16. Time is proportional to the size of the answer,
and is exponential in *p*. The tri mode is slower per *p*-mino counted because of
the increased depth of recursion with larger *p*.

Tables III–V show our results, with the square *p*-mino totals for complete-
ness. Table III corrects Klarner [3] and Read [7], who both (!) found
$HE(6) = 83$. $HX(p)$ has been confirmed for $p \leqslant 10$ by M. F. Sykes, by hand, in
a single afternoon (!!). Table IV gives $TE(9)$ and $TE(10)$ larger by 1 and 4 than
Read [7], because he is counting only simply connected objects.

Future computations ought to produce more detailed results, breaking
the totals down by symmetry and bh as in [1], by number of free edges (of
interest to theoretical physicists) or by connectivity (numbers of internal holes
and boundary loops).

6. Asymptotic Behavior

In Tables III–V, we have included the ratios of successive fixed totals. The
conjecture is irresistible that these approach limits, and the same for the free.

TABLE III

Hex p-mino Totals, Free and Fixed

p	$HE(p)$	$HX(p)$	Ratio	p	$HE(p)$	$HX(p)$	Ratio
1	1	1	1.0000	7	333	3652	4.4865
2	1	3	3.0000	8	1448	16,689	4.5698
3	3	11	3.6667	9	6572	77,359	4.6353
4	7	44	4.0000	10	30,490	362,671	4.6882
5	22	186	4.2273	11	143,552	1,716,033	4.7317
6	82	814	4.3763	12	683,101	8,182,213	4.7681

TABLE IV

Tri p-mino Totals, Free and Fixed

p	$TE(p)$	$TX(p)$	Ratio	p	$TE(p)$	$TX(p)$	Ratio
1	1	2	1.0000	9	160	1838	2.7230
2	1	3	1.5000	10	448	5053	2.7492
3	1	6	2.0000	11	1186	14,016	2.7738
4	3	14	2.3333	12	3334	39,169	2.7946
5	4	36	2.5714	13	9235	110,194	2.8133
6	12	94	2.6111	14	26,166	311,751	2.8291
7	24	250	2.6596	15	73,983	886,160	2.8425
8	66	675	2.7000	16	211,297	2,529,260	2.8542

TABLE V

Square p-mino Totals, Free and Fixed

p	$PE(p)$	$PX(p)$	Ratio	p	$PE(p)$	$PX(p)$	Ratio
1	1	1	1.0000	10	4655	36,446	3.6777
2	1	2	2.0000	11	17,073	135,268	3.7115
3	2	6	3.0000	12	63,600	505,861	3.7397
4	5	19	3.1667	13	238,591	1,903,890	3.7637
5	12	63	3.3158	14	901,971	7,204,874	3.7843
6	35	216	3.4286	15	3,426,576	27,394,666	3.8022
7	108	760	3.5185	16	13,079,255	104,592,937	3.8180
8	369	2725	3.5855	17	50,107,911	400,795,860	3.8320
9	1285	9910	3.6367	18	192,622,052	1,540,820,542	3.8444

Several authors, notably Klarner [3], have investigated this question theoretically. All that has been proved is that, by subadditivity, there exists

(14) $$FL = \lim_{p \to \infty} (FX(p))^{1/p},$$

where F stands for P, H, or T. If the fixed ratio limit exists, then the two are equal:

$$FL = \lim_{p \to \infty} FX(p)/FX(p-1), \qquad \text{if it exists.}$$

As to the free ratio, it is shown with some difficulty in [5] that, for square p-minos, the proportion of p-minos with any symmetry (other than I) approaches zero (but rather weakly), whence PE and PX behave the same in the limit but for a factor of 8. Following a general principle, the fixed totals are a little more tractable theoretically and numerically, which is why we choose to work with them.

The known bounds on the limits FL—or *critical points*, to borrow a sonorous phrase from theoretical physics—are

$$4 < \text{HL} \leqslant 6.75,$$

(15) $$2.13 < \text{TL} \leqslant 4,$$

$$3.72 < \text{PL} \leqslant 4.65.$$

These are mainly due to Eden [4] Klarner [3, 10]; however, the upper bound on HL seems to have escaped previous notice, so here is the proof.

Let each cell have r edges, $r = 3, 4, 6$. We shall construct an encoding of fixed p-minos; suppose we are given a particular one. Starting from some fixed edge (the *root*) of some fixed cell (cell 1) of the p-mino, perform the following process on cell i for $i = 1, 2, \ldots$. Make up a string of binary digits $d_{ij} = 1$ if the jth edge from the root around the perimeter of cell i has an unnumbered cell of the p-mino adjoining, otherwise $d_{ij} = 0$. In the former case, give the cell the next unused number, and mark the common edge as its root. Omit the root edge of cell $i > 1$, since it must adjoin a numbered cell $j < i$. This procedure maps the p-minos one to one into the sequences with roughly p ones and $(r-2)p$ zeros. Hence

$$FX(p) \leqslant \binom{(r-1)p}{p} \simeq ((r-1)^{r-1}/(r-2)^{r-2})^p, \qquad \text{by Stirling's approximation,}$$

and

(16) $$FL \leqslant (r-1)^{r-1}/(r-2)^{r-2}.$$

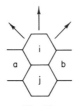

Fig. II

For hexagons we can omit not only the root edge but also the two adjacent, since the cells a, b adjoining them also adjoin the root cell j, and must already have been investigated (see Fig. 11). So now the string has length $(r - 3)p$, and instead of (16) we have

$$(17) \qquad HL \leqslant (6-3)^{6-3}/(6-4)^{6-4} = 6.75.$$

Our own investigation will be empirical. We assume that the functions can be approximated by exponentials, and try to estimate the base γ and index β in the relation

$$(18) \qquad FX(p) \sim \gamma^p p^\beta \times \text{constant}.$$

Then of course $\gamma = FL$. We hope shortly to describe our methods in a separate paper [9]. Our values are

$$(19) \qquad HL = \gamma_H \simeq 5.181 \pm 0.002,$$

$$\beta_H \simeq -0.98 \pm 0.01,$$

$$TL = \gamma_T \simeq 3.02 \pm 0.05,$$

$$\beta_T \simeq -0.8 \pm 0.5,$$

$$PL = \gamma_P \simeq 4.061 \pm 0.001,$$

$$\beta_P \simeq -0.98 > 0.02.$$

The triangular data behaves rather badly, and not just because of the small numbers involved. It is quite possible that actually $\beta = -1$ in all cases and there is an extra logarithmic factor. It is less likely that the γ are integers; γ_H is clearly not an integer.

References

1. Lunnon, W. F., Counting polynominoes, *in* "Computers in Number Theory" (A. O. L. Atkin and B. J. Birch, eds.), pp. 347–372, Academic Press, London, 1971.
2. Golomb, S. W., "Polyominoes." Scribner, New York, 1965.
3. Klarner, D. A., Cell growth problems, *Canad. J. Math.* **19**, 851–863 (1967).

4. Eden, M., A two dimensional growth process, *Proc. Berkeley Symp. Math. Statistics Probability, 4th*, 223–239 (1961).
5. Lunnon, W. F., Three combinatorial problems, Doctoral thesis. University of Manchester, 1969.
6. Coxeter, H. S. M., "Regular Polytopes," 2nd ed. Macmillan, New York, 1963.
7. Read, R. C., A census of triangular-celled animals, *Sci. Rep. UWI/CC 12*, United States AFOSR Project 1026–66 (1968). (Obtainable from University of the West Indies.)
8. Harary, F., and Read, R. C., The enumeration of tree-like polyhexes, *Proc. Edinburgh Math. Soc.* **17**, 1–13 (1970).
9. Lunnon, W. F., Asymptotic estimates of exponential functions (to appear).
10. Klarner, D. A., and Rivest, R., A procedure for improving the upper bound for the number of *n*-ominoes, *Canad. J. Math.* in press (1972).

SYMMETRY OF CUBICAL AND GENERAL POLYOMINOES

W. F. Lunnon[†]

Atlas Computer Laboratory
Chilton, Didcot
Berkshire, England

1. Hypercubic Polyominoes and Their Symmetry

We have investigated square polyominoes [1], remarking that the symmetry possessed by such an object is one of 8 types possible, which were cataloged. Here we make some observations about symmetry of more general configurations, in particular of cubical polyominoes. The group theory we shall use is mostly elementary and may be found in Ledermann [8]; the application of groups to symmetry is simply expounded by Weyl [9].

† Present address: Department of Computing in Mathematics, University College, Cardiff, Wales.

101

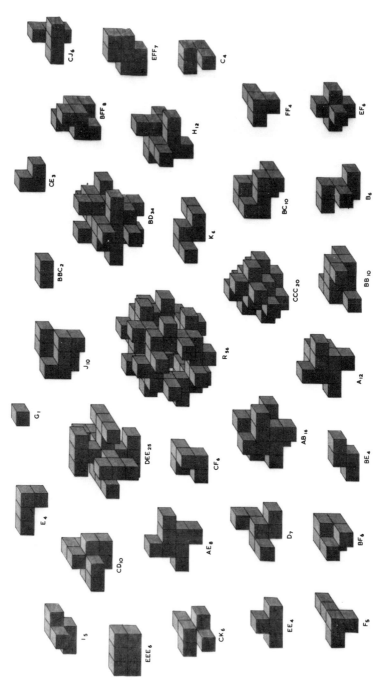

Fig. 1. Models of cubical symmetry types.

Suppose we are given a discrete tessellation in d-dimensional Euclidean space. A *Euclidean polyomino* is a connected set of e-dimensional cells of the tessellation, where two cells are *connected* if they have at least an f-dimensional cell in common, $0 \leqslant f \leqslant e \leqslant d$, all fixed constants. For example, the tessellation of cubes in ordinary space with $e = d$, $f = d-1$, and $d = 3$ yields polyominoes of cubes connected by their faces, of which some examples are shown in Fig. 1. Similarly the hypercubic tessellation for arbitrary d yields hypercubic polyominoes.

The tessellation has a symmetry group G_l and a normal subgroup G_t comprising all translations. *Fixed* polyominoes are equivalence classes under G_t; *free* polyominoes are classes under the whole G_l. Let $G = G_l/G_t$, the *special group* of the tessellation. A fixed polyomino P will be invariant under some subgroup H of G, its *symmetry*. Operating on P by an arbitrary member of G will yield another fixed polyomino P', the symmetry group H' of which is conjugate to H in G, and which corresponds to the same free polyomino as P. So the symmetry of a free polyomino is described by a set of conjugate subgroups of G; we shall prove later that every such set is the symmetry type of some polyomino.

Turning now to the hypercubic case, the group of the tessellation is what Coxeter calls $\mathbf{R}_{d+1}$ [3], and G is the group of the hypercube $\mathbf{O}_d$. To calculate the symmetry types of hypercubic polyominoes we need to know something about the latter group.

2.   The Hyperoctahedral Group $\mathbf{O}_d$

The general element of $\mathbf{O}_d$, the symmetry of a hypercube drawn in the natural orientation in Cartesian d space with its center at the origin, is a combination of axis reversals and axis permutations. For example, a rotation through $\pi/2$ about the z axis is

(1) $(x, y, z) \rightarrow (+y, -x, +z).$

Regarding the right-hand side as a signed permutation on the d symbols $x, y, z, \ldots$, we see an immediate analogy with ordinary permutations. Multiplication is analogous, the signs being multiplied on the way. For example, (1) repeated is a rotation through π about the z axis, or

(2) $(+y - x + z) \times (+y - x + z) = (-x - y + z).$

Conjugacy is also analogous [7]: two elements are conjugate and geometrically similar if their cycles correspond in pairs of the same type. Two cycles are of the same *type* when they are of the same length and the products $(\pm)$ of their signs are the same. So, for example, breaking (1) into cycles gives

$$(3) \qquad (+y - x + z) = (+y - x)(+z);$$

we say that its cycle type is $-2 + 1$, one negative 2 cycle and one positive 1 cycle. All six symmetries of this type will be found to rotate through $\pm \pi/2$ about some axis.

Every element of $\mathbf{O}_d$ leaves some subspace fixed, called its *center*: for example, the $(d-1)$-dimensional mirror of a reflection, or the $(d-2)$-dimensional axis of a rotation. The dimension c of the center is simply the number of positive cycles in the element. For each positive cycle, in the subspace it spans, leaves fixed the line $(\pm t, \pm t, \ldots)$, where $+$ changes to $-$ once for every $-$ in the cycle. Each negative cycle does the same but imposes the condition $t = -t$, which leaves only the point $(0, 0, \ldots)$. So each positive cycle adds 1 to c and each negative cycle adds 0.

Elements for which $d - c$ is odd are *improper*, that is, they interchange left and right. For it is easily seen that an improper element factors into an odd number of reflections, the primitive reflections being axis transpositions and axis reversals, and a simple manipulation shows that this number is congruent to $d - c$ modulo 2. These are the analogs of odd permutations. For example, (3) has one positive cycle. Its center is therefore a line, $c = 1$. Since $d - c = 2$ is even, it is proper—as befits a rotation.

Table I lists the 10 conjugacy classes of $\mathbf{O}_3$, showing their cycle types and geometrical effect. Notice that the "symmetry types" of Slepian [7] are conjugacy classes; we use the term to mean a set of conjugate subgroups of $\mathbf{O}_d$. To catalog the latter we wrote an uninteresting and inefficient subgroup

TABLE I

Elements of $\mathbf{O}_3$ by Conjugacy Class: 48 Operations in 10 Classes

Name	Cycle structure	Order	No. of conjugates	Description
I	$+1+1+1$	1	1	No effect
A	$+1-2$	4	6	$\pi/2$ rotation face axis
B	$+1-1-1$	2	3	π rotation face axis
C	$-1+2$	2	6	π rotation edge axis
D	$+3$	3	8	$2\pi/3$ rotation vertex axis
E	$-1+1+1$	2	3	Reflection in face plane
F	$+1+2$	2	6	Reflection in edge plane
H	-3	6	8	$D \times K$
J	$-1-2$	4	6	$A \times E$
K	$-1-1-1$	2	1	Reflection in center point

TABLE II

Subgroups of O_3 by Conjugacy Class: 98 Subgroups in 33 Classes

Name	Order	No. of conjugates	Class structure	Name	Order	No. of conjugates	Class structure
I	1	1	I	AB	8	3	I+2A+2C+3B
A	4	3	I+B+2A	AE	8	3	I+B+E+K+2A+2J
B	2	3	I+B	BFF	8	3	I+2F+2J+3B
C	2	6	I+C	CJ	8	3	I+B+2C+2E+2J
D	3	4	I+2D	EEE	8	1	I+K+3B+3E
E	2	3	I+E	EF	8	3	I+B+2A+2E+2F
F	2	6	I+F	EFF	8	3	I+B+E+K+2C+2F
H	6	4	I+K+2D+2H				
J	4	3	I+B+2J	BD	12	1	I+3B+8D
K	2	1	I+K	CF	12	4	I+K+2D+2H+3C+3F
				BBC	16	3	I+K+2A+2C+2F+2J+3B+3E
BB	4	1	I+3B	CCC	24	1	I+3B+6F+6J+8D
BC	4	3	I+B+2C	DEE	24	1	I+K+3B+3E+8D+8H
BE	4	3	I+B+E+K	R	24	1	I+3B+6A+6C+8D
BF	4	3	I+B+2F	G	48	1	I+K+3B+3E+6A+6C+6F+6J+8D+8H
CE	4	6	I+C+E+F				
CK	4	6	I+C+F+K				
EE	4	3	I+B+2E				
CD	6	4	I+2D+3C				
FF	6	4	I+2D+3F				

program, which shall not be described. Cannon surveys appropriate methods in his thesis [5], mentioning in particular the work of Neubüser (see Bülow and Neubüser [6]).

We find that O_3 has 98 subgroups falling into 33 conjugacy sets (symmetry types), including the 8 square ($d = 2$) types. Table II summarizes these subgroups. As the subgroups themselves are of no particular interest, we just show their breakdowns by conjugacy classes of their elements. For example, a type BB polyomino is invariant under the identity, class I, and the three rotations through π about an axis, class B. So the structure of type BB is shown as $I + 3B$.

3. The Existence of Models

While we have shown that any symmetry type of a free polyomino is a set of conjugate subgroups of the special group G of the tessellation, we have not shown the converse, that every such set is the type of some polyomino. For cubical polyominoes this is assured by our painful hand construction of models of each symmetry type (see Fig. 1). Below each figure is its symmetry type, defined by Table II, and the number of cubes composing it. Notice that CD_{10} and DEE_{25} possess a hidden central cell whereas BD_{34} does not. They are intended to be minimal in their number of cells, but it is quite possible that some of them can be further reduced.

For one-dimensional polyominoes the converse is false, however. There are two symmetry types, I (no symmetry) and G (reflection in the center point). Polyominoes consist of connected segments of integral length, all of which are clearly of type G.

Notice incidentally that we do not distinguish between various possible positions of the center relative to the cells, where the *center* of a polyomino with given symmetry is the intersection of the centers of its symmetries. So, for example, if a cubical is invariant under a single reflection of class E, its symmetry type is E whether the mirror lies on the faces of cubes or on the midplanes of cubes.

That the converse is true provided $d > 1$ was suggested by the model R_{56} of Fig. 1. The idea is to construct a large enough shell P with full symmetry G, then to reduce the symmetry to given $H < G$ by sticking on an asymmetric knob Q together with its images under H only. For R_{56}, P is a skeletal $4 \times 4 \times 4$ cube of edges and vertices only (32 cells), and Q is a cell projecting from its edge. Q has 24 images under the group R of all rotations of the cube, and $32 + 24 = 56$. Of course, such a model is not necessarily minimal.

THEOREM. Given a tessellation and d, e, and f, as in Section 1, with G, etc. as defined there and $d > 1$, then for any $H < G$ there is a fixed Euclidean polyomino with precisely that symmetry. This implies the free converse referred to above.

We need some facts about Euclidean tessellations, as discussed by Coxeter [3]. The full symmetry group G_t is generated by reflections in finitely many $(d-1)$-dimensional hyperplanes or *primes*, where 4 will do for the cubical case. The primes that act as mirrors for the pure reflections of G_t fall into finitely many families of equally spaced parallel primes, on which the center of any symmetry not equal to I must lie. There exist points lying simultaneously on one member of every family, 9 families for the cubical case with such points at centers and vertices of cubes. Let S be such a point. Then G is generated by reflections in the mirrors through S.

Define the *k-ball* to be the set of all cells whose distance is $\leqslant k$. The *distance* of a cell adjacent to S is defined to be zero; and of a cell adjacent to a cell of distance $k-1$, but to no nearer cell, is defined to be k. The k ball possesses a surface *k-sphere*, the cells of which are at distance k.

Let P be a k-ball. Then P has order k^d cells. Its surface has order k^{d-1}, and the mirrors through S contain order k^{d-2} of the latter. So for large enough k there exists a cell Q at distance $k+1$, adjacent to the surface, not lying on any mirror through S. Attaching this to P shifts the center of gravity away from S by a small length of order

(4) $k/(1+k^d) \simeq k^{1-d}$

in a direction away from all the mirrors through S. Since $d > 1$ we can choose k large enough to make (4) smaller than the distance between any two parallel mirrors.

The symmetry of $P \cup Q$ is I, since its center of gravity lies on no mirror. The symmetry of P on the other hand is G, since any element of G applied about center S will move one cell into another of the same distance. Let us now apply H to $P \cup Q$ with center S (or containing S if it is not a point). P remains invariant while Q traces out a system of cells $Q', Q'',...$, all distinct, and of distance $k+1$, so attached to P. Clearly $P \cup Q \cup Q' \cup ...$ has the required symmetry H.

4. Cubical Counts

Having spent so much effort in discussing cubical polyominoes, we felt constrained to sit down and enumerate them for small numbers p of cells. This is really a job for a computer, and a program using the ideas of Lunnon [1] and this paper would be straightforward. We present our hand-calculated counts in Table III, together with parallel counts of square ($d = 2$) polyominoes in Table IV for comparison, $1 \leqslant p \leqslant 6$.

TABLE III

Cubical Polyominoes

p	Free	Fixed	Real
1	1	1	1
2	1	3	1
3	2	15	2
4	7	86	8
5	23	534	29
6	112	3481	166

TABLE IV

Square Polyominoes

p	Free	Fixed	Real
1	1	1	1
2	1	2	1
3	2	6	2
4	5	19	7
5	12	63	18
6	35	216	60

Free polyominoes whose symmetry groups contain no improper elements (classes E, F, H, J, and K for the cubical case) are *enantiomorphic*: that is, if no rotations outside d space are permitted, they exist in distinct left and right forms. (Reflections in d space are rotations in $(d+1)$ space.) Under "real" we give the counts when each enantiomorphic is counted twice, as would be natural for the cubical case in ordinary space.

A trivial branching argument shows that as $p \to \infty$ there are at least 4^p p-celled cubical polyominoes. A method due to Eden [10, 11] shows that there are at most $(5^5/4^4)^p \simeq 12.21^p$. Extrapolating from our totally inadequate data, we estimate that there are very roughly 8^p.

References

1. Lunnon, W. F., Counting polyominoes, *in* "Computers in Number Theory" (A. O. L. Atkin, and B. J. Birch, eds.), pp. 347–352. Academic Press, London, 1971.
2. Golomb, S. W., "Polyominoes." Scribner, New York, 1965.
3. Coxeter, H. S. M., "Regular Polytopes," 2nd ed., Chapter 11. Macmillan, New York, 1963.
4. Coxeter, H. S. M., and Moser, W. O. J., "Generators and Relations for Discrete Groups," 2nd ed. Springer–Verlag, Berlin and New York, 1965.
5. Cannon, J. J., Computation in finite algebraic structures. Doctoral thesis, University of Sydney, 1969.
6. Bülow, R., and Neubüser, J., On some applications of group-theoretical programs to the derivation of the crystal classes of R_4, *in* "Computational Problems in Abstract Algebra" (J. Leech, ed.), pp. 131–135. Pergamon, Oxford, 1969.
7. Slepian, D., On the number of symmetry types of Boolean functions of n variables, *Canad. J. Math.* **5**, 185–193 (1953).
8. Ledermann, W., "Introduction to the Theory of Groups." Oliver & Boyd, Edinburgh, 1964.
9. Weyl, H., "Symmetry." Princeton Univ. Press, Princeton, New Jersey, 1952.
10. Lunnon, W. F., Counting hexagonal and triangular polyominoes, *in* "Graph Theory and Computing" (R. Read, ed.). Academic Press, New York, 1972. (this volume).
11. Klarner, D. A., Cell growth problems, *Canad. J. Math.* **19**, 851–863 (1967).
12. Klarner, D. A., Methods for the general cell growth problem, *in* "Combinatorial Theory and Its Applications," pp. 705–720. Balatonfüred (Hungary), 1969.

GRAPH COLORING ALGORITHMS†

David W. Matula   George Marble   Joel D. Isaacson

Department of Applied Mathematics   Department of Mathematical Studies
  and Computer Science   Southern Illinois University
Washington University   at Edwardsville
St. Louis, Missouri   Edwardsville, Illinois

1. Introduction

Considerable literature in the field of graph theory has dealt with the coloring of graphs, a fact which is quite apparent from Ore's extensive book *The Four-Color Problem* [8]. The majority of this effort has been devoted to

† This research was supported in part by the Advanced Research Projects Agency of the Department of Defense under contract SD-302 and by the National Science Foundation under contract GJ-446.

the theory of graph coloring, and relatively little study has been directed towards the design of efficient graph coloring procedures. Since numerous proofs of properties relevant to graph coloring are constructive, many coloring procedures are at least implicit in the theoretical development.

In this paper we focus attention on sequential vertex colorings, where vertices are sequentially added to the portion of the graph already colored and new colorings are determined to include each newly adjoined vertex. At each step an attempt is made to keep the total number of colors necessitated relatively small without an undue amount of computation being expended.

In Section 2 the concept of sequential colorings is formalized and certain upper bounds on the minimum number of colors needed to color a graph, the *chromatic number* $\chi(G)$, are described. It is noted that while sequentially coloring the vertices with highest degrees first appears reasonable and leads to an upper bound on $\chi(G)$, the sequential coloring determined by recursively adding vertices so that the last vertex added has minimum degree in the graph so far colored leads to a tighter bound. The notion of a bichromatic interchange is discussed, and efficient sequential coloring algorithms utilizing bichromatic interchange are formulated.

Our main result occurs in Section 3, where it is shown that the recursive-smallest-vertex-degree-last-ordering-with-interchange coloring algorithm will color any planar graph in five or fewer colors. The algorithm is evidently quite efficient even on large planar graphs.

The various algorithms have been programmed and applied to a selection of random graphs. The computed bounds on $\chi(G)$ and the number of colors used in the effected colorings are tabulated and compared in Section 4. The bounds are seen to vary considerably, with even the best bound being far from tight. Practically, the addition of the bichromatic interchange step to the sequential coloring procedure is shown to provide a significant improvement in reducing the number of colors utilized closer to the chromatic number, while still allowing for a reasonably fast computation time.

2. Sequential Vertex Colorings

A *graph* $G = (V, E)$ with vertex set V and edge set E will herein be assumed to have no loops or multiple edges. For $A \subset V$, the induced *subgraph* $\langle A \rangle = (A, E')$ of G will be the subgraph of G, where E' contains all edges of E, both end points of which are in A. Also $\langle v_1, v_2, ..., v_j \rangle$ will denote $\langle \{v_1, ..., v_j\} \rangle$. A k *coloring* of G is an assignment of colors to the vertices of G using no more than k colors and such that adjacent vertices have different colors. For an ordering $v_1, v_2, ..., v_n$ of the vertices V of G, a *sequential coloring* of G corresponding to this order is a k coloring of G utilizing each of the colors $1, 2, ..., k$ determined recursively as follows:

(1) v_1 is assigned color 1, thus 1 coloring $\langle v_1 \rangle$;
(2) if $\langle v_1, v_2, ..., v_{i-1} \rangle$ has been j colored, then $v_1, ..., v_{i-1}$ are assigned the same colors in $\langle v_1, ..., v_i \rangle$, and v_i is assigned color m, where $m \leqslant j+1$ is the minimum positive integer not occurring on adjacent vertices in $\langle v_1, ..., v_i \rangle$. Thus, $\langle v_1, ..., v_i \rangle$ is j colored for $m \leqslant j$, and $(j+1)$ colored otherwise.

A *complete graph* G has an edge for every distinct pair of vertices, and a *complete k-partite graph* $G = (V, E)$ has a vertex partition $A_1, A_2, ..., A_k$, such that each edge with an end point in A_i and A_j is in E for $i \neq j$ and is not in E for $i = j$.

THEOREM 1. Any sequential coloring of a complete k-partite graph G is a k coloring of G.

Proof: Let $G = (V, E)$ be a complete k-partite graph where all edges of E have end points in different parts of the vertex partition $A_1, A_2, ..., A_k$. For $v \in A_i$, $w \in A_j$, $i \neq j$, v and w must be colored different by any sequential coloring since v and w are adjacent in G. Suppose $v_1, v_2, ..., v_{|V(G)|}$ is any particular ordering of the vertices of G, and let $v_i, v_j \in A_p$, $i < j$. Now in $\langle v_1, v_2, ..., v_j \rangle$, v_j will have precisely the same neighbors as v_i. So the minimum color value m not occurring on its neighbors in $\langle v_1, v_2, ..., v_{j-1} \rangle$ must be the color value of v_i. Hence, v_i and v_j are assigned the same color by the sequential coloring algorithm corresponding to the order $v_1, v_2, ..., v_{|V(G)|}$. Then all vertices of each A_p, $1 \leqslant p \leqslant k$, will be assigned the same color, so G is k colored. Since the ordering of the vertices was arbitrary, the theorem is proved.

For any graph G, the smallest k such that G can be k colored is termed the *chromatic number*, $\chi(G)$, of G. In general not all sequential colorings of a graph G will yield $\chi(G)$ colorings. For the graph of Fig. 1, the sequential

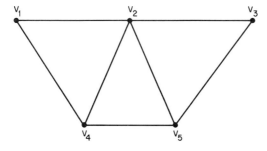

Fig. 1

coloring corresponding to v_1, v_2, v_3, v_4, v_5 utilizes 4 colors, whereas the order v_5, v_4, v_3, v_2, v_1 yields a 3 coloring. In particular some sequential coloring must yield a $\chi(G)$ coloring. To see this let A_i be the vertices colored i by a $\chi(G)$ coloring of G in the colors $1, 2, ..., \chi(G)$. Then for any ordering of the vertices $V(G)$, which has all members of A_i before any member of A_j for $1 \leqslant i < j \leqslant \chi(G)$, the corresponding sequential coloring will be a $\chi(G)$ coloring.

It is not easy to determine in general if a particular sequential coloring is a $\chi(G)$ coloring of G. No efficient general procedure for determining $\chi(G)$ is known, and even the known bounds on $\chi(G)$ are not always very sharp. Let the *degree of v* in the graph G, $\deg_G(v)$ ($\deg(v)$ when G is understood), be the number of adjacent vertices of v in G. It is evident from the sequential coloring procedure that $\chi(G) \leqslant 1 + \max_{v \in V(G)} \{\deg(v)\}$. Brooks [2] has improved upon this.

THEOREM 2 (BROOKS [2]). Let G be a connected graph with

$$\max_{v \in V(G)} \{\deg(v)\} \geqslant 3$$

where G is not a complete graph. Then

(1) $$\chi(G) \leqslant \max_{v \in V(G)} \{\deg(v)\}.$$

We shall term this inequality the *max-degree* bound on $\chi(G)$.

For graphs with only a few vertices of large degree, it is evident from the sequential coloring procedure that coloring these vertices first will generally avoid the need for as many as $\max_{v \in V(G)} \{\deg(v)\}$ colors. By ordering the vertices $v_1, v_2, ..., v_{|V(G)|}$ such that $\deg(v_i) \geqslant \deg(v_{i+1})$ for $1 \leqslant i < |V(G)|$ and considering the corresponding sequential coloring, the following bound of Welsh and Powell [10] is obtained.

THEOREM 3. Let G be a graph with $V(G) = \{v_1, v_2, ..., v_n\}$, where $\deg(v_i) \geqslant \deg(v_{i+1})$ for $i = 1, ..., n - 1$. Then

(2) $$\chi(G) \leqslant \max_{1 \leqslant i \leqslant n} \min\{i, 1 + \deg(v_i)\}.$$

Inequality (2) will be termed the *truncated-max-degree* bound on $\chi(G)$. The proof of Theorem 3 given by Welsh and Powell [10] is essentially the algorithmic proof we have sketched. Bondy [1] has given a shorter existential proof.

An ordering of the vertices of a graph G such that $\deg(v_i) \geqslant \deg(v_{i+1})$ for $1 \leqslant i < |V(G)|$ will be called a *largest-first* (LF) ordering of the vertices. Determination of a sequential coloring corresponding to such an ordering will be termed the *largest-first* algorithm (LF algorithm). The sequential coloring corresponding to a given LF ordering will effect the same coloring as described by the algorithm of Welsh and Powell [10] and will utilize no more than

$\max_i \min \{i, 1 + \deg(v_i)\}$ colors. Note that since the LF ordering of the vertices is not necessarily unique, the number of colors utilized in the coloring provided by the LF algorithm can vary depending on the particular LF ordering chosen. Application of the LF algorithm will mean its application for a particular largest-first ordering.

Cole [3] discusses a procedure for scheduling subject examinations utilizing graph-coloring techniques. If each subject corresponds to a vertex and any pair of subjects that must be taken by the same student corresponds to an edge of a graph G, then $\chi(G)$ is the minimum number of examination periods needed to avoid scheduling conflicts. Cole [3] gives the subject incompatibility data for 34 examination subjects to be offered for the First Year General Degree at Leicester University for June 1963, which is reproduced in Fig. 2 as the adjacency matrix of a graph. There are further refinements in Cole's model as some subjects require multiple papers that may need to be in sequential periods.

Cole describes an algorithm ordering the 34 subjects in an LF ordering with further subordering being determined on the basis of the multiple paper conditions. He then generates a period assignment table utilizing 14 periods. It is evident from Cole's solution that fewer periods would be needed if the multiple papers were not required, and let us consider how many periods will be needed in this case.

The graph G with the adjacency matrix of Fig. 2 must have $\chi(G) \leqslant 20$ from the max-degree bound, and $\chi(G) \leqslant 14$ is confirmed by the truncated-max-degree bound as cited by Welsh and Powell [10]. Actually the sequential coloring utilizing the ordering $v_1, v_2, \ldots, v_{34}$ yields a coloring using only nine colors. The LF ordering with subordering by label value yields an 8 coloring, where $\chi(G) = 8$ can be verified since $\langle v_7, v_8, v_9, v_{26}, v_{27}, v_{28}, v_{29}, v_{30} \rangle$ is a complete subgraph. The sequential coloring determined by this LF ordering is indicated in Fig. 2.

The previous example suggests that sequential coloring algorithms may perform considerably better than the two bounds (1) and (2) suggest. A closer inspection of the sequential coloring procedure shows that for a given ordering $v_1, v_2, \ldots, v_n$ of the vertices of a graph G, the corresponding sequential coloring algorithm could never require more than k colors where

$$(3) \qquad k = \max_{1 \leqslant i \leqslant n} \{1 + \deg_{\langle v_1, v_2, \ldots, v_i \rangle}(v_i)\}.$$

The determination of a vertex ordering minimizing k in (3) was derived earlier by Matula [5], and can be found by the following procedure:

(1) for $n = |V(G)|$, let v_n be chosen to have minimum degree in G;
(2) for $i = n - 1, n - 2, \ldots, 2, 1$, let v_i be chosen to have minimum degree in $\langle V(G) - \{v_n, v_{n-1}, \ldots, v_{i+1}\} \rangle$.

```
      1  2  3  4  5  6  7  8  9 10 11 12 13 14 15 16 17 18 19 20 21 22 23 24 25 26 27 28 29 30 31 32 33 34
  1 [ 4  1  1  1     1     1  1  1        1     1  1     1  1     1        1     1        1  1  1
  2 [ 1  1  1     1  1  1  1     1  1     1     1  1  1  1                 1           1  1
  3 [ 1  1  2     1  1  1        1  1     1     1  1  1  1                 1           1  1
  4 [ 1        8        1        1        1  1     1           1              1        1  1
  5 [ 1  1  1     4  1           1  1     1     1  1  1              1
  6 [ 1  1  1     1  3  1        1  1     1           1
  7 [ 1  1        1  4  1  1  1                 1           1        1  1  1  1  1
  8 [ 1              1  5  1              1              1        1  1  1  1  1  1  1  1  1
  9 [ 1           1  1  1  1     1           1  1     1  1        1     1  1  1  1  1  1  1  1  1  1
 10 [ 1  1  1     1  1  1        5  1     1                 1              1
 11 [    1  1     1  1           1  7     1     1  1  1              1
 12 [       1              1           4                                         1  1
 13 [ 1  1  1     1  1           1  1     6           1  1
 14 [             1                 1  1  1  1     1  1  1     1  1  1           1
 15 [ 1  1  1  1  1              1     1     1  5  1  1  1                 1  1  1        1  1  1
 16 [ 1  1  1  1  1              1     1     1  1  6  1  1                 1  1  1        1  1  1
 17 [    1  1  1                 1     1     1  1  1  3
 18 [ 1  1  1  1     1     1  1  1        1     1  1  7  1        1     1        1  1  1
 19 [ 1  1  1        1     1        1        1  1        1  5        1  1     1  1     1  1
 20 [                               1                 2  1     1  1
 21 [                               1                 1  5     1  1
 22 [             1     1                                         5              1  1
 23 [ 1  1  1     1              1           1     1  1  1  1     3  1
 24 [                   1  1        1                 1  1  1     1  4  1        1  1  1
 25 [ 1        1                    1  1  1     1              1  2  1  1
 26 [ 1  1        1  1  1           1  1        1           1        1  7  1  1  1  1
 27 [ 1  1        1  1  1           1  1        1           1        1  1  8  1  1  1
 28 [ 1           1  1  1           1           1                 1        1  1  6  1  1  1  1  1  1
 29 [ 1     1     1  1  1     1        1  1     1  1     1     1        1  1  1  1  2  1  1  1
 30 [ 1     1     1  1  1     1        1  1     1  1     1     1        1  1  1  1  3  1  1
 31 [             1  1                 1  1                                   1  1  1  4  1
 32 [             1  1                                                         1  1  1  1  8  1  1
 33 [             1  1                                                                     1  2  1
 34 [             1  1                                                                  1     1  1  3
```

Fig. 2. The adjacency matrix for the subject incompatibility data for 34 examination subjects offered for the First Year General Degree at Leicester University for June 1963 from Cole [3]. The color values determined by the largest-first algorithm corresponding to the largest-first ordering with subordering by subject number are given on the main diagonal.

For any vertex ordering $v_1, \ldots, v_n$ determined in this manner, we must have

$$(4) \qquad \deg_{\langle v_1, v_2, \ldots, v_i \rangle}(v_i) = \min_{1 \leqslant j \leqslant i} \deg_{\langle v_1, v_2, \ldots, v_i \rangle}(v_j), \qquad 1 \leqslant i \leqslant n,$$

so that such an ordering will be termed a *smallest-last* (SL) vertex ordering. The fact that any smallest-last vertex ordering minimizes k in (3) over the $n!$ possible orderings is shown by Matula [6].

Note that the determination of a smallest-last vertex ordering has a feature of recursiveness not shared by the largest-first ordering procedure. The degrees of vertices computed in determining a smallest-last ordering are over subgraphs, whereas determination of the largest-first ordering utilizes only the degrees of vertices in the whole graph. Thus, the orderings are not necessarily equivalent.

The procedure of determining a smallest-last ordering of the vertices and then determining the corresponding sequential coloring will be termed the *smallest-last* coloring algorithm (SL algorithm). It is evident from the construction that the SL algorithm will always determine a coloring requiring no more than $1 + \max_H \min_{v \in V(H)} \{\deg_H(v)| H$ a subgraph of $G\}$ colors. This provides an alternate proof of the following bound on $\chi(G)$ derived earlier and independently by Szekeres and Wilf [9] from consideration of the eigenvalues of the adjacency matrix of G.

THEOREM 4.   For any graph G,

(5) $$\chi(G) \leqslant 1 + \max_H \min_{v \in V(H)} \{\deg_H(v)| H \text{ a subgraph of } G\}.$$

We shall refer to inequality (5) as the *max-subgraph min-degree* bound on $\chi(G)$. Szekeres and Wilf [9] give both an existential and a constructive (algorithmic) proof of Theorem 4. Their algorithmic proof yields an ordering of the vertices, not necessarily an SL ordering, minimizing k in (3). Matula [6] has shown that the bound (5) can be sharpened to

(6) $$\chi(G) \leqslant 1 + \max_H \{\lambda(H)| H \text{ a subgraph of } G\},$$

where $\lambda(H) \leqslant \min_{v \in V(H)} \{\deg_H(v)\}$ is the edge connectivity of H. However, this improved bound does not pertain to sequential colorings and will not be utilized herein.

It is evident that the max-subgraph min-degree bound (5) is always sharper than or equal to the truncated-max degree bound (2). Applying the SL algorithm to the graph, G, with the adjacency matrix of Fig. 2 taken from Cole's problem, the max-subgraph min-degree upper bound on $\chi(G)$ is determined to be 10 and again an 8 coloring is achieved. This bound compares favorably with the truncated-max degree bound of 14 for this graph. Although not sharp, the max-subgraph min-degree bound appears quite superior to the max-degree and truncated-max-degree bounds for such graphs having a variety of vertex degrees.

The SL algorithm does not always effect a $\chi(G)$ coloring of G. Some instances of this will be discussed in Section 4. In attempting to improve the sequential coloring procedure previously described note that when the vertex v_i is adjoined to the $(k-1)$ colored subgraph $\langle v_1, v_2, \ldots, v_{i-1} \rangle$, a kth color is needed only if v_i is adjacent to vertices with colors $1, 2, ..., k-1$. Now if a complete graph on $k-1$ vertices exists among the neighbors of v_i, then $\chi(\langle v_1, v_2, ..., v_i \rangle) = k$ and the new color is necessary. Otherwise, it may be possible to change the colors on some neighbors of v_i so as to preserve the $(k-1)$ coloring of $\langle v_1, v_2, ..., v_{i-1} \rangle$ while leaving at most $k-2$ colors on neighbors of v_i, thus freeing a color for v_i.

Given a graph G with a k coloring in the colors $1, 2, ..., k$, let A_i be the set of vertices of G colored i. For $i \neq j$, the i, j *bichromatic subgraph* of G is the subgraph $\langle A_i \cup A_j \rangle$, and a component of $\langle A_i \cup A_j \rangle$ is an i, j *component*. If the distinct vertex colors i and j are interchanged on an i, j component of the k colored graph G, then another k coloring of G is obtained. This procedure is termed an $i \leftrightarrow j$ *interchange* on the k colored graph G. A *bichromatic interchange* on the k-colored graph G is an $i \leftrightarrow j$ interchange on G for some $i \neq j$.

A search for bichromatic interchanges at critical points in the coloring process is introduced in the following coloring algorithm.

For an ordering $v_1, v_2, ..., v_n$ of the vertices V of G, a *sequential-with-interchange* coloring of G corresponding to this ordering is a k coloring of G utilizing each of the colors $1, 2, ..., k$ determined recursively as follows:

(1) v_1 is assigned color 1, thus 1 coloring $\langle v_1 \rangle$;

(2) if $\langle v_1, v_2, ..., v_{i-1} \rangle$ has been j colored using each of the colors $1, 2, ..., j$, and if m is the minimum positive integer not occurring on vertices of $\langle v_1, v_2, ..., v_{i-1} \rangle$ adjacent to v in G, then

(a) for $m \leqslant j$, we assign each vertex of $\langle v_1, ..., v_{i-1} \rangle$ the same color in $\langle v_1, ..., v_i \rangle$, and v_i is assigned color m, thus j coloring $\langle v_1, ..., v_i \rangle$;

(b) for $m = j + 1$, let $K \subset \{1, 2, ..., j\}$ be the set of color values such that $\alpha \in K$ implies exactly one vertex adjacent to v_i in $\langle v_1, ..., v_i \rangle$ has color α in $\langle v_1, ..., v_{i-1} \rangle$. If for some $\alpha, \beta \in K$, $\alpha \neq \beta$, an α, β component of $\langle v_1, ..., v_{i-1} \rangle$ has only one vertex adjacent to v_i in $\langle v_1, ..., v_i \rangle$, then perform one α, β interchange on one such α, β component of $\langle v_1, ..., v_{i-1} \rangle$. Now color the vertices $v_1, ..., v_{i-1}$ the same as in this new coloring of $\langle v_1, ..., v_{i-1} \rangle$, and v_i with the available color, either α or β, and a j coloring of $\langle v_1, ..., v_i \rangle$ is obtained; otherwise if no such interchange is possible, color $v_1, ..., v_{i-1}$ the same as in $\langle v_1, ..., v_{i-1} \rangle$, and color v_i with color $j + 1$, thus $(j+1)$ coloring $\langle v_1, v_2, ..., v_i \rangle$.

The *largest-first-with-interchange* coloring algorithm (LFI algorithm) will refer to the sequential-with-interchange coloring algorithm applied to a vertex sequence in largest-first order. The *recursive-smallest-vertex-degree-last-with-interchange* coloring algorithm (SLI algorithm) will refer to the sequential-with-interchange coloring algorithm applied to a vertex sequence in smallest-last order. It should be noted that both of these algorithms depend on the particular LF or SL ordering used and on the particular bichromatic interchange made in Step 2b when more than one suitable interchange is available.

We shall now investigate some properties of the bounds developed and the algorithms proposed, first on planar and then on random graphs.

3. 5 Coloring Planar Graphs

A tree having at least n vertices of degree n is easily constructed for any $n \geqslant 1$. A tree is a planar graph, so the max-degree (1) and truncated-max-degree (2) bounds can give arbitrarily high bounds on $\chi(G)$ for some planar graphs G, while it is known [8, p. 84] that $\chi(G) \leqslant 5$ for all planar graphs G. We assume that planar graphs are defined to exclude loops and multiple edges. Szekeres and Wilf [9] point out that the max-subgraph min-degree bound (5) is always less than or equal to 6 for any planar graph. In this section we shall show that the recursive-smallest-vertex-degree-last-ordering-with-interchange coloring algorithm (SLI algorithm) will utilize at most five colors in coloring any planar graph.

LEMMA 1. Let H be a planar graph where $H - v$ is k colored for some $v \in V(H)$ with $\deg(v) \geqslant 4$. Then for any four vertices adjacent to v, either two of these are of the same color or there exists a bichromatic interchange on $H - v$ which will yield two similarly colored vertices among the four.

Proof: Given the planar graph H with $H - v_0$ k colored, where $\deg(v_0) \geqslant 4$, choose any four adjacent vertices of v_0. If no two of these have the same color then we may assume these four neighbors, v_1, v_2, v_3, and v_4, occur in clockwise order around v_0 and have colors 1, 2, 3, and 4, respectively (see Fig. 3). Now if v_1 and v_3 are not in the same component of the 1, 3 bichromatic subgraph, then the $1 \leftrightarrow 3$ interchange on the 1, 3 component containing v_1 will yield a coloring that verifies the lemma in this case. Otherwise, consider the 2, 4 bichromatic subgraph and assume v_2 and v_4 are in the same component. Then there is a path in H from v_2 to v_4 and a disjoint path in H from v_1 to v_3 since v_1 and v_3 are in the same 1, 3 component. From the planarity of H and the clockwise order of v_1, v_2, v_3, and v_4 about v_0 in H, it is evident that the vertex v_0 may be added to the graph $H - v_0$ along with edges $v_0 v_1$, $v_0 v_2$, $v_0 v_3$, $v_0 v_4$, $v_1 v_2$, $v_2 v_3$, $v_3 v_4$, $v_4 v_1$, and the resulting graph H' will be planar (see Fig. 3). But H' then has vertex disjoint paths between any two of the vertices $\{v_0, v_1, v_2, v_3, v_4\}$, and by Kurtowski's theorem [8, p. 22], H' can not be planar. Thus, v_2 and v_4 are not in the same 2, 4 component, so the

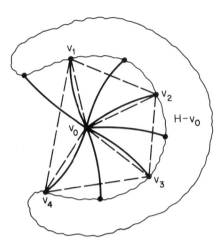

Fig. 3. The planar graph H includes $H - v_0$, the vertex v_0, and the solid lines from v_0 to vertices of $H - v_0$. The graph H' includes $H - v_0$, the vertex v_0, and the dashed lines shown. It is evident that H' is also planar.

$2 \leftrightarrow 4$ interchange on the $2, 4$ component containing v_2 is an interchange proving the lemma.

Now we show the main theorem.

THEOREM 5. The recursive-smallest-vertex-degree-last-ordering-with-interchange coloring algorithm (SLI algorithm) will color any planar graph in five or fewer colors.

Proof: Given the planar graph G, we may assume that the vertices are ordered in a smallest-last ordering so that v_i has minimum degree in $\langle v_1, v_2, \ldots, v_i \rangle$, $i = 1, 2, \ldots, |V(G)|$.

Proceeding by induction, $\langle v_1 \rangle$ is 1 colorable. Assume that $\langle v_1, \ldots, v_{i-1} \rangle$ is 5 colored by SLI for some i. Now $\langle v_1, \ldots, v_i \rangle$ is planar so $\deg_{\langle v_1, v_2, \ldots, v_i \rangle}(v_i) \leq 5$ [8, p. 51]. If fewer than five distinct colors occur on the vertices of $\langle v_1, \ldots, v_{i-1} \rangle$ adjacent to v_i in $\langle v_1, \ldots, v_i \rangle$, then clearly v_i will be colored by SLI without resorting to a sixth color. If exactly five distinct colors appear, then by Lemma 1 some bichromatic interchange in $\langle v_1, \ldots, v_{i-1} \rangle$ including a vertex adjacent to v_i in G exists which will cause fewer than five distinct colors to occur on the neighbors of v_i among $v_1, v_2, \ldots, v_{i-1}$. Since the SLI algorithm searches for all possible interchanges on the adjacent vertices of v_i in this case, an appropriate interchange will be made and v_i will be colored by SLI without recourse to a sixth color. Thus, $\langle v_1, \ldots, v_i \rangle$ is 5 colored by SLI and by the induction assumption $G = \langle v_1, v_2, \ldots, v_{|V(G)|} \rangle$ is then colored in five or fewer colors.

If a graph G contains no subgraph of minimum degree five or greater, then for a smallest-last ordering of the vertices,

$$\deg_{\langle v_1,...,v_i\rangle}(v_i) = \min_{1\leqslant j\leqslant i} \deg_{\langle v_i,v_2,...,v_i\rangle}(v_j) \leqslant 4, \qquad i = 1,2,...,|V(G)|.$$

By methods analogous to the proof of Theorem 5 we can readily prove Theorem 6.

THEOREM 6. Let G be a planar graph with no subgraph of minimum degree five. Then the SLI algorithm will color G in four or fewer colors.

4. Coloring Random Graphs

The various sequential coloring algorithms were applied to a collection of random graphs and the results are tabulated in Table I. The random graphs were computer generated utilizing pseudo random number generators starting with a fixed set of vertices and adding edges chosen uniformly from the remaining possible edges until a specified average degree was obtained.

A total of twenty random graphs were investigated. Five of these graphs had order twenty-five and average degree eight; fifteen had order one hundred, five each having average degree ten, twenty, and forty. For each graph the max-degree (1), truncated-max-degree (2), and max-subgraph min-degree (5) upper bounds on the chromatic number were determined. For each of the random graphs 1–15, the largest complete subgraph was determined, thus providing a lower bound on the chromatic number. For random graphs on 100 edges with average degree 40, such as random graphs 16–20, it can be estimated [7] that the probability that the largest complete subgraph is of order nine or greater is only a few percent.

The graphs were each colored by each of the four previously described algorithms, LF, LFI, SL, and SLI, via programs prepared in the PL/I language and executed on an IBM 360/50. The execution time in seconds for each coloring procedure is tabulated. In addition, for 10 of the 20 problems the vertices were randomly ordered in five ways and the number of colors used in the corresponding sequential vertex coloring was determined in each case.

Statistically, it is quite evident that the five graphs in each of the four blocks (1–5, 6–10, 11–15, 16–20) exhibit only a small variation in each of the parameters measured. Regarding the number of colors utilized by the LF algorithm, other results on coloring random graphs having approximately the same density and order, given by Wood [11], are in close agreement. This suggests that global properties of random graphs such as we have measured can be quite sharply determined (see also Holgate [4] and Matula [7]), despite the variability in local structure inherent in random graphs, a result akin to the situation in statistical thermodynamics.

TABLE I

Some Results From the Application of the Coloring Algorithms to Several Random Graphs

Graph number	Number of vertices	Average degree	$U_1{}^a$	$U_2{}^b$	$U_3{}^c$	L^d	Largest first Colors used	Time^e	Largest first with interchange Colors used	Time^e	Smallest last Colors used	Time^e	Smallest last with interchange Colors used	Time^e	Colors used	Colors used	Colors used	Colors used
1	25	8	14	10	7	5	6	0.44	5	1.40	5	1.38	5	1.44	7	6	6	7
2	25	8	13	10	7	5	6	0.38	5	1.40	6	1.38	5	4.12	6	6	6	6
3	25	8	14	10	7	5	6	0.38	6	1.47	6	2.00	5	6.31	7	7	7	7
4	25	8	13	10	7	4	6	0.44	5	1.45	5	1.25	5	1.44	6	6	6	5
5	25	8	12	10	7	4	6	0.44	5	0.94	5	1.31	5	2.50	6	6	5	6
6	100	10	18	14	7	4	6	2.25	6	5.90	6	3.41	6	8.50				
7	100	10	17	14	8	4	7	2.30	5	4.19	7	3.35	6	7.81				
8	100	10	16	14	8	4	6	2.26	6	4.30	6	3.31	6	7.81				
9	100	10	20	14	8	4	7	2.27	6	3.35	6	3.55	5	6.12				
10	100	10	16	14	8	4	6	2.29	6	4.90	7	3.12	6	7.75				
11	100	20	32	23	15	5	9	2.55	9	7.81	10	5.94	9	16.25	11	11	12	11
12	100	20	32	24	15	6	10	2.54	9	4.88	11	6.06	9	13.69	10	11	11	12
13	100	20	31	24	15	5	9	2.42	9	8.19	10	5.94	10	16.07	11	12	11	11
14	100	20	30	23	16	5	10	2.45	9	11.44	10	5.94	9	13.67	11	10	11	10
15	100	20	31	24	16	5	10	2.62	9	5.12	10	6.00	9	11.88	11	10	11	11
16	100	40	53	43	32		16	2.81	16	19.81	17	15.19	15	26.88				
17	100	40	51	42	32		15	2.75	15	11.31	16	15.12	14	27.81				
18	100	40	51	42	32		16	2.81	15	12.53	16	15.06	15	34.38				
19	100	40	52	42	32		17	2.88	15	15.79	18	15.31	14	29.00				
20	100	40	52	42	32		17	3.26	16	13.37	17	15.06	16	28.94				

(Column groups: **Random graphs** — Graph number, Number of vertices, Average degree. **Bounds on colorings** — Upper ($U_1{}^a$, $U_2{}^b$, $U_3{}^c$), Lower (L^d). **Colorings** — Largest first; Largest first with interchange; Smallest last; Smallest last with interchange; Sequential colorings—Random orderings, each "Colors used".)

^{a} $U_1 = \max_{v \in V(G)}\{\deg(v)\}$.

^{b} $U_2 = \max_{1 \le i \le n} \min\{i, 1 + \deg(v_i)\}$, where $\deg(v_{i-1}) \ge \deg(v_i)$ for $2 \le i \le |V(G)|$.

^{c} $U_3 = 1 + \max_H \min_{v \in V(H)}\{\deg_H(v)\}$ | H a subgraph of G.

^{d} L is the largest number of vertices in any complete subgraph of G.

^{e} Time (sec) on an IBM 360/50.

A dramatic variation in the value of the upper bounds for $\chi(G)$ is evident in Table I, with the max-subgraph min-degree bound (U_3) giving consistent significant improvements over the truncated-max-degree (U_2) and max-degree (U_1) bounds. The U_3 bound still must be considered quite poor for random graphs. Note that U_3 gives a value at least twice the actual value of $\chi(G)$ for the larger denser graphs 16–20. It is significant that the colorings obtained by the naïve sequential colorings, corresponding to the random orderings, gave a coloring in a number of colors less than or equal to the best upper bound (U_3) in every case.

Practically, for a random graph G, it appears that the simple procedure of sequentially coloring G gives a better upper bound on $\chi(G)$ than the best known upper bounds based on other graph theoretic properties of G. Despite the superior bound attendant to the smallest-last ordering as compared to the largest-first ordering, the colorings effected by the SL and LF algorithms were almost equivalent for these random graphs. In addition the number of colors needed by both the SL and LF algorithms tended to be only slightly less than the average number of colors needed for sequential colorings based on random orderings. The effect of adding the interchange step to the sequential coloring algorithms was significantly beneficial. In 60% of the graphs in Table I the interchange step gave a more efficient coloring for the largest-first ordering, and in 70% of the graphs the interchange step improved the smallest-last ordering. Thus the use of bichromatic interchange would seem to represent an advancement in the state of the art for practical graph-coloring procedures.

Timing considerations show that the SL algorithm takes considerably more time than the LF algorithm and the interchange step adds precipitously in both cases. Yet the largest times indicated for the random graphs of order 100 with average degree 40, hence 2000 edges, still are only of the order of 0.5 min. Thus, considerably larger problems should be economically viable by any of these algorithms.

The behavior of these sequential coloring algorithms in theory on planar graphs and empirically on certain random graphs is hopefully indicative of their general behavior. Yet acceptable performance of these algorithms on other classes of structured graphs that might arise in practical applications can not be predicted with any certainty. Investigations to further the understanding of the performance of these graph-coloring algorithms on other classes of graphs is to be encouraged.

Theoretically it would be helpful to know to what extent the number of colors used in the smallest-last-with-interchange algorithm could exceed the chromatic number for other classes of graphs, similar to the results established here for planar graphs. A concurrent study into sharper lower bounds on $\chi(G)$ could be beneficial in this process.

References

1. Bondy, J. A., Bounds for the chromatic number of a graph, *J. Combinatorial Theory* **7**, 96–98 (1969).
2. Brooks, R. L., On coloring the nodes of a network, *Proc. Cambridge Philos. Soc.* **37**, 194–197 (1941).
3. Cole, A. J., The preparation of examination time-tables using a small-store computer, *Comput. J.* **7**, 117–121 (1964).
4. Holgate, P., Majorants of the chromatic number of a random graph, *J. Roy. Statist. Soc. Ser. B* **31**, 303–309 (1969).
5. Matula, D. W., A min–max theorem for graphs with application to graph coloring, *SIAM Rev.* **10**, 481–482 (1968).
6. Matula, D. W., *k*-components, clusters and slicings in graphs, *SIAM J. Appl. Math.* **22**, 1972 (in press).
7. Matula, D. W., On the complete subgraphs of a random graph, *Proc. Conf. Combinatorial Math. and Its Applications*, 2nd, Chapel Hill, 356–369 (1970).
8. Ore, O., "The Four Color Problem." Academic Press, New York, 1967.
9. Szekeres, G., and Wilf, H. S., An inequality for the chromatic number of a graph, *J. Combinatorial Theory* **4**, 1–3 (1968).
10. Welsh, D. J. A., and Powell, M. B., An upper bound for the chromatic number of a graph and its application to timetabling problems, *Comput. J.* **10**, 85–86 (1967).
11. Wood, D. C., A Technique for coloring a graph applicable to large scale timetabling problems, *Comput. J.* **12**, 317–319 (1969).

ALGEBRAIC ISOMORPHISM INVARIANTS FOR GRAPHS OF AUTOMATA†

John F. Meyer

Department of Electrical and Computer Engineering
The University of Michigan
Ann Arbor, Michigan

1. Introduction

Finite automata, which are mathematical models of discrete-time finite-state systems, can be represented by a finite sequence of directed graphs called *transition graphs*. In the discussion that follows, such graphs are studied from an algebraic point of view in terms of a natural representation of the graphs by linear transformations. The representation is *natural* in the sense that its matrix equivalent coincides with the usual representation of graphs by

† This work was supported by the U.S. Air Force, Rome Air Development Center under Contract AF30(602)–3546. It is derived from a disertation submitted in partial fulfillment of the requirements for the degree of Doctor of Philosophy at The University of Michigan.

123

adjacency matrices. Under this representation, the classical invariants of linear transformation similarity become invariants of graphical isomorphism. The principal objective of the investigation is to determine the extent to which these algebraic invariants specify the structure (isomorphism class) of an arbitrary transition graph.

In the past, efforts to relate similarity invariants to graphical structure (for example, see Collatz and Sinogowitz [2] and Harary *et al.* [8]) have focused on a single invariant, namely, the characteristic polynomial of the adjacency matrix or, equivalently, the eigenvalue spectrum, if the characteristic polynomial has all its roots in the representation field. Since the characteristic polynomial does not, in general, characterize similarity, this approach is generalized in the present investigation by considering complete sets of similarity invariants, such as, all the invariant factors or all the elementary divisors of the representing transformation, or matrix.

It is shown first that such invariants still fall short of characterizing transition graphs, up to isomorphism, by exhibiting two nonisomorphic transition graphs having similar representations. The question remains, however, why this is so or, more precisely, what structural invariants correspond to a complete set of similarity invariants.

This problem is studied first for weakly connected transition graphs by examining structural invariants that suffice to determine a complete set of similarity invariants. This results in a procedure by which the invariant factors of a representing linear transformation can be determined directly from the structure of the corresponding graph. Moreover, the procedure is independent of the choice of the representation field. Multicomponent graphs are then considered. Here the problem decomposes rather naturally into a study of two extreme cases: permutation graphs and forests. Regarding permutation graphs, it is shown that the elementary divisors can be formulated in terms of the cycle structure, where these formulas depend on the characteristic of the representation field. Regarding forests, on the other hand, there is no such dependence on the nature of the field. Based on the solution obtained for connected graphs, it is shown that the elementary divisors can be formulated in terms of the depths of the points of a forest. By combining these results and solving for the graphical invariants in terms of the elementary divisors, one is able to determine, for an arbitrary transition graph, the precise extent of the structural information conveyed by these algebraic isomorphism invariants.

2. Finite Automata and Transition Graphs

Since several types of finite-state automata, alternatively referred to as sequential machines, have been distinguished in the literature (for example, see Arbib [1], Ginsburg [4], or Hartmanis and Stearns [9]), we begin by making precise the class under discussion here.

DEFINITION 1.   A *finite automaton* is a system

$$M = (I, Q, \delta)$$

where

(1) I is a finite, nonempty set of *inputs*;
(2) Q is a finite, nonempty set of *states*;
(3) δ is a partial function from $Q \times I$ into Q, where Q is the *transition function*, or *next-state function*.

Since no output function is specified, automata of this type are sometimes referred to as state machines [9]. Since δ is a partial function, if we let $D(\delta)$ denote the *domain of δ*, that is

$$D(\delta) = \{(q, a) \mid \delta(q, a) \text{ is defined}\},$$

then $D(\delta)$ is a subset of $Q \times I$ and need not include every state-input pair. Thus, interpreting $q \in Q$ as the present state of M, and $a \in I$ as the present input to M, if $(q, a) \in D(\delta)$, then $\delta(q, a)$ is the next state of M. If $(q, a) \notin D(\delta)$, then the next state is unspecified. Consequently, such automata are usually qualified as being incompletely specified, or simply incomplete [4]. As we have chosen to dispense with this terminology in the above definition, we will say instead that a finite automaton $M = (I, Q, \delta)$ is *complete*, if $D(\delta) = Q \times I$.

Let us now consider a graphical representation of finite automata, henceforth referred to simply as automata, that is conceptually the same as the usual representation of automata by transition diagrams, or state graphs, but avoids explicit labeling of the lines of the graph. We assume a familiarity with basic concepts of graph theory and, in particular, directed graphs (for example, see Harary *et al.* [6]). The graphical terminology used here will follow that of the reference just cited, unless otherwise specified. We begin, however, with a slightly more general notion of a directed graph by allowing loops, that is, by defining a *directed graph* as an ordered pair

$$G = (X, \gamma),$$

where X is a finite, nonempty set of *points* of G, and γ is a relation on X ($\gamma \subseteq X \times X$), which is the directed *lines* of G. The class of graphs of interest here is a special class of directed graphs defined as follows. The name is due to Yoeli [13].

DEFINITION 2.   A directed graph G is a *transition graph* if every point of G has outdegree 0 or 1. Thus, in relational terms, a directed graph $G = (X, \gamma)$ is

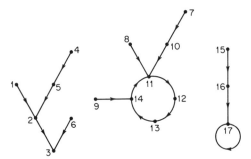

Fig. 1. A 17-point transition graph.

a transition graph, if and only if γ is a single-valued relation (partial function) on X (see Fig. 1)

Suppose now that $M = (I, Q, \delta)$ is an automaton, and for each $a \in I$, let δ_a be a relation on Q, defined as follows:

$$(1) \qquad\qquad (q, r) \in \delta_a \qquad \text{iff} \quad \delta(q, a) = r.$$

Then for all $a \in I$ it is immediate from the definitions that (Q, δ_a) is a transition graph, thereby yielding the following graphical representation of M.

DEFINITION 3. If $M = (I, Q, \delta)$ is an automaton where $I = \{a_1, a_2, ..., a_k\}$, the *graph sequence* of M is the sequence of transition graphs

$$S(M) = ((Q, \delta_{a_1}), (Q, \delta_{a_2}), ..., (Q, \delta_{a_k})).$$

The graph sequence of a 2-input, 4-state automaton is illustrated in Table I and Fig. 2.

TABLE I

An Automaton M

a q	a_1	a_2
q_0	q_2	q_3
q_1	—	q_3
q_2	q_3	—
q_3	q_0	q_2

$$S(M) = \left(\quad\right.$$

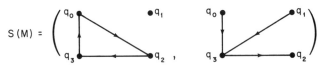

Fig. 2. The graph sequence $S(M)$ of automaton M.

To state the above representation in somewhat more general terms let

$$\mathscr{M}(I) = \{M \mid M \text{ a finite automation with input set } I\}$$

and

$$\mathscr{G} = \{G \mid G \text{ a transition graph}\}.$$

Then the above representation can be regarded as a function

$$\phi: \mathscr{M}(I) \to \mathscr{G}^{|I|}$$

where

$$\phi(M) = S(M).$$

From this viewpoint, it is easily verified that the representation is faithful in the sense that ϕ is one-to-one. Moreover, the range of ϕ is obviously the set of all sequences $S \in \mathscr{G}^{|I|}$, such that all transition graphs in S have the same set of points. Consequently, one can alternatively regard an automaton as a finite sequence of transition graphs having a common set of points. Moreover, if $M = (I, Q, \delta)$ and $M' = (I, Q', \delta')$ are *isomorphic*, as automata, that is, there is a one-to-one correspondence η between Q and Q', such that

(2) $\qquad\qquad (q, a) \in D(\delta) \qquad \text{iff} \qquad (\eta(q), a) \in D(\delta')$

and

(3) $\qquad\qquad \eta(\delta(q, a)) = \delta'(\eta(q), a), \qquad (q, a) \in D(\delta),$

then isomorphic automata can be characterized in terms of their sequence graphs. If G and G' are directed graphs, and η is a graph *isomorphism from G to G'*, then let

$$G \underset{\eta}{\simeq} G'$$

denote that G *is isomorphic to G' under η.* If there is no need to refer explicitly to an isomorphism,

$$G \simeq G'$$

will denote that G *is isomorphic to G'.*

THEOREM 1. If $M, M' \in \mathscr{M}(I)$ with $\phi(M) = (G_1, G_2, ..., G_k)$ and $\phi(M') = (G_1', G_2', ..., G_k')$, then M is isomorphic to M', if and only if there is a graph isomorphism η such that

$$G_i \underset{\eta}{\simeq} G_i', \qquad i = 1, 2, ..., k.$$

Proof: Since $\phi(M)$ is the graph sequence of M, the condition

$$G_i \simeq G_i', \qquad i = 1, 2, ..., k$$

holds, if and only if

$$(Q, \delta_a) \underset{\eta}{\simeq} (Q, \delta_a'),$$

for all $a \in I$, or equivalently, by definition of a graph isomorphism,

(4) $(q, r) \in \delta_a$ iff $(\eta(q), \eta(r)) \in \delta_a'$, $a \in I$.

If η is a one-to-one correspondence between Q and Q', it follows from the definition of δ_a, see (1), that (4) is equivalent to (2) and (3). As the latter are the defining conditions for an automaton isomorphism from M to M', the theorem is proved.

In interpreting the conditions of Theorem 1, one should be careful not to paraphrase the result by saying automata are isomorphic, if and only if corresponding transition graphs in their graph sequences are pairwise isomorphic. Being pairwise isomorphic means that

(5) $G_i \simeq G_i'$, $i = 1, 2, ..., k,$

and indeed, by Theorem 1, this is a necessary condition. On the other hand, it is not sufficient when $k > 1$. One requires, in addition, that among all the isomorphisms existing between the various isomorphic pairs, at least one of these is common to all pairs. However, Theorem 1 does have the following important consequence. By the necessity of (5), any isomorphism invariant determined for transition graphs immediately yields a set of k isomorphism invariants for k-input automata. It is this observation that motivates the following study of algebraic isomorphism invariants for transition graphs.

3. Algebraic Isomorphism Invariants

The algebraic invariants we wish to consider are induced by a natural linear representation of directed graphs. It is *natural* in the sense that it is the linear transformation equivalent of the usual representation of graphs by adjacency matrices. To be more precise, let $\mathcal{G}_n$ denote the set of all digraphs on points $N_n = \{1, 2, ..., n\}$, and let $\mathcal{T}_n(F)$ denote the set of all linear transformations on an n-dimensional vector space V over a field F. Then the *natural representation* of $\mathcal{G}_n$ relative to F and some basis $\mathcal{A}$ for V, called the *representation field* and *representation basis*, respectively, is the function

$$\rho : \mathcal{G}_n \to \mathcal{T}_n(F),$$

where, if $G = (N_n, \gamma)$, the representing transformation $T_G = \rho(G)$ is defined as follows for all $\alpha_i \in \mathcal{A}$:

(6) $$T_G(\alpha_i) = \sum_{=1}^{n} a_{ij} \alpha_j,$$

where

$$a_{ij} = \begin{cases} 1, & (i,j) \in \gamma, \\ 0, & \text{otherwise.} \end{cases}$$

Since $\mathscr{A}$ is a basis for V, it follows that T_G is uniquely specified for each $G \in \mathscr{G}_n$. Moreover, it should be obvious from the definition that the matrix of T_G with respect to $\mathscr{A}$ is simply the *adjacency matrix of G* regarded as a matrix over the representation field F. Thus, we will sometimes refer to T_G as the *adjacency transformation of G*, relative to F and $\mathscr{A}$. The reason for defining a linear representation in terms of transformations rather than matrices is that we find transformations to be more suggestive of structural interpretation. However, all the results that are obtained concerning T_G can be stated equivalently in terms of the adjacency matrix A_G.

In the terminology of the representation set $\mathscr{T}_n(F)$, an *isomorphism* is a *nonsingular* linear transformation. Two linear transformations $T, T' \in \mathscr{T}_n(F)$ are linearly isomorphic if they are *similar*, that is, there exists a nonsingular linear transformation $S \in \mathscr{T}_n(F)$ such that $T = STS^{-1}$. A fundamental observation regarding the natural representation is that it preserves isomorphism in the sense that isomorphic graphs have similar adjacency transformations. As this fact has been previously observed for adjacency matrices over the real numbers [5], we will simply state the result, without proof, in terms of the natural representation.

THEOREM 2. If G, $G' \in \mathscr{G}_n$ and G is isomorphic to G' $(G \simeq G')$, then, under the natural representation, relative to any choice of representation field F and basis $\mathscr{A}$, T_G is similar to $T_{G'}$ $(T_G \sim T_{G'})$.

Hence, by Theorem 2, every invariant of similarity for linear transformations is, under the inverse of the natural representation, an invariant of isomorphism for directed graphs. It is important to note, however, that a complete set of similarity invariants does not yield a complete set of isomorphism invariants, that is, the converse of Theorem 2 does not hold. This was first revealed by Collatz and Sinogowitz [2] for undirected graphs represented by adjacency matrices over the field of real numbers. In their tabulation of characteristic polynomials for trees, the adjacency matrices of two nonisomorphic 8-point trees are shown to have the same characteristic polynomial. Since these adjacency matrices are symmetric (because the graphs are undirected) and real, the adjacency matrices are therefore, similar.

Of interest here, of course, is whether the same is true for transition graphs. In other words, can nonisomorphic transition graphs be represented by similar linear transformations? Indeed, it would be fortunate if this were not the case for then a complete set of similarity invariants would yield a complete set of isomorphism invariants. However, transition graphs are no exception.

THEOREM 3.   There exist nonisomorphic transition graphs G and G' such that $T_G \sim T_{G'}$.

Proof: Consider the two transition graphs

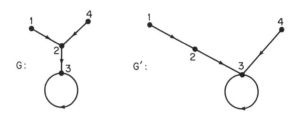

and suppose that the representation field is the reals. On computing the invariant factors of T_G and $T_{G'}$ both have the nontrivial invariant factors $x^3 - x^2$ and x and so $T_G \sim T_{G'}$. As G and G' are obviously nonisomorphic, the theorem holds.

Thus, even for this relatively restricted class of graphs, a complete set of similarity invariants for the representing linear transformations, for example, their invariant factors or elementary divisors, fails to yield a complete set of isomorphism invariants. The investigation that follows is concerned with the discovery of just why this is so, and more specifically, the extent to which this is so. The main result is a graphical characterization of the structural information conveyed by any complete set of similarity invariants.

In summarizing some terminology and results concerning the general structure of transition graphs [14], we see that a weakly *connected* transition graph G is

(1)   a *flower* if every point of G has out degree 1;
(2)   an *in tree* [7] if exactly one point of G has out degree 0.

Since a connected transition graph can have at most one point with out degree 0, it follows that every weak *component* of a transition graph is either a flower or an in-tree, subsequently referred to as a tree. Also, transition graphs are obviously *unipathic*, and, consequently, we can use the notation $[x, y]$ to denote a path from x to y. $l[x, y]$ will denote the length of path $[x, y]$. A *cycle point* of a transition graph is any point that lies in a cycle. A *tree point* is any point that is not a cycle point. The *period* of a flower G is the number of cycle points of G, that is, the length of its unique cycle. The *root* of a tree G is the unique point of G having out degree 0.

If G is a transition graph, and x is a point of G, let $C(x)$ denote the unique component of G containing x. Then the notion of height, as usually defined for trees, can be extended to transition graphs.

DEFINITION 4.   If $G = (X, \gamma)$ is a transition graph and $x \in X$, then the height $h(x)$ of x is defined as follows:

(1)   if $C(x)$ is a flower, then $h(x) = \min\{l[x, y] \,|\, y$ a cycle point of $C(x)\}$;
(2)   if $C(x)$ is a tree, then $h(x) = l[x, x_0]$, where x_0 is the root of $C(x)$.

The *height $h(G)$ of a transition graph* G is the maximum height of any point of G. Note that $h(x) = 0$, if and only if x is either a cycle point or a root.

For connected transition graphs, that is, trees and flowers, we find that the invariant factors of the representing transformations are intimately related to the heights of certain points in the corresponding graphs. This important relationship can be expressed in the form of an algorithm for computing the invariant polynomials of T_G directly from the structure of G. If $G = (X, \gamma)$ is a transition graph, let $R(G, x)$ denote the *reachable set of x*, that is

$$R(G, x) = \{y \,|\, [x, y] \text{ a path of } G\}.$$

If Y is a proper subset of X, let $G - Y$ denote the *removal of Y from G*, that is, $G - Y = G$ restricted to the set of points $X - Y$. Then, given any transition graph G, we define a sequence of subgraphs

$$G_1, G_2, \ldots, G_m$$

as follows:

(1) $G_1 = G$;
(2)   if x_i is a point of maximum height in $G_i = (X_i, \gamma_i)$ and

(7) $R(G_i, x_i) \neq X_i,$

then

$$G_{i+1} = G_i - R(G_i, x_i).$$

Otherwise the sequence terminates, that is, $G_m = G_i$.

We say that such a sequence is *derived from* G and, although a derived sequence is not necessarily unique, even up to isomorphism, we obtain an important result.

THEOREM 4.   If G is a connected transition graph, and $G_1, G_2, \ldots, G_m$ is a sequence of subgraphs derived from G, then the representing linear transformation T_G, relative to any choice of representation field F and basis $\mathscr{A}$, has m nontrivial invariant factors $\psi_i(x)$, $i = 1, 2, \ldots, m$, which can be graphically determined as follows:

(1)   if G is a flower of period r, then

$$\psi_1(x) = x^{h(G_1)+r} - x^{h(G_1)}.$$

If G is a tree, then

$$\psi_1(x) = x^{h(G_1)+1}.$$

(2) if $m > 1$, then

$$\psi_i(x) = x^{h(G_i)+1}, \qquad i = 2, 3, \ldots, m.$$

The proof of the theorem is based on the classical decomposition of a vector space V, relative to a linear transformation T on V, into cyclic subspaces $V_1, V_2, \ldots, V_m$ such that the minimum polynomial of V_i coincides with the ith nontrivial invariant factor of T (see Gantmacher [3]). The process of forming a derived sequence of subgraphs parallels this decomposition process where points of maximum height correspond to vectors which generate the various cyclic subspaces. The graph obtained on removing a maximum reachable set $R(G_i, x_i)$ corresponds to the linear transformation $\overline{T}$ induced by T on the quotient space $V/V_1 + V_2 + \cdots + V_i$. Therefore, although a somewhat lengthy proof is required to take care of all the details (see Meyer [11]), the verification is conceptually rather straightforward.

To illustrate the theorem, consider the following transition graph, which is a flower of period 2 on 12 points:

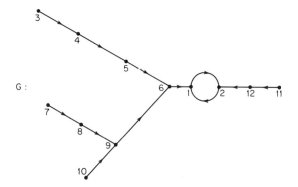

Forming a derived sequence of subgraphs we have

$$G_1 = G$$

Since $h(3) = h(G_1)$, (we could have also chosen point 7), and $R(G_{1,3} = \{1, 2, 3, 4, 5, 6\}$, we have

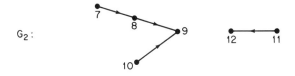

Since $h(7) = h(G_2)$

$$G_3: \quad \underset{10}{\bullet} \qquad \underset{12}{\bullet}\!\longleftarrow\!\underset{11}{\bullet}$$

Removing $R(G_{3,11})$

$$G_4: \quad \underset{10}{\bullet}$$

and as $R(G_4, 10) = \{10\}$, the process terminates. Accordingly, T_G has four nontrivial invariant factors, namely

$$\psi_1(x) = x^6 - x^4, \qquad \psi_2(x) = x^3, \qquad \psi_3(x) = x^2, \qquad \psi^4(x) = x.$$

Thus, given any connected transition graph G, that is, a flower or a tree, the invariant factors of T_G can be computed by inspection of the sequence derived from G by peeling away points reachable from a point of maximum height. This provides insight to how the structure of a connected transition graph relates to the invariant factors of T_G. Moreover, the observations made by Theorem 4 are fundamental to solving the more general problem of characterizing a complete set of similarity invariants for the class of all transition graphs.

4. Disconnected Graphs and Elementary Divisors

If G is a disconnected transition graph, then, in general, the invariant factors of the adjacency transformation T_G cannot be determined by simply applying Theorem 4 to each component of G. To illustrate, consider the transition graph G of Fig. 3. Applying Theorem 4 to each of the components

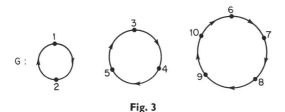

Fig. 3

$C(1)$, $C(3)$, and $C(6)$, the invariant factors so determined would be $x^2 - 1$, $x^3 - 1$, and $x^5 - 1$. However, assuming a representation field of characteristic 0, or as some prefer, ∞, direct computation of the first invariant factor of T_G, for example, shows that

$$\psi_1(x) = x^8 + 2x^7 + 2x^6 + x^5 - x^3 - 2x^2 - 2x - 1.$$

Moreover, we observe that, unlike the representation of connected transition graphs, the invariant factors can depend on the nature of the representation field. For example, if the representation field has characteristic 2, instead of 0 as assumed above, then the first invariant factor of T_G for G of Fig. 3 is

$$\psi_1(x) = x^8 + x^5 + x^3 + 1.$$

This dependency on the representation field F obviously complicates the process of relating algebraic invariants to graphical structure. For if the invariant factors depend on F, then so must their graphical determination. It is apparent, therefore, that some additional effort is required in order to appropriately generalize the results of the previous section. In particular, given some transition graph G, we want to be able to determine the values of a complete set of similarity invariants for T_G directly from the structure of G (compare with Theorem 4). Furthermore, we want to be able to do this relative to any given representation field F, where different fields, of course, may require different procedures. This is the intent of the development that follows, and we will find that such procedures can indeed be formulated.

We begin by considering a more refined description of the invariant factors usually referred to as the elementary divisors of a linear transformation, or matrix. To quickly review this well known concept, suppose that T is a linear transformation on an n-dimensional vector space over F with invariant factors

$$\psi_1(x), \psi_2(x),..., \psi_n(x).$$

Then each polynomial $\psi_i(x)$ can be factored uniquely, except for order, as a product of powers of monic, prime in $F[x]$ polynomials, that is

(8) $\psi_i(x) = \psi_1(x)^{l_{i1}} \psi_2(x)^{l_{i2}} ... \psi_m(x)^{l_{im}}, \qquad i = 1, 2,..., n.$

Here, for notational convenience, we assume that $\psi_1(x), \psi_2(x),..., \psi_m(x)$ are all the distinct prime factors of the invariant factors and, hence, some of the integers l_{ij} may be equal to 0. Each of the factors

(9) $\psi_j(x)^{l_{ij}}$

for which

$$l_{ij} > 0, \qquad 1 \leqslant i \leqslant n, \quad 1 \leqslant j \leqslant m$$

is called an *elementary divisor of T*. All such factors, including repetitions, are collectively referred to as the *elementary divisors of T*. As each invariant factor $\psi_i(x)$ divides its predecessor, $1 < i \leqslant n$, it is important to note that

$$l_{(i+1)j} \leqslant l_{ij}, \qquad 1 \leqslant i < n, \quad 1 \leqslant j \leqslant m.$$

For this reason, not only do the invariant factors uniquely determine the elementary divisors of T, but also, conversely, the elementary divisors uniquely determine the invariant factors. However, elementary divisors have an important property not generally shared by invariant factors (see Gantmacher [3]).

LEMMA 1. If $T \in \mathcal{M}_n(F)$ and $V = V_1 + V_2$, the *direct sum* of V_1 and V_2, where V_1 and V_2 are invariant subspaces of V, relative to T, then the elementary divisors of T are obtained by combining the elementary divisors T restricted to V_1 with those of T restricted to V_2.

To apply Lemma 1, we make use of what is usually meant by the direct sum of two digraphs, that is, if $G_1 = (X_1, \gamma_1)$ and $G_2 = (X_2, \gamma_2)$ are *disjoint* $(X_1 \cap X_2 = \varnothing)$, their direct *sum* is the graph

(10) $$G_1 + G_2 = (X_1 \cup X_2, \gamma_1 \cup \gamma_2).$$

Suppose now that $G \in \mathcal{G}_n$ is represented by the adjacency transformation

$$T_G : V - V$$

under the natural representation, with representation basis $\mathcal{A} = \{\alpha_1, \alpha_2, \ldots, \alpha_n\}$, and there exist subgraphs $G_1 = (X_1, \gamma_1)$ and $G_2 = (X_2, \gamma_2)$ of G, such that

$$G = G_1 + G_2.$$

Then one immediately obtains a corresponding direct sum decomposition of V. If we let

(11) $$V_1 = \langle \{\alpha_j | j \in X_1\} \rangle \quad \text{and} \quad V_2 = \langle \{\alpha_j | j \in X_2\} \rangle$$

where $\langle \rangle$ denotes *subspace spanned by*, then obviously

$$V = V_1 + V_2,$$

since $X_1 \cap X_2 = \varnothing$ and $X_1 \cup X_2 = N_n$. Moreover, we observe that each of these subspaces is invariant relative to the representing transformation T_G. In short,

(12) $$T_G(V_1) \subseteq V_1 \quad \text{and} \quad T_G(V_2) \subseteq V_2.$$

To verify this for V_1, it suffices to show that $T_G(\alpha_i) \in V_1$ for all $i \in X_1$. However, by the definition of the natural representation (6),

$$T_G(\alpha_i) = \sum_{i=1}^n a_{ij} \alpha_j,$$

where, since G is the direct sum of G_1 and G_2, $a_{ij} = 1$ implies $j \in X_1$. Thus, $T_G(\alpha_i)$ is a linear combination of the vectors $\{\alpha_j | j \in X_1\}$ which, by (11), says $T_G(\alpha_i) \in V_1$. The same argument applies to V_2, thereby proving (12).

Accordingly, if T_{G_1} and T_{G_2} denote the restrictions of T to V_1 and V_2, respectively, then Lemma 1 applied to the above decomposition proves Lemma 2.

LEMMA 2. If $G \in \mathcal{G}_n$ and $G = G_1 + G_2$, then the elementary divisors of T_{G_1} and T_{G_2}, taken in their totality, are the elementary divisors of the adjacency transformation T_G.

Generalizing Lemma 2 to an arbitrary, finite number of summands, we obtain Theorem 5.

THEOREM 5. If $G \in \mathcal{G}_n$ and $G = G_1 + G_2 + \cdots + G_s$, then the elementary divisors of T_G are obtained by combining all the elementary divisors of the transformations $T_{G_1}, T_{G_2}, \ldots, T_{G_s}$, where T_{G_i} is the restriction to T_G to V_i, that is,

$$T_{G_i} = V_i \to V_i,$$

where

$$V_i = \langle \{\alpha_j | j \in X_i\} \rangle, \qquad i = 1, 2, \ldots, s.$$

Proof: The proof is by induction on the number summands ($s \geqslant 1$), using Lemma 2 to verify the induction step.

In particular, Theorem 5 applies to a direct sum where the summands are all the components of G. Thus, if we choose elementary divisors as the similarity invariants we wish to interpret, it follows that disconnected transition graphs can be analyzed in terms of their components, thereby simplifying the analysis. This then is the point we adopt and, in the investigation that follows, we will establish that the elementary divisors of T_G both determine, and are determined by, two particular sequences of numerical invariants of G.

The first of these sequences is a tabulation of the cycle lengths (periods) of the flowers of G.

DEFINITION 5. If G is a transition graph with n points, the *period sequence* of G is the sequence $\pi(G) = (r_1, r_2, \ldots, r_n)$, where r_j equals the number of components of G that are flowers of period j, $j = 1, 2, \ldots, n$.

The second sequence is a tabulation of the depths of tree points, where, if x is a point of G, the *depth* $d(x)$ *of* x is the length of the longest directed path to x, that is

$$d(x) = \max \{l[y, x] | [y, x] \text{ a path of } G\}.$$

DEFINITION 6. If G is a transition graph with n points, the *depth sequence* of G is the sequence

$$\delta(G) = (d_0, d_1, \ldots, d_{n-1}),$$

where d_j equals the number of tree points x of G such that $d(x) = j$, $j = 0, 1, \ldots, n-1$.

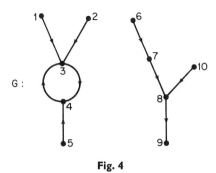

Fig. 4

To illustrate these concepts, if G is the transition graph of Fig. 4, then

$$\pi(G) = (0, 1, 0, 0, 0, 0, 0, 0, 0, 0)$$

and

$$\delta(G) = (5, 1, 1, 1, 0, 0, 0, 0, 0, 0).$$

To relate $\pi(G)$ and $\delta(G)$ to the elementary divisors of T_G, we first obtain solutions for two special cases: permutation graphs and forests. These two solutions will then be combined to obtain a general characterization.

5. Permutation Graphs

A directed graph $G = (X, \gamma)$ is a *permutation graph* if γ is a permutation on X. In terms of some other concepts defined earlier, it can easily be verified that the following statements are equivalent for any transition graph G:

(13) $\begin{cases} \text{(1)} & G \text{ is a permutation graph;} \\ \text{(2)} & \text{every component of } G \text{ is a flower of height 0, that is, a cycle;} \\ \text{(3)} & T_G \text{ is nonsingular;} \\ \text{(4)} & \text{if } \pi(G) = (r_1, r_2, \ldots, r_n), \text{ then } \sum_{j=1}^{n} j r_j = n; \\ \text{(5)} & \delta(G) = (0, 0, \ldots, 0). \end{cases}$

In particular, if G is a permutation graph, note that its period sequence $\pi(G)$ is just the cycle structure of the corresponding permutation. In using $\pi(G)$ to determine the elementary divisions of T_G, note also that our attention can be restricted to prime fields. A field is *prime* if it contains no proper subfields. This is possible, since T_G is defined in terms of the scalars 0 and 1. Consequently, if F is a representation field of characteristic k where $k = 0$ or some prime p, T_G is also over the prime subfield F_k of F. Since two linear transformations are similar over F_k, if and only if they are similar over any extension of F_k, that is, any field of characteristic k, no loss of generality will result from such a restriction.

If F_k is a prime field of characteristic k, let $\Phi_e(x)$ denote the eth cyclotomic polynomial (see Van der Waerden [12]), where e is any positive integer not divisible by k. If $k = 0$, $\Phi_e(x)$ is defined for all $e \geq 1$. $\Phi_e(x)$, by definition, is the polynomial whose roots are all the primitive eth roots of unity found in any extension of F_k. Since every eth root of unity is a primitive dth root of unity for some d such that $d \mid e$, d *divides* e, we have the well-known identity

$$(14) \qquad\qquad x^e - 1 = \prod_{d \ni d \mid e} \Phi_d(x).$$

Based on (14), if G is a permutation graph, and we consider first the case where $F = F_0 = Q$, the rational numbers, the elementary divisors of T_G can be graphically determined.

THEOREM 6a. If G is a permutation graph with period sequence $\pi(G) = (r_1, r_2, ..., r_n)$, and the natural representation is over Q, then for $1 \leq i \leq n$ the ith cyclotomic polynomial $\Phi_i(x)$ occurs exactly

$$m_i = \sum_{j \ni i \mid j} r_j$$

times as an elementary divisor of T_G. Moreover, when taken in their totality, these are all the elementary divisors of T_G.

Proof: If G is a permutation graph, G can be expressed as the sum

$$G = G_1 + G_2 + \cdots + G_s,$$

where each subgraph G_t is a component of G ($t = 1, 2, ..., s$). If we let T_{G_t} denote the restriction of T_G to V_t, then, by Theorem 5, the elementary divisors of T_G are obtained by combining all the elementary divisors of the transformations $T_{G_1}, T_{G_2}, ..., T_{G_s}$. Suppose then that G_t is a component of G. By (13), G_t is a flower of height $h(G_t) = 0$, and if G_t has period j, then, by Theorem 4,

$$(15) \qquad\qquad x^j - 1$$

is the only invariant factor of G_t. Accordingly, by (8) and (9), the elementary divisors of T_{G_t} are the primary factors in $Q[x]$ of $x^j - 1$. Hence, by (14),

$$(16) \qquad\qquad x^j - 1 = \prod_{i \ni i \mid j} \Phi_i(x).$$

Since cyclotomic polynomials are prime in $Q[x]$, we conclude that

$$(17) \qquad \Phi_i(x) = \text{an elementary divisor of } \; T_G \qquad \text{iff} \quad i \mid j(t),$$

where $j(t)$ is the period of G_t. To simplify the remainder of the argument, let $\delta_{i,j}$ denote the *Kronecker delta function*, and let

$$(18) \qquad \xi_{i,j} = \begin{cases} 1, & i \mid j, \\ 0, & \text{otherwise.} \end{cases}$$

If as in the statement of the theorem, we let m_i denote the total number of occurrences of $\Phi_i(x)$ as an elementary divisor of T_G, then, by (17) above and Theorem 5, we have

$$m_i = \sum_{t=1}^{s} \xi_{i,j(t)}, \qquad i = 1, 2, \ldots, n.$$

Since $\xi_{i,j(t)}$ can be rewritten as

$$\xi_{i,j(t)} = \sum_{j=1}^{n} \delta_{j,j(t)} \, \xi_{i,j},$$

we have

$$m_i = \sum_{t=1}^{s} \sum_{j=1}^{n} \delta_{j,j(t)} \, \xi_{i,j},$$

and changing the order of summation,

$$m_i = \sum_{j=1}^{n} \xi_{i,j} \left(\sum_{t=1}^{s} \delta_{j,j(t)} \right).$$

However,

$$\sum_{t=1}^{s} \delta_{j,j(t)}$$

is just the number of components of G having period j. Thus,

$$m_i = \sum_{j=1}^{n} \xi_{i,j} r_j, \qquad i = 1, 2, \ldots, n,$$

thereby concluding the proof of Theorem 6a.

Thus, over the field of rational numbers, the elementary divisors of T_G can be determined directly from the period sequence of a permutation graph G. That this can be done for fields of characteristic 0 is not too surprising, since others (see Marcus and Minc [10]), have shown how to determine the characteristic roots in the complex number field of a permutation matrix. The merit of the present approach, however, is that it can be extended to fields of prime characteristic.

Suppose F_k is a prime field with k equal to some prime integer p. Then, by

(15), we must be able to determine the primary factors of polynomials of the form

$$x^j - 1 \in F_p[x],$$

where j is a positive integer. As j may be a multiple of p, in which case there are no roots of order j, let $\varepsilon(j)$ denote the *exponent of the highest power of p that divides j* (if $p \nmid j$, $\varepsilon(j) = 0$), that is,

$$j = ep^{\varepsilon(j)}, \qquad \text{where} \quad p \nmid e.$$

Then letting $k = \varepsilon(j)$,

$$x^j - 1 = (x^e)^{p^k} - 1,$$

and since $(a^{p^k} - b^{p^k}) = (a - b)^{p^k}$ in any field of characteristic p, we observe the important fact that

(19) $$x^j - 1 = (x^e - 1)^{p^k}.$$

Since $p \nmid e$, the roots of $x^e - 1$ are eth roots of unity. By (14) and (19), we conclude

(20) $$x^j - 1 = \prod_{i \ni i | j | p^{\varepsilon(j)}} \Phi_i(x)^{p^{\varepsilon(j)}}.$$

Therefore, using an argument similar to the proof of Theorem 6a with (20) replacing (16), one obtains Theorem 6b.

THEOREM 6b.   If G is a permutation graph with period sequence $\pi(G) = (r_1, r_2, \ldots, r_n)$, and the natural representation is over F_p, where p is a prime, then for all i such that $1 \leqslant i \leqslant n$ and $p \nmid i$, and all k such that $0 \leqslant k \leqslant [\log_p n]$, each primary factor of the polynomial $\Phi_i(x)^{p^k}$ occurs exactly

$$m_{i,k} = \sum_{j \ni i | j} \delta_{k, \varepsilon(j)} r_j$$

times as an elementary divisor of T_G. Moreover, when taken in their totality, these are all the elementary divisors of T_G.

Thus, by Theorems 6a and 6b, the elementary divisors of T_G can be computed directly from $\pi(G)$ if G is a permutation graph. Conversely, given the elementary divisors of T_G, we find that $\pi(G)$ is uniquely determined.

THEOREM 7.   If G is a permutation graph and the natural representation is over a prime field F_k, then the elementary divisors of T_G uniquely determine the period sequence $\pi(G)$.

Proof: We will consider the case where $k = 0$, that is, the rational numbers. The proof for $k = p$ is similar. By Theorem 6a, if $\pi(G) = (r_1, r_2, \ldots, r_n)$, then

$$\sum_{j \ni i | j} r_j = m_i, \qquad i = 1, 2, \ldots, n$$

or equivalently, in terms of the divisibility function $\xi_{i,j}$, Eq. (18),

$$\sum_{j=1}^{n} \xi_{i,j} r_j = m_i, \qquad i = 1, 2, \ldots, n,$$

where m_i is the number of occurrences of the elementary divisor $\Phi_i(x)$.

To prove, conversely, that the integers $m_1, m_2, \ldots, m_n$ uniquely determine $\pi(G)$, the above system of equations can alternatively be expressed as the matrix equation

$$Dr = m,$$

where

$$D = [d_{ij}]_{n \times n}, \qquad \text{with} \quad d_{ij} = \xi_{i,j}, \qquad r = \begin{bmatrix} r_1 \\ r_2 \\ \vdots \\ r_n \end{bmatrix}, \qquad m = \begin{bmatrix} m_1 \\ m_2 \\ \vdots \\ m_n \end{bmatrix}.$$

On closer examination of the matrix D, since

$$d_{ij} = \begin{cases} 1, & i \text{ divides } j, \\ 0, & \text{otherwise,} \end{cases}$$

D is upper triangular with all its diagonal elements equal to 1 and, therefore, D is a nonsingular matrix. Consequently,

$$r = D^{-1} m$$

and, as m and r describe the elementary divisors and period sequence, respectively, we obtain the desired result.

Combining Theorems 6 and 7 with the fact that two linear transformations are similar over F_k, iff they are similar over any extension of F_k, we have Theorem 8.

THEOREM 8. If G and G' are permutation graphs, then, under the natural representation relative to any choice of representation field F and basis $\mathscr{A}$, the adjacency transformations T_G and $T_{G'}$ are similar, if and only if $\pi(G) = \pi(G')$.

Since $\pi(G)$ determines G up to isomorphism, we conclude that a complete set of similarity invariants yields a complete set of isomorphism invariants for permutation graphs.

COROLLARY 1. If G and G' are permutation graphs, then $T_G \sim T_{G'}$ iff $G \cong G'$.

Thus, the question of how similarity invariants relate to isomorphism invariants is settled for permutation graphs relative to all possible representation fields. The fact that Corollary 1 holds for fields of prime characteristic is rather surprising, since in this case it is possible for nonisomorphic permutation graphs to have adjacency transformations with the same characteristic polynomial. For example,

are both represented over F_2 by transformations with the characteristic polynomial $x^2 + 1$. Accordingly, it is only when we consider additional similarity invariants that we are able to distinguish such graphs.

6. Forests

A directed graph G is a *forest* if G has no semicycles. In terms of other concepts defined earlier, it can be verified that the following statements are equivalent for any transition graph G:

$$(21)\begin{cases} (1) & G \text{ is a forest;} \\ (2) & \text{every component of } G \text{ is an in-tree;} \\ (3) & T_G \text{ is nilpotent;} \\ (4) & \text{if } \delta(G) = (d_0, d_1, \ldots, d_{n-1}), \text{ then } \sum_{j=0}^{n-1} d_j = n; \\ (5) & \pi(G) = (0, 0, \ldots, 0). \end{cases}$$

Thus, in the context of transition graphs, we will take *forest* to mean a *forest of in trees* and, as earlier, *tree* will mean *in tree*.

Comparing the above characterizations with those given for permutation graphs (13), forests are at the opposite extreme. This applies to their representations as well, and, unlike what was observed for permutation graphs, we find that results obtained for connected graphs are more easily generalized. This is due to the fundamental fact that the elementary divisors of a nilpotent transformation coincide with its invariant factors. If T is nilpotent of index q, $\psi_1(x) = x^q$. Hence, each nontrivial invariant factor of T has the form x^k, where $1 \leqslant k \leqslant q$. As x is prime over any field, these polynomials, by definition, are the elementary divisors of T. Consequently, if G is a forest with components $G_1, G_2, \ldots, G_s$, one can apply Theorem 4 to determine the elementary divisors of T_{G_i} for each tree G_i where $i = 1, 2, \ldots, s$. One thereby determines the elementary divisors of T_G, as in Theorem 5. Moreover, using Theorem 4, we find that the elementary divisors of T_G can be related directly to the depth sequence $\delta(G)$ by Definition 6, once we establish that depth is invariant under removal of a reachable set $R(G, x)$.

LEMMA 3.   If y is a tree point of $G' = G - R(G, x)$ then the depth of y in G' is equal to the depth of y in G.

Proof: Let y be a tree point of $G' = G - R(G, x)$, and suppose the conclusion is false, that is,

$$d(y, G) \neq d(y, G'),$$

where $d(y, G)$ and $d(y, G')$ denote the depths of point y in G and G', respectively. Then it must be the case that

$$d(y, G') < d(y, G),$$

since, obviously, a tree point of G cannot have greater depth in a subgraph of G. This says, in turn, that if x_0 is a point of G such that

$$l[x_0, y] = d(y, G),$$

then the sequence of points $[x_0, y] = (x_0, x_1, \ldots, y)$ cannot be a path of $G - R(G, x)$. In other words, for some point x_k in $[x_0, y]$, $x_k \in R(G, x)$. Thus, both $[x, x_k]$ and $[x_k, y]$ are paths of G which implies $y \in R(G, x)$, and contradicts the assumption that y is a point of G'. Hence, $d(y, G) = d(y, G')$, thereby proving the lemma.

In view of this fact, we can establish an important connection between the depth sequence of a forest and the elementary divisors of its adjacency transformation.

THEOREM 9.   If G is a forest with depth sequence $\delta(G) = (d_0, d_1, \ldots, d_{n-1})$, then, under the natural representation relative to any choice of representation field F and basis $\mathscr{A}$, d_j is the number of elementary divisors of T_G having degree greater than j, where $j = 0, 1, \ldots, n-1$.

Proof: Without loss of generality, we can suppose G is connected, that is, a tree. If the theorem holds for every component of a forest H, then, by Theorem 5 and the definition of δ, it must hold for H. Suppose then that G is a tree with derived sequence (7) $G_1, G_2, \ldots, G_m$. Then

$$G_1 = G;$$

$$G_{i+1} = G_i - R(G_i, x_i), \qquad i = 1, 2, \ldots, m - 1,$$

where x_i is a point of maximum height in G_i. By Theorem 4.

$$\psi_i(x) = x^{h(G_i) + 1}$$

is the ith nontrivial invariant factor or, since T_G is nilpotent, the ith elementary divisor of T_G. Thus, the degree of $\psi_i(x)$ is greater than j, if and only if $h(G_i) \geq j$. To prove the theorem, therefore, it suffices to prove

$$(22) \qquad d_j = |\{i \mid h(G_i) \geq j\}|, \qquad j = 0, 1, \ldots, n - 1.$$

To verify the above, consider now the collection of reachable sets

(23) $\{R(G_1, x_1), R(G_2, x_2),..., R(G_m, x_m)\},$

where the first $m-1$ sets are as above and the point x_m has maximum height in G_m, that is, $R(G_m, x_m)$ is the set of points of G_m (otherwise the removal process would not have terminated). Since G_i is a forest, the component $C(x_i)$ is a tree, and if y_i denotes to root of $C(x_i)$, then $R(G_i, x_i)$ is just the set of points of the path

$$[x_i, y_i] = (z_0, z_1,..., z_l),$$

where

$$z_0 = x_i, \quad z_l = y_i \quad \text{and} \quad l = h(x_i) = h(G_i).$$

Since the depth $d(z_j, G_i)$ of z_j in G_i is the length of the longest path of G_i to z_j, and since $z_0 = x_i$ is a point of maximum height in G_i, we have

$$d(z_j, G_i) = l[z_0, z_j] = j$$

for $j = 0, 1,..., h(G_i)$. Moreover, by Lemma 3, z_j also has depth j in G_{i-1}, if $i > 1$, and, by repeated applications of this fact, z_j has depth j in $G_1 = G$.

The result we seek is now immediate for, by its definition, the collection (23) of reachable sets is obviously a partition of the points of G. Hence, for each integer j, $0 \leqslant j < n$, the number of points of G of depth j is just the number of sets $R(G_i, x_i)$ such that $h(G_i) \geqslant j$. In short

$$d_j = |\{i | h(G_i) \geqslant j\}|,$$

which proves (22) and, therefore, concludes the proof of Theorem 9.

Thus, in case G is a forest, the elementary divisors of T_G uniquely determine the depth sequence $\delta(G)$ of G. The relationship just established says even more, for, turning it around, we find that the depth sequence of a forest is enough to uniquely determine the elementary divisors of its adjacency transformation. This important consequence can be precisely stated.

THEOREM 10.   If G is a forest with depth sequence

$$\delta(G) = (d_0, d_1,..., d_{n-1}),$$

then, under the natural representation relative to any choice of representation field F and basis $\mathscr{A}$, the polynomial

$$x^i, \quad i = 1, 2,..., n$$

occurs exactly

$$m_i = d_{i-1} - d_i, \quad d_n \equiv 0,$$

times as an elementary divisor of T_G. Moreover, when taken in their totality, these are all the elementary divisors of T_G.

Proof: Suppose G is a forest. Then T_G is nilpotent, and any elementary divisor of T_G is of the form x^k, $0 < k \leqslant n$. In particular, if we let m_i denote the number of occurrences of x^i as an elementary divisor, then m_i is just the number of elementary divisors having degree equal to i. By Theorem 9, therefore,

$$\sum_{i > j} m_i = d_j$$

or, equivalently

$$\sum_{i=j+1}^{n} m_i = d_j, \qquad j = 0, 1, \dots, n - 1.$$

Solving these equations for m_i, if $i = n$, then, by the last equation,

$$m_n = d_{n-1}.$$

If $1 \leqslant i < n$, then

$$m_i = m_i + \sum_{j=i+1}^{n} m_j - \sum_{j=i+1}^{n} m_j$$

$$= \sum_{j=i}^{n} m_j - \sum_{j=i+1}^{n} m_j.$$

Thus, $m_i = d_{i-1} - d_i$, thereby proving the theorem.

As elementary divisors are a complete set of similarity invariants, by combining Theorems 9 and 10 we have proved Theorem 11.

THEOREM 11. If G and G' are forests, then, under the natural representation relative to any choice of representation field F and basis $\mathscr{A}$, the adjacency transformations T_G and $T_{G'}$ are similar, if and only if $\delta(G) = \delta(G')$.

Thus, the question of how similarity invariants relate to isomorphism invariants is settled for forests as well as permutation graphs (compare with Theorem 8). We now show how these two special cases can be combined to obtain a general solution.

7. Arbitrary Transition Graphs

The ability to graphically characterize similarity under the natural representation in case G is a permutation graph (see Theorem 8) or a forest (see Theorem 11) is sufficiently general, as we now show. If G is an arbitrary transition graph, then there exists a suitably derived graph $\bar{G}$ consisting of a permutation graph, and/or a forest, such that T_G is similar to $T_{\bar{G}}$. More precisely, if $G = (X, \gamma)$ is a transition graph, let X_F and X_P denote its cycle points and tree points respectively. That is,

$$X_P = \{x \mid x \text{ a cycle point of } G\}$$

and

$$X_F = \{x \mid x \text{ a tree point of } G\},$$

where we observe that $X_P \cap X_F = \varnothing$ and $X_P \cup X_F = X$. By removing the lines of G that are from a tree point to a cycle point, we obtain a special subgraph of G.

DEFINITION 7. The reduction $\bar{G}$ of a transition graph $G = (X, \gamma)$ is the transition graph $\bar{G} = (X, \bar{\gamma})$, where $\bar{\gamma} = \gamma - \{(x, y) \mid x \in X_F \text{ and } y \in X_P\}$.

To obtain an alternative and, perhaps, more vivid description of the reduction, if $G = (X, \gamma)$ and $X_P \neq \varnothing$, let G_P denote the restriction of G to X_P, that is,

$$G_P = (X_P, \gamma_P),$$

where $\gamma_P = (X_P \times X_P) \cap \gamma$. If $X_F \neq \varnothing$, let

$$G_F = (X_F, \gamma_F),$$

where γ_F is similarly defined. Thus, if defined, G_P is a permutation graph and G_F is a forest. We note also that $G_P(G_F)$ is just the removal of $X_F(X_P)$ from G, that is,

(24) $G_P = G - X_F$ and $G_F = G - X_P$.

From the definitions, it is immediate that the reduction of G can be alternatively described as follows:

$$\bar{G} = \begin{cases} G_P, & G \text{ is a permutation graph,} \\ G_F, & G \text{ is a forest,} \\ G_P + G_F, & \text{otherwise.} \end{cases}$$

The last case, of course, is the interesting one where, if G contains both tree points and cycle points, its reduction can be expressed as the sum of a permutation graph and a forest. To illustrate, if G is the transition graph G, (a) of Fig. 5, then $X_P = \{1, 2, \ldots, 6\}$, $X_F = \{7, 8, \ldots, 18\}$, and $\bar{G}$ is obtained from G be removing lines

$$\{(7, 1), (8, 2), (9, 2), (10, 5), (11, 5)\}.$$

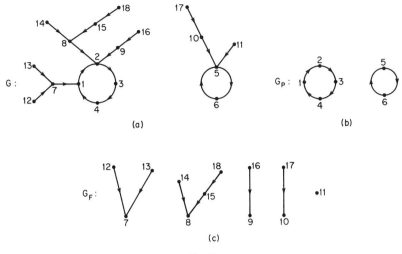

Fig. 5

Restricting G to X_P, we have the permutation graph G_P, (b) of Fig. 5, and restricting G to X_F, the forest G_F, (c) of Fig. 5.

The justification of this reduction is complete once we observe the basic property of Lemma 4.

LEMMA 4. If G is a transition graph and $\bar{G}$ is the reduction of G, then, under the natural representation relative to any choice of representation field F and basis $\mathscr{A}$, T_G is similar to $T_{\bar{G}}$.

Proof: It suffices to prove the lemma for connected graphs, since G can always be expressed as a sum $G = G_1 + G_2 + \cdots + G_s$ of components where, obviously, $\bar{G} = \bar{G}_1 + \bar{G}_2 + \cdots + \bar{G}_s$. Consequently, if the lemma holds for connected graphs, we have

$$T_{G_i} \sim T_{\bar{G}_i}, \qquad i = 1, 2, \ldots, s.$$

Applying Theorem 5, we conclude that $T_G \sim T_{\bar{G}}$.

Suppose, therefore, that G is a connected transition graph with reduction $\bar{G}$. If G is a cycle or a tree we are through for in either case $\bar{G} = G$. We can suppose further, therefore, that G is a flower of period r and height $h > 0$. To determine the elementary divisors of T_G, let $G_1, G_2, \ldots, G_m$ be a sequence derived from G (7). Then, by Theorem 4, the first invariant factor of T_G is

$$\psi_1(x) = x^{h+r} - x^h = x^h(x^r - 1).$$

If x_1 is the tree point chosen in deriving G_2, assuming $m > 1$, that is,

(25) $h(x_1, G) = h$

and

(26) $G_2 = G - R(G, x_1),$

then the remaining invariant factors of T_G are all the invariant factors of T_{G_2}. Accordingly, the elementary divisors of T_G are x^h, the primary factors of $x^r - 1$, and, if $m > 1$, the elementary divisors of T_{G_2}.

If we now consider the reduction $\bar{G} = G_P + G_F$, the elementary divisors of $T_{\bar{G}}$ are the elementary divisors of T_{G_P} along with those of T_{G_F} (see Theorem 5). Since G_P is a cycle of length r, the elementary divisors of T_{G_P} are just the primary factors of $x^r - 1$. As for G_F, we note first that the height $h(x, G_F)$ of a point x in G_F is exactly one less than its height in G. Suppose that $[x, z]$ is the shortest path in G to a cycle point z, and $(y, z) \in \gamma$, where y is a tree point, then y precedes z in the path $[x, z]$ and on removal of z and (y, z) in forming $G_F = G - X_P$, y has out degree 0 in G_F. Thus, for all $x \in X_F$,

$$h(x, G_F) = l[x, y] = l[x, z] - 1$$
$$= h(x, G) - 1.$$

In particular, for the point x_1 (25), we have

$$h(x_1, G_F) = h - 1,$$

and if we begin determination of the elementary divisors of T_{G_F} with the tree $C(x_1)$, the first invariant factor of this component is

$$\psi_1(x) = x^{(h-1)+1} = x^h.$$

Hence, x^h is an elementary divisor of T_{G_F}. The remaining elementary divisors, if $m > 1$, are determined by $C(x_1) - R(C(x_1), x_1)$ and the remaining components of G_F, that is, the forest

$$G_F - R(C(x_1), x_1) = G_F - R(G_F, x_1).$$

However, by the choice of x_1 and the definition of $\bar{G}$, this is just the second graph G_2 (26) in the sequence derived from G, that is,

$$G_F - R(G_F, x_1) = G - R(G, x_1) = G_2.$$

In summary, therefore, the elementary divisors of $T_{\bar{G}}$ are the primary factors of $x^r - 1$, x^h, and the elementary divisors of T_G. As these coincide with the elementary divisors of T_G, we conclude that $T_G \sim T_{\bar{G}}$, thereby proving the lemma.

Combining this important observation with the characterizations of similarity obtained earlier for permutation graphs and forests, we are now prepared to establish the main result of the investigation.

THEOREM 12. If G and G' are transition graphs, then, under the natural representation relative to any choice of representation field F and basis $\mathscr{A}$, the adjacency transformations T_G and $T_{G'}$ are similar, if and only if $\pi(G) = \pi(G')$ and $\delta(G) = \delta(G')$.

Proof: By Lemma 4, $T_G \sim T_{G'}$, if and only if $T_{\bar{G}} \sim T_{\bar{G}'}$. Also, by definition of the reduction, it should be obvious that period sequences and depth sequences are preserved, that is, $\pi(G) = \pi(\bar{G})$ and $\delta(G) = \delta(\bar{G})$, for any transition graph G. Thus, it suffices to prove

(27) $T_{\bar{G}} \sim T_{\bar{G}'}$,   iff $\pi(\bar{G}) = \pi(\bar{G}')$ and $\delta(\bar{G}) = \delta(\bar{G}')$.

Suppose then that $T_{\bar{G}} \sim T_{\bar{G}}$ where $\bar{G} = G_P + G_F$ and $\bar{G}' = G_P' + G_F'$. Since T_{G_P} and $T_{G_{P'}}$ are both nonsingular, and T_{G_F} and $T_{G_{F'}}$ are both nilpotent, the elementary divisors of T_{G_P} must be those of $T_{G_{P'}}$. Similarly, the elementary divisors of T_{G_F} must be those of $T_{G_{F'}}$, that is

$$T_{G_P} \sim T_{G_{P'}}   \text{and}   T_{G_F} \sim T_{G_{P'}}.$$

Conversely, the above implies $T_{\bar{G}} \sim T_{\bar{G}'}$ and so

(28) $T_{\bar{G}} \sim T_{\bar{G}'}$ iff $T_{G_F} \sim T_{G_{F'}}$ and $T_{G_P} \sim T_{G_{P'}}$

If we now apply the characterizations obtained earlier for permutation graphs and forests, by Theorem 8,

(29) $T_{G_P} \sim T_{G_{P'}}$ iff $\pi(G_P) = \pi(G_P')$

and, by Theorem 11,

(30) $T_{G_F} \sim T_{G_{F'}}$ iff $\delta(G_F) = \delta(G_F')$

Moreover, for any reduction $\bar{G} = G_P + G_F$ with c cycle points and t tree points, it follows from the definitions of π and δ that

$$\pi(\bar{G}) = \big(\pi(G_P), \underbrace{0, 0, ..., 0}_{t}\big).$$

and

$$\delta(\bar{G}) = \big(\delta(G_F), \underbrace{0, 0, ..., 0}_{c}\big).$$

Thus,

(31) $\pi(G_P) = \pi(G_P')$ iff $\pi(\bar{G}) = \pi(\bar{G}')$

and

(32) $\delta(G_F) = \delta(G_F')$ iff $\delta(\bar{G}) = \delta(\bar{G}')$.

By linking (28)–(32), we establish (27) and thereby prove the theorem.

Thus, for the class of transition graphs, we have obtained a complete characterization of what similarity invariants of adjacency transformations or, equivalently adjacency matrices, have to say about graphical structure. That is, two transition graphs on n points will have similar adjacency transformations, if and only if they agree both in the number of components that are flowers of period j, $1 \leqslant j \leqslant n$ and in the number of points that are tree points of depth k, $0 \leqslant k \leqslant n-1$. Moreover, given the elementary divisors of T_G, those of T_{G_F}, that is, those of the form x^i, can be used to compute $\delta(G_F)$, by Theorem 9, thereby determining $\delta(G) = (\delta(G_F), 0, 0, ..., 0)$. Those of T_{G_P}, that is, those which remain, can be used to compute $\pi(G_P)$, as in proof of Theorem 7, thereby determining $\pi(G) = (\pi(G_P), 0, 0, ..., 0)$. Conversely, given $\delta(G)$ and $\pi(G)$, the elementary divisors of T_{G_F} can be computed from $\delta(G_F)$, by Theorem 10, and the elementary divisors of T_{G_P} from $\pi(G_P)$, by Theorem 8, thereby determining all the elementary divisors of T_G.

Given this characterization of the graphical information that is conveyed by a complete set of similarity invariants, one can now examine the reasons for incompleteness (see Theorem 3) that is, the information that is not conveyed by these invariants. Considering first the connected case, if G is a tree, then the depth sequence, in general, fails to provide information as to where maximal paths of G, that is, paths from points of depth 0 to the root of G, intersect. Thus, for example, although the trees

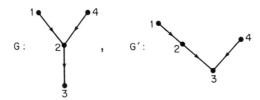

are represented by similar linear transformations, G and G' are obviously not isomorphic. If we consider next the case where G is a flower such that $h(G) > 0$, then, relative to the tree points of G, we have the same kind of information loss noted above. In addition, since G and its reduction $\bar{G}$ are similarly represented, we lose information as to where tree points of height 1 attach to the cycle. Thus, for example, the flowers

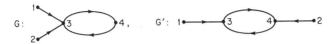

are similarly represented and yet nonisomorphic. Finally, if we consider graphs that are not necessarily connected, then, in addition to what we have just observed, $\pi(G)$ and $\delta(G)$ fail to provide any information as to which tree points belong to components that are flowers and which belong to components that are trees, assuming of course that G is neither a transition graph or a forest. Consequently, for example, each of the transition graphs

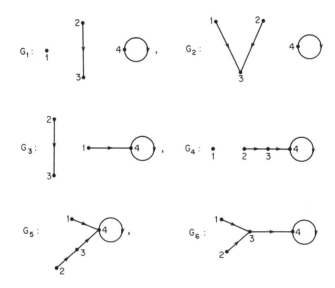

has period sequence $(1, 0, 0, 0)$ and depth sequence $(2, 1, 0, 0)$ yet no one graph is isomorphic to any of the others.

In conclusion, we remark that all that has been said in this investigation of transition graphs applies dually to the directionally dual class of graphs

$$\mathcal{H} = \left\{ G \,\middle|\, \begin{array}{l} G \text{ a digraph such that} \\ \text{every point of } G \text{ has in degree 0 or 1} \end{array} \right\},$$

for if G^d is the dual (converse) of G, then T_{G^d} is the transpose T_G^T of T_G. Thus, $T_G \sim T_{G^d}$ and, accordingly, the theorems established here for transition graphs apply dually, for example, with *out tree* replacing *in tree*, *height sequence* replacing *depth sequence*, etc., to the class $\mathcal{H}$.

References

1. Arbib, M. A., "Theories of Abstract Automata." Prentice Hall, Englewood Cliffs, New Jersey, 1969.
2. Collatz, L., and Sinogowitz, U., Spektren Endlicher Grafen, *Abh. Math. Sem. Univ. Hamburg* **21**, 63–77 (1957).
3. Gantmacher, F. R., "Matrix Theory," Vol. 1. Chelsea, Bronx, New York, 1959.
4. Ginsburg, S., "An Introduction to Mathematical Machine Theory." Addison-Wesley, Reading, Massachusetts, 1962.
5. Harary, F., The determinant of the adjacency matrix of a graph, *SIAM Rev.* 202–210 (1962).
6. Harary, F., Norman, R. Z., and Cartwright, D., "Structural Models: An Introduction to the Theory of Directed Graphs." Wiley, New York, 1965.
7. Harary, F., "Graph Theory." Addison-Wesley, Reading, Massachusetts, 1969.
8. Harary, F., King, C., and Read, R. C., Cospectral graphs and diagraphs, unpublished manuscript.
9. Hartmanis, J., and Stearns, R. E., "Algebraic Structure Theory of Sequential Machines." Prentice-Hall, Englewood Cliffs, New Jersey, 1966.
10. Marcus, M., and Minc, H., "A Survey of Matrix Theory and Matrix Inequalities." Allyn & Bacon, Rockleigh, New Jersey, 1964.
11. Meyer, J. F., Algebraic isomorphism invariants for transition graphs, *Rome Air Development Center Rep. RADC-TR-68-351* (1968).
12. Van der Waerden, B. L., "Modern Algebra," Vol. I. Ungar, New York, 1949. (Translated from the second revised German edition, Springer-Verlag, Berlin and New York, 1940.)
13. Yoeli, M., and Ginzburg, A., On homomorphic images of transition graphs, *J. Franklin Inst.* **278**, 291–296 (1964).
14. Yoeli, M., and Ablow, C. M., Representations and products of transition graphs, *Air Force Cambridge Research Laboratories Rep. AFCRL-65-400* (1965).

THE CODING OF VARIOUS KINDS OF
UNLABELED TREES

Ronald C. Read

Department of Combinatorics and Optimization
University of Waterloo
Waterloo, Ontario
Canada

1. Introduction: Coding in General

Let $\mathscr{G}$ be a collection of graphs of some specified kind, and let X be a specified set of objects. A *coding procedure* is a mapping $c: \mathscr{G} \to X$ such that two graphs in $\mathscr{G}$ map onto the same element of X, if and only if they are

153

isomorphic. In practice the set X is usually the set of all strings of symbols of some kind. The definition of *isomorphic* will depend on the kind of graphs under discussion.

The image under c of a graph G in $\mathscr{G}$ will be called the *code* of G. For emphasis, we may refer to an element of X as a *valid code* if it belongs to the image of c. Clearly there is a one-to-one correspondence between the isomorphism classes in $\mathscr{G}$ and the set of valid codes. Any process for obtaining from a valid code x a graph whose code is x, that is, a representative of the corresponding isomorphism class, will be called a *decoding procedure*.

It is worth remarking, in passing, that if our set of symbols has finite cardinality m, we can interpret a string of these symbols as an integer in the scale of m, using the symbols as the digits $0, 1, ..., m - 1$. Thus, for theoretical purposes, we can, without any loss of generality, take X to be the set of positive integers. We shall see later that, under certain circumstances, this is also a practical procedure.

Isomorphic graphs have the same code, and conversely. Hence, the *coding problem*—the problem of devising a coding procedure for a given set $\mathscr{G}$—is effectively the same as the *isomorphism problem*, which is the problem of devising an algorithm to test whether two graphs in $\mathscr{G}$ are isomorphic or not. Both problems are notoriously intractable when $\mathscr{G}$ is the set of all unlabeled graphs. We shall consider here only the somewhat easier problems that arise when $\mathscr{G}$ is taken to be sets of trees of various kinds. Before we do this, however, it will be worth while to look at the two problems just mentioned, while still in the context of graphs in general.

If we have a one-shot problem of determining whether two graphs are isomorphic or not, then a computer program that tests this directly, that is, whose input is the pair of graphs and whose output is "yes" or "no," will be quite satisfactory. The alternative of using a coding program, whose input is a single graph and whose output is the code, on each of the graphs and then comparing the two codes, might well be roundabout. If the task in hand is to search a list of graphs to find if a particular graph is present in the list or not, then, by coding the graphs, we can replace, let us say, N applications of the isomorphism program by one application of the coding program and N comparisons between the codes in the list and the code of the given graph. If the isomorphism and coding programs are of comparable complexity, as they probably would be, then this will result in a great saving in time.

As an illustration of this point, consider the problem of constructing all trees on a given number p of nodes. A method that produces no duplicates has been described [13], but is complicated. It is more straightforward to derive these trees from those on $p - 1$ nodes, which we shall suppose we have constructed already. To each of the trees on $p - 1$ nodes we add, in every possible way, an extra edge, one node of which is already in the tree, while the other is a new node of valency 1. A node of valency 1 in a tree will be called an *end node*. It is clear that we shall get all trees on p nodes in this way, but that they will be produced many times over. Thus, each time we produce a tree we must look to see whether it is one that has occurred before, discard it if it has, and add it to the list if it has not. This is precisely the kind of application in which a coding procedure can be advantageously used. A catalog of all unlabeled trees on up to 13 nodes has been produced in just this sort of way by Morris [9].

There are many kinds of trees that one might like to catalog, and, of course, there are many other applications, other than mere cataloging, where similar techniques could be employed. This paper will be a discursive survey of several methods, some old and some new, for coding trees of various kinds. We shall consider planted plane trees, rooted plane trees, rooted trees, and plane trees, as well as common-or-garden trees, those which the computer man calls "free" trees. All our trees will be given as unlabeled. We shall not mention methods for coding labeled trees except in so far as they can be subverted to serve our unlabeled purposes.

It is unlikely that readers of this book will need to be told that the study of tree structures is of importance in the theory of computing. Knuth [7, Chapter 2.3] gives further enlightenment, should this be necessary. Accordingly, much of the background of this paper, such as the concept of walking around, or traversing, a tree, will be old hat to the computer man. Yet this paper is not written from the computer man's viewpoint. These trees are not thought of as related to, or arising from, computer problems, but are regarded as the primary objects of study, and of interest in themselves, rather than as a means to an end. Thus, this paper is basically theoretical. The only reason it gets in on the "graph theory and computing" act at all is that the sort of applications to which tree coding can be applied are liable to require the handling of a great number of trees, possibly large ones. Thus, the computer's help is needed, and it becomes necessary to consider questions of economical storage, efficient algorithms, and various kinds of programming tricks from time to time.

2. Definitions

A *tree* is a connected graph with no circuits. Two trees are *isomorphic* if there exists a one-to-one correspondence between their nodes which preserves adjacency.

A *rooted tree* is a tree in which one node, the root, has been distinguished from the others. Two rooted trees are isomorphic if there is a one-to-one adjacency-preserving correspondence between them, which maps the root of one on to the root of the other.

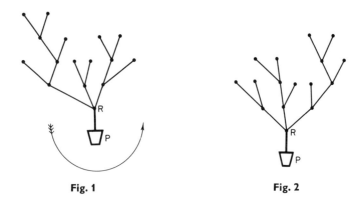

Fig. 1   Fig. 2

A *plane tree* is one that has been imbedded in the plane. Two plane trees are isomorphic if there is an orientation-preserving homeomorphism of the plane onto itself, which maps one tree onto the other.

A *planted plane tree* is usually defined [5, 17] as a rooted plane tree for which the root is an end node. This root is then the "pot" in which the tree has been planted, and the edge incident with it is the "trunk" of the tree. It will be convenient for our purposes not to regard this node and this edge as being part of the tree, and we modify our definition accordingly. Moreover, we shall disregard the howls of the botanists and give the name "root" to the node that is adjacent to the pot, and hence incident with the trunk. Thus, instead of regarding Fig. 1 as a planted plane tree on 17 nodes of which the pot P is the root, we shall regard it as having only 16 nodes, and being rooted at R. The function of the pot and the trunk is to prevent us from, for example, swiveling the left-hand branch of the tree in Fig. 1 in a counter clockwise direction, as indicated by the arrow, to obtain the tree in Fig. 2. This restriction can be indicated just as well by reducing the pot and the edge PR to a short vertical line, or simply by agreeing to draw the root at the bottom of the figure, as will be done for all planted trees in this paper.[†] Using this convention we can readily see that the planted plane trees of Fig. 1 and Fig. 2 are different, even if we leave out the pot and the trunk. These can be added in a unique way if considered necessary, but they will not form part of the tree, and, in particular, they will not contribute to the node and edge counts of the tree.

Consider a tree T which is rooted at a node R, and consider a particular node X of it. Consider the set of nodes A with the property that the unique path from A to R contains the node X. This set of nodes, which includes X, defines a subtree of T which we shall call the *branch* at X. Clearly the branch at R is the whole rooted tree.

[†] Unlike Knuth [7, p. 307], I have not (yet) been converted to the bat's-eye view of graph theory which requires trees to be drawn with their roots at the top of the page.

3. Binary Codes for Planted Plane Trees

We shall first describe a method for coding planted plane trees, the code being a string of 0's and 1's, which can, if we wish, be interpreted as a binary integer. This coding procedure will be the prototype for several others that will be discussed in this paper for different kinds of trees. The general idea of this coding algorithm is not new. It has been described by Edmonds [1, p. 196], and has been discovered independently by others [3, 16]. Indeed, the fundamental idea on which the method rests was used by Cayley [2] in 1875 in his study of trees of various kinds, and so has quite a venerable history. The fundamental idea is that if we remove the root from a planted plane tree, together with the incident edges, we obtain an *ordered* set of planted plane trees. These trees will be the branches at the nodes that are adjacent to the root of the original tree. We shall refer to them as the subtrees adjacent to the root. Figure 3 illustrates their genesis.

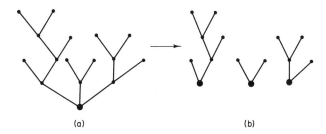

(a)   (b)

Fig. 3

The fact that these trees are ordered is a consequence of the fact that the original tree was planted, and that the edges incident with the root therefore occurred in a specific order, which could not be altered (by virtue of the remark made earlier about not being able to swivel edges round the root). The fact that they are planted trees follows from the fact that each edge incident with the root of the original tree will be the "trunk" for the corresponding subtree.

Suppose we have defined a binary code for any planted plane tree on p nodes or less, and wish to define a code for a tree T on $p+1$ nodes. Removal of the root gives a number, let us say k, of planted plane trees, each of which will have a code, since it has p nodes or fewer. Let $C_1, C_2, ..., C_k$ be their binary codes. We then define the code of T to be

$$0, C_1, C_2, ... \quad ..., C_k, 1$$

where the commas denote catenation of the 0–1 strings.

To complete this recursive definition we specify the code of the planted

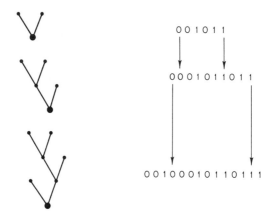

Fig. 4

plane tree on 1 node, necessarily the root. It is 01. Figure 4 shows how the codes for the trees of Fig. 3b can be derived. From this we deduce that the code for the tree of Fig. 3a is

$$0001000101101110010110001011011 1$$

It is easily proved by mathematical induction that the code of a planted plane tree satisfies the following two conditions:

(1) the number of 0's equals the number of 1's;

(2) if the code is scanned from left to right, then at each stage the number of 0's is greater than the number of 1's, except when the end of the code is reached, when they are equal.

A string of 0's and 1's which satisfies these two conditions will be said to have the *level property*, a term taken from de Bruijn and Morselt [3].

The recursive definition just given is the simplest way of defining the code, but gives rise to computational difficulties. Starting at the root, we know how to construct the code of the whole tree from those of the subtrees at the root, but we do not yet know what these codes are. Thus, we need to work upward from the root until we come to subtrees that are small enough to code, that is, until we come to an end node.

This process can be systematized in an obvious way by defining the *height* of a node, a term which is almost self-explanatory. The root is of height 1. All nodes adjacent to the root are of height 2. All nodes adjacent to these, except the root, are of height 3, and so on.[†] This progressive manner of defining the

---

† The height is sometimes defined by letting the root be of height zero, thus giving $h-1$ where we have h. Our definition has some advantages in the present context.

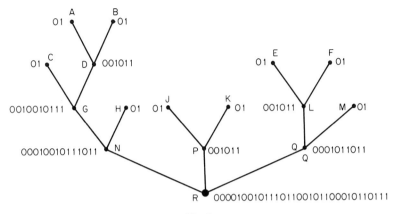

Fig. 5

height of a node indicates an obvious algorithm for finding the heights of all the nodes of a given rooted tree. In drawing planted trees, nodes of the same height will be drawn on the same level, with higher nodes further up the page than lower ones. This definition of height applies also to trees that are rooted but not planted.

The maximum height h of the nodes of a rooted tree is called the *height* of the tree. Having classified the nodes of a tree according to their height, we can easily implement the coding procedure. We first code the branches on the nodes of height $h-1$, since the subtrees, if any, formed by the removal of these nodes are single nodes. We then code the branches on nodes of height $h-2$, and so on. This is illustrated in Fig. 5, where the code for the branch at each node is indicated next to that node. These branches are coded in the alphabetical order of the letters attached to the nodes.

4. Binary Codes for Plane Rooted Trees

The difference between a plane tree and a planted tree is that the swiveling about the root mentioned in Section 2 *is* allowed. If the valency of the root is k, then, as with planted plane trees, the removal of the root gives rise to k subtrees, each of which is a planted plane tree. However, whereas with planted trees we could recognize one of these subtrees as being the left-most and arrange them from left to right uniquely, we can now only recognize a cyclic order in which these branches occur around the root.

Thus, we can code each of the branches as before. However, when we attempt to combine the codes into a code for the whole tree, we do not know where to start. We can overcome this difficulty by taking each branch in turn, regarding it as the left-most and taking the others in cyclic order, thus treating the tree as if it were a planted plane tree. This gives us k possible codes, which may not be all distinct, from which we need to choose one, in some unique way, to be the desired code for the tree. There are several ways in which one can do this, and for the time being we shall mention the most straightforward of these. We regard each of the k codes as being an integer expressed in binary notation, and choose the code which gives the smallest such integer. It is easily seen that two isomorphic plane rooted trees will give the same code, and conversely.

Another way of looking at this process is to observe that to any plane rooted tree there correspond k planted plane trees, one for each of the k spaces between the edges round the root in which a trunk for the planted tree can be drawn. We choose the planted tree with minimum code to represent the plane rooted tree.

5. Binary Codes for Rooted Trees

When we consider trees that are merely rooted, so that the manner in which they are imbedded in the plane, if they are, is not important, we see that the order of the subtrees at the root is immaterial. If we permute them in any way we merely get another imbedding of the same, that is, an isomorphic, tree. Moreover, the same remark applies to these subtrees, regarded as rooted trees, and so on. Thus we must recast the recursive definition. Let $C_1, C_2, \ldots, C_k$ be the codes of the branches at the nodes adjacent to the root. Let these codes be written in some preferred order, to be defined, given by a permutation $(i_1, i_2, i_3, \ldots, i_k)$ of the integers $1, 2, 3, \ldots, k$. Then the code of the tree is defined to be

$$0, C_{i_1}, C_{i_2}, \ldots \quad \ldots, C_{i_k}, 1.$$

As before, the code of the tree consisting of a single node is defined to be 01.

For practical computation of the code we can first determine the heights of the nodes of the tree, and code the branches at the nodes in order of the heights, with highest nodes first. The order in which we do this for the nodes of the same height will not matter. These codes will be sorted when they are incorporated into the codes of branches at lower nodes.

We have left unspecified the particular preferred order in which we will take the codes $C_1, C_2, \ldots \ldots, C_k$. The most straightforward choice would be to arrange these codes in nondescending order of the binary integers that they represent. Thus, if the removal of the root resulted in trees the codes of which were

$$00001111, \quad 0001011011, \quad 00000101101111,$$

then we would take them in the order in which they have just been given, since they correspond to the integers 15, 91, and 367 in decimal notation. This method would certainly serve. In fact, *any* method of ordering the codes will serve, provided it is unique.

There are some theoretical advantages, however, in using a rather less direct method of ordering. Instead of regarding the codes as binary integers, we shall regard them as binary fractions, following a *binary point*, which is the binary analogue of the decimal point. Thus, we would take the above codes in the order

$$00000101101111, \qquad 00001111, \qquad 0001011011$$

by virtue of the inequalities

$$.00000101101111 < .00001111 < .0001011011$$

We shall see later the advantages of this method of ordering.

6. The Decoding Algorithm

In all the codes defined so far we have ended with what is, in effect, the code of a planted plane tree. We have used an ordering procedure to establish a preferred order for the subtrees rooted at any given node. The planted tree whose code we obtain, therefore provides a canonical way of drawing the given tree, be it rooted and plane or just rooted.

It follows that any method of decoding the code of a planted tree to obtain a drawing of the tree, or other information from which we could draw the tree, such as the adjacency matrix or a list of edges, will apply equally well to all the kinds of trees so far considered.

The decoding procedure is straightforward. It is easily verified that the symbols 0 and 1 appearing in the code of a tree behave like left and right parentheses, respectively. Let us therefore write the code of the tree of Fig. 5, using parentheses. We get

$$((((\,)((\,)(\,)))(\,)((\,)(\,))(((\,)(\,))(\,))).$$

Each left parenthesis has a matching right parenthesis. Let us regard each such pair of parentheses as the left and right portions of a circle, or other closed curve, and fill in the missing parts of this curve. We obtain Fig. 6.

If we now regard the circles as nodes, and define adjacency to mean immediate inclusion of one circle inside another, we arrive at the original planted plane tree. The fact that this decoding procedure is effective can be proved by mathematical induction. The removal of the first and last symbols of the code, which correspond to the root, from a string of 0's and 1's having the level

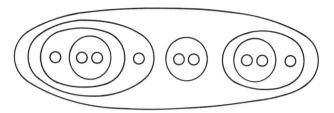

Fig. 6

property results in a succession of substrings, each having that property. The breaks between one substring and the next occur where the number of 0's equals the number of 1's, reading from left to right. These substrings are the codes of the subtrees that result when the root is removed. Thus, this procedure simply reverses the coding procedure.

This geometric method of drawing circles, which has been rather loosely described above, is equivalent to the following more formal algorithm, which produces as its output a list of those pairs of nodes that make up the edges of the tree. It also produces a canonical way of labeling the nodes of the tree with any convenient ordered set of labels. We shall always take our set of labels to be the integers $1, 2, 3, ..., p$.

Algorithm 1. Decoding Algorithm.

Step 1: Associate a label with each 0 occurring in the code, by numbering them, in order, from left to right.

Step 2: Scan the code from left to right until the configuration 001 is found. Note the pair of labels associated with the two 0's in this configuration, and then delete the second 0 and the 1. Note that the two labels noted will define an edge of the tree

Step 3: (1) If the resulting string has more than 2 symbols, repeat from Step 2.

(2) Otherwise the string is just 01, the label associated with this 0 is that of the root, and the algorithm terminates.

The connection between this algorithm and the geometrical approach can be seen as follows. The sequence 01 occurring in a code gives the smallest size circle and therefore represents an end node. It is adjacent only to the node represented by the circle S which contains it. If it is left-most inside this circle, then it will produce the configuration 001. If not, then it will eventually become the left-most circle inside S when the other circles, also in S which lie to its left, have been deleted. This decoding process is illustrated in Table I, which gives the decoding of the string

00001001011101100101100010110111

In Table I the digits making up the 001 configuration have been underlined in each line. The pairs of nodes that are the edges of the tree are written on the right as they are found. The digits that are deleted at each stage of the algorithm have not been physically deleted, but merely struck through. In implementing

TABLE I

Decoding of the String 00001001011101100101100010110111

1 2 3 4	5 6	7	8	9 10	11	121314	15	16	Edges
0 0 0 0 1 0 0 1	0 1 1	1 0	1 1 0 0	1 0 1	1 0 0 0	1 0 1	1 0 1 1 1	(3, 4)
0 0 0 0 1 0 0 1	0 1 1	1 0	1 1 0 0	1 0 1	1 0 0 0	1 0 1	1 0 1 1 1	(5, 6)
0 0 0 0 1 0 0 1	0 1	1 0	1 1 0 0	1 0 1	1 0 0 0	1 0 1	1 0 1 1 1	(5, 7)
0 0 0 0 1 0 0 1	1	1 0	1 1 0 0	1 0 1	1 0 0 0	1 0 1	1 0 1 1 1	(3, 5)
0 0 0 0 1 0 0 1	1 1	1 0	1 1 0 0	1 0 1	1 0 0 0	1 0 1	1 0 1 1 1	(2, 3)
0 0 0 0 1 0 0 1	1 1 1	0 1	1 0 0 1	0 1 1	0 0 0 1	0 1 1	0 1 1 1	(2, 8)
0 0 0 0 1 0 0 1	1 1 1	1 0	0 1 0 1	1 0 0 0	1 0 1	1 0 1 1 1	(1, 2)
0 0 0 0 1 0 0 1	1 1	0 1	1 0 0 0	1 0 1	1 0 1 1 1	(9, 10)
0 0 0 0 1 0 0 1	1 1	0 1	0 0 1 0	1 1 0 0 0	1 0 1	1 0 1 1 1	(9, 11)
0 0 0 0 1 0 0 1	1 1	0 1	0 0 0 1	0 1 1 0 0 0	1 0 1	1 0 1 1 1	(1, 9)
0 0 0 0 1 0 0 1	1 1	1 0 0 0	1 0 1	1 0 1 1 1	(13, 14)
0 0 0 0 1 0 0 1	1 1	0 0	1 0 1	1 0 1 1 1	(13, 15)
0 0 0 0 1 0 0 1	1 1	0 0 0 1	0 1 1 0 1 1 1	(12, 13)
0 0 0 0 1 0 0 1	1 1	0 0 0 1	0 1 1 0 1 1 1	(12, 16)
0 0 0 0 1 0 0 1	1 1	0 0 0 1	0 1 1 0 1 1 1	(1, 12)
0 0 0 0 1 0 0 1	1 1	0 1 1 0 0 0 1 0 1 1 0 1 1 1	end.

this algorithm on a computer, this is a possible way of coping with these deletions. Some kind of tag is assigned to each digit to show whether it has been deleted or whether it has still to be considered. This has the disadvantage that at each iteration one has to wade through a lot of garbage to find the next 001 configuration. The alternative is to delete the digits literally and close up the rest of the code, so that the code to be searched gets shorter as the algorithm progresses. This could lead to problems of keeping track of the labels of the 0's as they move around in storage. However, it is not, in fact, necessary to label these 0's all at the beginning as, for simplicity, we have stated above. It is sufficient to allocate labels only until we come to a 001 configuration, and resume the labeling after the appropriate deletions have been made. In this way no 0 receives a label until it has reached its final resting place from which, ultimately, it will be deleted.

7. Binary Codes for Unrooted Trees

The problem of coding unrooted trees can be reduced to that of coding rooted trees by means of the center–bicenter theorem for trees, which we now briefly review.

Let T_1 be a tree. Note all the end nodes of T_1. If there are more than 2 of them, simultaneously delete them all from T_1, together with their incident edges. Call the resulting tree T_2. In a similar way, form a new tree T_3 from T_2, and so on. We obtain a sequence

$$T_1, T_2, T_3, \ldots$$

of trees. Each of these trees has at least 2 nodes fewer than its predecessor and, hence, this sequence must terminate. It will do so when a tree is reached that has either one node, which is called the *center* of the original tree T_1, or two nodes and the edge joining them, called the *bicenter* of T_1.

If T_1 has a center, we can associate with it a unique rooted tree T_1^* by making the center the root. T_1^* can be coded by the method of Section 3, and we can take this code to be the code of the unrooted tree T_1.

If T_1 has a bicenter, the removal of the bicentral edge will result in two rooted trees. These can be coded as before. We then arrange these codes in the preferred order, and insert the first code after the initial 0 of the second code. Thus, if the codes, in the preferred order, are $0A1$ and $0B1$, where A and B are binary strings, the code of the tree becomes $00A1B1$. It will be seen that this is the same as truncating each code by removing the initial 0, catenating the two resulting strings, and placing 00 at the beginning. In terms of the tree, the effect of this is to obtain the code of a rooted tree in which the second of the two bicentral nodes is the root. Thus, in either case, the code obtained is that for a planted plane tree, and it follows that it can be decoded by the decoding algorithm already described.

We note that by a method essentially the same as that just described, the coding of unrooted plane trees can be made to depend on the coding procedure for rooted plane trees that was described in Section 4.

8.   A Streamlined Algorithm for Coding Unrooted Trees

It will be clear from the last section that the algorithm described there for coding unrooted trees requires three passes through the tree. First, we need to work from the outside inward, deleting end nodes, in order to find the center or bicenter. Then we must work outward from the center or bicenter in order to find the heights of the nodes. Finally we work inward again, using the coding algorithm on the branches at nodes of decreasing height.

The speed with which a tree could be coded would be increased if we could replace this three-pass process by a single-pass process. It turns out that this is possible. We combine the detection of the center or bicenter with the outside-inward coding process, to form a sort of "code-as-you-go" algorithm. Since we shall not know the heights of the nodes at any stage, and shall not be aware of where the center or bicenter is until we come to it, it is clear that this streamlined algorithm must differ in some ways from those already considered. As it happens the differences are slight.

At each stage of the algorithm we shall consider a subtree of the tree T_1, obtained from it by deletions of end nodes, as described at the beginning of Section 8. In this subtree each end node will bear a tag, which is the binary code of the branch at that node in the original tree. Initially each end node of T_1 bears the tag 01. The transition from one subtree to the next is effected in the following way. Look at those nodes that are adjacent to an end node and pick out those for which all but one of the adjacent nodes are end nodes. It will be convenient to call such a node a *ripe node*. For each ripe node X, construct the code of the branch at X from the tags of the adjacent end nodes as described in Section 5. Having done this, delete the end nodes adjacent to X. When this has been done for all ripe nodes, we have a tree in which all the end nodes, and only they, bear a tag. This is the next tree in the sequence, and we repeat the process if possible. The algorithm ends when the resulting tree has only one or two nodes. If it has one node, this will be the center of T_1, and its tag will be the code of the tree. If it has two nodes, these will be the bicentral nodes of T_1. We form the code by catenating their tags in the preferred order, having first dropped their initial zeros, and then inserting 00 in front, as described in Section 7.

We now formally present this algorithm.

ALGORITHM 2.   STREAMLINED CODING ALGORITHM.   At each stage of this algorithm the nodes will be partitioned into two sets. A is the set of tagged nodes, and B is the set of untagged nodes. With each element of A there is associated a tag, that is, a binary string. Initially A consists of the end nodes of T_1, each having the tag 01, and B contains all the other nodes of T_1.

Step 1:   Construct the set R of those nodes of B that are adjacent to at most one other node of B.

Step 2:   For each node $X \in R$ consider the tags of the adjacent nodes that are in A. By catenating these tags in the preferred order, and enclosing them between a 0 and a 1, construct the tag for the node X.

Step 3:   (1)   If B has only one element X, then the tag of X is the required code for T_1, and the algorithm terminates;

(2)   if B has exactly two elements, then the code of T_1 is obtained by catenating the truncated tags of these two nodes in the preferred order, and prefixing 00;

(3)   otherwise put $A = A \cup R$, $B = B - R$, and go back to Step 1.

Note that in this more formal algorithm we have made no provision for deleting the end nodes from the tree, as we did in the informal presentation.

This is not necessary, since once the tag of an end node P in T_1 has been incorporated into the tag of an adjacent node, all nodes adjacent to P are then tagged. Hence, P cannot subsequently be adjacent to a node X of a set R, since X must be in B and, hence, is untagged. Thus, P plays no further part in the algorithm and is as good as deleted. It follows that all we need to know about a node is whether it is tagged, and if so, with what.

It is easily verified that if, in the above algorithm, we forget about the nature of the tags and merely note which nodes are tagged and which are not, then we obtain an algorithm for determining the center or bicenter of the tree. It will differ from that described at the beginning of Section 7 only in the timing of the deletions of the nodes. Thus, if we apply to the tree of Fig. 5 the usual method for finding centers or bicenters, we would delete nodes A, B, C, E, F, G, I, J, and K in the first iteration. In the streamlined coding algorithm only nodes A, B, C, E, F, G, J, and K would be deleted, since the nodes adjacent to them are "ripe." On the other hand, node C, for example, would be ear-marked for deletion at this stage, but would not be actually deleted, in the sense that its adjacent node H would become tagged, until a later iteration when H became ripe. The same applies to node I. This clearly makes no difference to the final result, and the node or nodes last tagged when the streamlined algorithm terminates will be the center or bicenter of the tree, as the case may be. Thus, the code that results is the same as that given in Section 7.

This algorithm will not work for a rooted tree, since it, so to speak, generates its own root. However, by a species of low cunning, we can trick the algorithm into coding rooted trees should the occasion arise. The natural way of determining whether a node is ripe is to subtract the number of adjacent tagged nodes from its valency. The node is ripe if the result, r say, is 1 or 0. However, note that $r = 0$ only when the algorithm is about to terminate and the node in question is the center of the tree. The valencies of the nodes can be computed at the beginning of the algorithm. All we have to do then, is to fudge the initial list of valencies by increasing the listed valency of the intended root by 1. This will prevent the value of r for that node from ever being 1, except when all other nodes have been tagged. It will then be exactly 1, instead of 0. In this way, the algorithm is fooled into taking the required node to be what it fondly imagines to be the center of the tree.

9. Some Properties of Tree Codes

In this section we discuss some of the properties of the binary codes so far defined for trees.

(1)   The code for a tree on p nodes consists of p zeros and p ones having the level property. This property of tree codes has already been discussed.

(2)   The difference between the number of 0's and 1's in a portion of the code, reading from left to right, up to, and including a particular zero, is the height of the node associated with that zero in the manner of Algorithm 1. The proof of this will be given in the next section.

The remaining properties are for trees not imbedded in the plane.

(3)   The code of a rooted tree of height h starts with h zeros.

The proof is by mathematical induction. A rooted tree of height 2 is a star in which the root is adjacent to each of the other nodes of the tree. Clearly its code is

$$0010101 \cdots 01011$$

and since this starts with 2 zeros, the result is true when $h = 2$. Assume that it is true for trees of height $2, 3, ..., h - 1$, and consider a tree of height h. The code of such a tree is of the form

$$c(T) = 0, c(T_1), c(T_2), ..., c(T_k), 1$$

where at least one of the subtrees $T_1, T_2, ..., T_k$ is of height $h - 1$.

Now, since we are ordering the codes of these subtrees by nondescending magnitude of the corresponding binary fraction, the codes with the most leading zeros will come first. Thus, T_1 will certainly be of height $h - 1$, and this means that the code $c(T)$ of T will start with h leading zeros. The result then follows by mathematical induction.

(4)   The code of an unrooted tree of diameter D starts with $[(D + 3)/2]$ zeros.

The *diameter* of a tree is defined as the longest distance, counted by the number of edges, between any two nodes. It is easily verified that if the tree has a center, then D is even, and that there are at least two end nodes at distance $D/2$ from the center and none at a greater distance. Therefore, when the tree is rooted at the center, its height is $D/2 + 1$. The present result then follows from property (3).

If the tree has a bicenter, then D is odd and the two rooted trees that are joined by the bicentral edge, and whose codes make up the code of the tree, are both of height $(D + 1)/2$. Since the code begins with a zero followed by the code of one of these rooted trees, it will begin with $(D + 3)/2$ zeros, again by property (3). Since D is odd, this agrees with the result given.

We see, in properties (3) and (4), the reason why the ordering for codes was based on the interpretation of the codes as binary fractions rather than as binary integers. This is also the reason why the code for the tree on just one node was chosen to be 01, rather than 10, which is what has been used in some other discussions of this topic [12, 16]. The whole object has been to concentrate as many zeros on the left of the code as possible, thus making its integer or fractional equivalent as small as possible. It is of interest to ask how small this can be.

When we are talking about single codes, or comparing codes of the same length, it makes little difference whether we interpret the code as a fraction or as an integer. In the rest of this section, we shall think of the code of a tree as representing an integer, and call this the *integer code* of the tree. Clearly, we lose no information by thus ignoring the leading zeros, since their number can be deduced from the level property. We now consider the values between which the integer codes of trees must lie.

It would be tedious to recount in detail how the maximum and minimum integer codes were calculated for various kinds of trees. It will suffice to summarize the results. For this purpose, we let $i(T)$ denote the integer code of a tree T.

(5) If T is a rooted tree, then

$$2^{p-1} \leqslant i(T) \leqslant \tfrac{1}{3}(2^{2p-1}+1).$$

(6) If T is unrooted and is central, then

$$f(p) \leqslant i(T) \leqslant \tfrac{1}{3}(2^{2p-1}+1),$$

where

$$f(2k+1) = 2^{3k+1} - 2^{2k+1} + 2^{k+1} - 1$$

and

$$f(2k) = 2^{3k-1} + 2^{3k-2} + 2^{2k-1} + 2^k - 1.$$

(7) If T is unrooted and is bicentral, then

$$g(p) \leqslant i(T) \leqslant h(p),$$

where

$$g(2k) = (2^k-1)(2^{2k-1}+1),$$

$$g(2k+1) = g(2k) + 2^{3k},$$

$$h(2k) = \tfrac{1}{3}(2^{2k-1}+1)^2,$$

and

$$h(2k+1) = \tfrac{1}{3}(2^{4k}+2^{2k}+1).$$

We give, in Table II, some actual values for these bounds between which $i(T)$ must lie.

It is of interest to note that, although these numbers increase exponentially with p, nevertheless, even for trees on what, for many applications, is a fairly large number of nodes, $p = 20$ or so, these integer codes still lie within the

TABLE II

Bounds for $i(T)$

p	Rooted trees		Unrooted central		Unrooted bicentral	
	minimum	maximum	minimum	maximum	minimum	maximum
3	7	11	11	11	–	–
4	15	43	43	43	27	27
5	31	171	103	171	91	91
6	63	683	359	683	231	363
7	127	2731	911	2731	743	1387
8	255	10923	2959	10923	1935	5547
9	511	43691	7711	43691	6031	21931
10	1023	174763	24095	174763	15903	87723
11	2047	699051	63551	699051	48671	349867
12	4095	2796203	194623	2796203	129087	1399467
13	8191	11184811	516223	11184811	391231	5593771
14	16383	44739243	1564799	44739243	1040511	22375083
15	32767	178956971	4161791	178956971	3137663	89483947
16	65535	715827883	12550399	715827883	8356095	357935787
17	131071	2863311531	33423871	2863311531	25133311	1431677611
18	262143	11453246123	100532735	11453246123	66978303	5726710443
19	524287	45812984491	267912191	45812984491	201196031	22906579627
20	1048575	183325937963	804783103	183325937963	1536347647	91626318507

range of values which can be handled, as integers, by modern computers and present-day programming languages. Thus, for example, the 16-digit integer allowance in APL will enable us to handle, as integers, the codes of trees with up to 27 nodes. Actually, the result will be a little better than this if we use property (9) below.

This suggests that the coding algorithm could be carried out using integer arithmetic instead of the manipulation of binary strings. At each stage of the algorithm the tags would be integers, and each end node would initially bear as its tag the integer 1. In Step 2 of the algorithm the tags $N_1, N_2, \ldots, N_k$ would be combined to give the tag

$$2 \times \left(\left(\cdots \left(\left(\left(N_1\, 2^{l_2} + N_2 \right) 2^{l_3} + N_3 \right) 2^{l_4} + N_4 \right) \cdots \right) 2^{l_k} + N_k \right) + 1$$

for the appropriate ripe node, where l_i is the length of the string corresponding to the integer N_i. Although these lengths can be deduced from the tags, this would be time-consuming. They are best stored along with the tags.

Two further properties of tree codes are worth a brief mention. It is sometimes of interest to know whether a tree code is that of a central or bicentral tree, without going to the trouble of decoding the whole tree. This question is settled by the following property.

(8)   If a binary string is known to be the code of some unlabeled tree, then this tree is central if and only if the first two substrings that result from the deletion of the extreme 0 and 1 have the same number of leading zeros.

Proof: If the tree is central, then at least two of the subtrees at the root have the same maximal height. Hence, by property (3), their codes have the same number of leading zeros. If the tree is bicentral, then the first substring denotes a tree which contains the bicentral edge, and, hence, will be higher by 1 than any other tree at the root. Thus, this substring will have one more leading zero than the second substring.

The other property enables us to shorten slightly the code of a tree.

(9)   The binary code of a tree ends with at least two 1's. The proof is obvious.

This means that all integer codes are congruent to 3, modulo 4. It also means that the binary code can be shortened, without any loss of information, by dropping these two 1's, all leading 0's, and the first 1. This will, in turn, give a smaller integer code.

10.   Canonical Labelings

Although the trees that we have been considering have all been unlabeled, it is clear that if we have a means of drawing a tree in the plane in some standard way (and Sections 6 and 7 provide such a way) then it is no problem to choose a method for labeling the nodes of the tree in a manner which depends only on the isomorphism class of the tree. There are many ways in which this choice can be made, and a canonical labeling obtained. We shall consider two of them. In this section and the next we shall be concerned entirely with planted plane trees and shall, for convenience, refer to these as just trees.

(1)   *The bottom-up labeling:*   We shall allocate integers $1, 2, 3, ..., p$ to the nodes, taking the nodes in the following order. Nodes of height r precede nodes

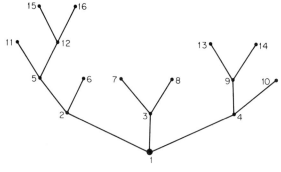

Fig. 7

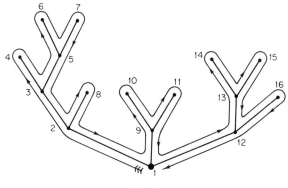

Fig. 8

of height $r+1$, and among the nodes of the same height the ordering is from left to right. This gives us a labeling like that of Fig. 7.

(2) *The walk-around labeling:* The notion of walking around, or through, a tree is a very well-known one in computer applications of tree structures. A fairly clear idea of what is meant can be gleaned from Fig. 8.

More formally, we can think of the walk, now strictly "on" the tree rather than "around" it, as being determined by a sequence of nodes in the order in which they are visited, starting at the root. This sequence is determined by the following maze-threading rules:

Rule 1: each edge is to be traversed exactly once in either direction;
Rule 2: if, subject to Rule 1, we can go upwards from the node where we are, we do so, and we take the left-most of the edges available to us;
Rule 3: if we cannot go up, we go down.

By walking around a given tree we can obtain a canonical labeling which is different from that given in labeling (1). As we walk around the tree, we give the next available label to a node when we meet it for the first time. This also is shown in Fig. 8.

A simple induction argument shows that the canonical labeling thus obtained is precisely that given by the decoding algorithm, Algorithm 1. This suggests that the code for a tree could be derived from a walk around the tree, and this is so. In fact, it is exactly the way in which de Bruijn and Morselt [3] derive a code that is essentially the same as that of Section 3. They observe that a walk around a tree is specified uniquely by stating, for each of the $2p-2$ steps of the walk, whether it goes up or down. The rule for taking the left-most of the available edges then ensures uniqueness. Thus, if we interpret 0 and 1 in the code to mean "up" and "down", respectively—

de Bruijn and Morselt use U and D—then the code gives us a recipe for performing the walk, and, hence, drawing the tree. Since going up (down) means going to a node of height greater (less) by 1, we immediately deduce Property (2) of Section 9.

De Bruijn and Morselt used this code to establish a simple one-to-one correspondence between the set of trees on p nodes and the set of binary trees on $2p-1$ nodes. Harary *et al.* [5] had previously observed that these two sets were equinumerous, and had established a one-to-one correspondence between them. However it was a somewhat complicated one.

11. Valency Codes

We now consider a rather different type of code, to which the name "valency code" is not inappropriate, though it is not the valency v_i of each node that is used but rather the number u_i of edges that go upwards from that node. For the root this is the same as the valency, while for every other node of the tree u_i is one less than the valency. A *valency code* is simply a list of these integers u_i in some specific order.

It is well known that the *set* $\{u_i\}$ will not, in general, determine the tree. However, it so happens that if these numbers are listed in the order of one or the other of the canonical labelings given in Section 10, then the tree *is* uniquely determined. This statement requires proof, and we consider first the simpler case.

(1) *The bottom-up valency code (BUV code) for a tree:* This code consists of a sequence $\{u_i\}$, where u_i relates to the node labeled i in the bottom-up labeling of the tree. Clearly, two isomorphic trees will give the same code.

To prove the converse we show how to draw the tree having a given BUV code. The method is so straightforward that a specific example will suffice to explain it. Consider the sequence

(1) $4, 2, 0, 2, 3, 0, 0, 2, 0, 0, 0, 0, 0, 0$

The first integer indicates that there are 4 edges going upwards from the root. We draw the root and these edges. This gives us 4 nodes of height 2. The next 4 terms 2, 0, 2, 3 give the numbers of edges going upwards from these 4 nodes, from left to right. We can therefore draw the tree up to and including the nodes of height 3. There are $2+0+2+3 = 7$ of these. The next 7 terms in the code enable us to construct the tree up to the nodes of height 4. There are two of these. They are end nodes, as the two remaining zeros in the code testify. We obtain the graph of Fig. 9. The method is clearly quite general.

This valency code consists of p integers. The number of consecutive zeros at the end will be the number of nodes of maximum height h, necessarily

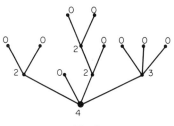

Fig. 9

end nodes, plus the number of end nodes of height $h-1$ to the right of the right-most node of height $h-1$, not an end node. This number could be as small as 1, so the only simple way of reducing the length of the code would be to omit the final zero. This gives a code consisting of $p-1$ terms.

A code essentially of this type has been described by Klarner [6], who pays particular attention to the special case of trees in which every node that is not an end node has the same valency $k+1$. Since Klarner includes the trunk in his planted plane trees, this means that in our notation $u_i = 0$ or k for all i. If it is understood that we are talking about trees of this kind, then it is not necessary to record the values of the u_i in the code. We need merely put a 1 in the ith position if $u_i = k$, and a 0 if $u_i = 0$. In this way, we obtain a string of p binary digits. Some economy is possible here, for we must have $u_1 = k$, so we can omit u_1. Further, there must be at least k consecutive zeros at the end of the sequence, and these can be omitted, too, without any loss of information. This reduces the code to a sequence of $p-k-1$ digits. It is easily proved that $p = kn+1$ for this kind of tree, where n is the number of nodes of valency greater than 1. Hence, the length of the code is $k(n-1)$ binary digits. This is a considerable saving over the $2p$ binary digits of the code of Section 3, but, of course, these trees are very special indeed.

Returning to the BUV code for general planted plane trees, we consider a way of expressing it in binary form. In place of each u_i, we write down a string consisting of u_i zeros followed by a one, and we catenate these to form the required binary code. Thus, from (1) we get

$$000010011001000111001111111$$

Clearly, the sequence of u_i's can be reconstructed from such a string. Each 1 signals the end of a substring for which the number, possibly 0, of 0's is the corresponding u_i. Since node i accounts for $u_i + 1 = v_i$ digits of this code, for $i > 1$, and node 1 accounts for $v_1 + 1$ digits, the total length of the code will be

$$\sum_{i=1}^{p} v_i + 1 = 2(p-1) + 1$$
$$= 2p - 1.$$

There will, therefore, be p ones (or $p-1$, if the economy of omitting the last u_i is used) and, hence, $p-1$ zeros. Klarner used this device, with 0 and 1 interchanged, to obtain yet another one-to-one correspondence between the two sets mentioned at the end of Section 10.

We note here, in passing, that the addition of a zero at the beginning of this binary version of a BUV code gives a binary string which has the level property. By regarding this as the binary code of a tree, as in Section 3, we can define an interesting one-to-one mapping of the set of all trees on p nodes onto itself.

(2) *The walk-around valency code* (*WAV code*) *for a tree:* The walk-around valency code is defined exactly like the bottom-up valency code except that the order of the nodes is taken to be that given by the walk-around labeling. Thus, the tree of Fig. 9 has the WAV code

$$4, 2, 0, 0, 0, 2, 2, 0, 0, 0, 3, 0, 0, 0.$$

Thus, the WAV code, like the BUV code, is a sequence of p integers $\{u_i\}$. We first demonstrate a property of WAV codes.

THEOREM. If the sequence $\{u_i\}$ is the WAV code of a tree, then for every $r \ (= 1, 2, ..., p)$

$$(2) \qquad \sum_{i=1}^{r} u_i \geqslant r - 1,$$

where the equality holds, if and only if, $r = p$.

Proof: When we code the tree T whose code is $\{u_i\}$, we start at the root and walk around the tree, noting the integer u_i for each new node we pass. Stop after noting the rth term. Some edges of T will have been traversed twice, once in each direction, some once, and some not at all. Delete from T all edges not traversed at all, thus obtaining a subtree T'. Now complete the walk around T'. We can add no more u_i's to the code, since we can reach a new node only along a path not previously traversed, and these have been eliminated. Thus, T' has r nodes.

In the special case when $T' = T$, that is, when every edge had been traversed in at least one direction, we now have the code of T. Thus, $r = p$ and

$$(3) \qquad \sum_{i=1}^{p} u_i = v_1 + \sum_{i=2}^{p} (v_i - 1)$$

$$= \sum_{i=1}^{p} v_i - (p - 1)$$

$$= 2(p - 1) - (p - 1)$$

$$= p - 1$$

so that (2) holds. If $T' \neq T$, then the truncated code that we have is not the code of T'. Some of the u_i's are too large because of the removal of edges. The fact that T is connected ensures that the removal of an edge not traversed at all must make at least one of the u_i smaller for T' than for T. Hence,

$$\sum_{i=1}^{r} u_i > \sum_{i=1}^{r} u_i'$$
$$= r - 1$$

by applying (3) to the tree T', where u_i' is defined for the tree T'. This proves the theorem.

This theorem gives us a decoding algorithm for WAV codes. If we remove the first term in the code, then the next so many terms will describe a walk around the left-most of the subtrees joined to the root. How many terms are required can be found by using (3). We take just so many as will make their sum one less than the number of terms. Thus, to decode

$$3, 3, 0, 2, 0, 0, 0, 0, 2, 0, 2, 0, 0,$$

we remove the first 3, and find that we must take the next 6 terms before (3) is satisfied, since $3+0+2+0+0+0 = 5$. This defines the left-most tree, and we delete these terms from the sequence. We now repeat this procedure with the rest of the sequence, and so on. We find two more subtrees, defined by the subsequences 0 and $2, 0, 2, 0, 0$.

Thus, by removing the first term of the code, we can break up the remainder into subsequences, each of which is the WAV code of a subtree joined to the root. The number of subtrees thus obtained must be u_1. It follows that this term is redundant and could be dropped from the code, if we wished, without loss of information.

As before, we can replace the term u_i by a string of u_i zeros and a 1. We catenate these strings to obtain a binary version of the WAV code. From (2) we deduce that, by putting a zero in front of this code, we obtain a binary string having the level property.

Before going further we give more formal algorithms for the decoding of BUV codes and WAV codes. These are equivalent to the informal descriptions given above, though we shall not take space to prove this.

ALGORITHM 3. ALGORITHM FOR DECODING BUV CODES. The terms of the code are numbered $1, 2, \ldots, p$ consecutively. There are two markers, a left marker and a right marker, which both point to the first term of the code at the start of the algorithm.

Step 1: If the right marker is pointing to the last term, the algorithm terminates. Otherwise, move the right marker to point to the next term.

Step 2: If the left marker points to a zero term, move it to the next nonzero term.

Step 3: Reduce by 1 the term to which the left marker is pointing, and note, as an edge of the tree, the pair of nodes corresponding to the present positions of the two markers. Repeat from Step 1.

TABLE III

Decoding Process for a BUV Code

1	2	3	4	5	6	7	8	9	10	11	12	13	14	Edge
4	2	0	2	3	0	0	2	0	0	0	0	0	0	(This is the code)
3	2	0	2	3	0	0	2	0	0	0	0	0	0	(1, 2)
2	2	0	2	3	0	0	2	0	0	0	0	0	0	(1, 3)
1	2	0	2	3	0	0	2	0	0	0	0	0	0	(1, 4)
0	2	0	2	3	0	0	2	0	0	0	0	0	0	(1, 5)
0	1	0	2	3	0	0	2	0	0	0	0	0	0	(2, 6)
0	0	0	2	3	0	0	2	0	0	0	0	0	0	(2, 7)
0	0	0	1	3	0	0	2	0	0	0	0	0	0	(4, 8)
0	0	0	0	3	0	0	2	0	0	0	0	0	0	(4, 9)
0	0	0	0	2	0	0	2	0	0	0	0	0	0	(5, 10)
0	0	0	0	1	0	0	2	0	0	0	0	0	0	(5, 11)
0	0	0	0	0	0	0	2	0	0	0	0	0	0	(5, 12)
0	0	0	0	0	0	0	1	0	0	0	0	0	0	(8, 13)
0	0	0	0	0	0	0	0	0	0	0	0	0	0	(8, 14)

If the sequence was a valid code, then all the terms will have been reduced to zero when the algorithm terminates. Table III gives an example of this decoding process. The positions of the markers are indicated by underlining.

ALGORITHM 4. ALGORITHM TO DECODE A WAV CODE. We number the terms of the WAV code from 1 to p. During the algorithm some terms will be altered, or deleted altogether, but the numbering remains the same. There is a marker, which initially points to the first term, and is moved around as the algorithm progresses according to the following two rules:

Rule 1: if the marker points to a nonzero term, say u_i, move it to the next term u_j, ignoring any deleted terms; note that nodes i and j form an edge of the tree.

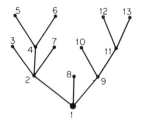

Fig. 10

TABLE IV

Decoding Process for a WAV Code

	1	2	3	4	5	6	7	8	9	10	11	12	13	Edge
Code	**3**	3	0	2	0	0	0	0	2	0	2	0	0	
	3	**3**	0	2	0	0	0	0	2	0	2	0	0	(1, 2)
	3	3	**0**	2	0	0	0	0	2	0	2	0	0	(2, 3)
	3	**2**		2	0	0	0	0	2	0	2	0	0	
	3	2		**2**	0	0	0	0	2	0	2	0	0	(2, 4)
	3	2		2	**0**	0	0	0	2	0	2	0	0	(4, 5)
	3	2		**1**		0	0	0	2	0	2	0	0	
	3	2		1		**0**	0	0	2	0	2	0	0	(4, 6)
	3	2		**0**			0	0	2	0	2	0	0	
	3	**1**					0	0	2	0	2	0	0	
	3	1					**0**	0	2	0	2	0	0	(2, 7)
	3	**0**						0	2	0	2	0	0	
	2							0	2	0	2	0	0	
	2							**0**	2	0	2	0	0	(1, 8)
	1								2	0	2	0	0	
	1								**2**	0	2	0	0	(1, 9)
	1								2	**0**	2	0	0	(9, 10)
	1								**1**		2	0	0	
	1								1		**2**	0	0	(9, 11)
	1								1		2	**0**	0	(11, 12)
	1								1		**1**		0	
	1								1		1		**0**	(11, 13)
	1								1		**0**			
	1								**0**					
	0													

Rule 2: If the marker points to a zero term, delete this term, move the marker back to the previous nondeleted term, which cannot be a zero, and decrease this term by 1.

If the code is valid, this algorithm will terminate when all terms have been deleted. The pairs of nodes noted under Rule 1 will give the edges of the tree.

A step-by-step example of this decoding procedure is given in Table IV. The term to which the marker is pointing is underlined. This gives us the tree of Fig. 10.

Returning to Table III, we see that the right-hand labels, in the pairs that make up the edges of the tree, are simply the integers from 2 to 14 (in general, 2 to p) in order. This is because the right-hand marker is moved along one place at each iteration of the algorithm. Since these integers can be supplied automatically, all the information about the structure of the tree is given by the left-hand labels of these pairs, namely,

$$1, 1, 1, 1, 2, 2, 4, 4, 5, 5, 5, 8, 8.$$

This sequence defines an even simpler code for a tree. We can call it a *tree-function code* (TF code) since the function $f(i)$, defined to be the *i*th term in the code, is a tree function as defined by, for example, Moon [8]. It is also known as a canonical representation, see [7]. This code hardly needs any decoding, for we need only write down the integers 2 to p underneath its successive terms to obtain the list of edges of the tree. For this reason a TF code is probably the most convenient way to store an unlabeled tree in a computer. The tree will be stored as a $(p-1)$-dimensional vector, say (a_i), and the edges will be all of the form $(a_i, i+1)$. If further economy is desired, the first term can be omitted, since it must be 1.

The above is not the only tree-function code. Any labeling of a rooted tree will give rise to a tree function. We put $f(i) = j$, if (i, j) is the first edge of the unique path from node i to the root. However, a tree function is of no use as a code for planted plane trees, and will not be called a TF code, unless we associate with it some convention that enables us to determine how the tree was imbedded in the plane. This is true of the tree function related to the bottom-up labeling since, with this labeling, the nodes at the same level are labeled consecutively from left to right. We need not be as specific as this, however. It is sufficient if the labels of the nodes at the same level increase monotonically from left to right. It is easily verified that this is true of the walk-around labeling. Hence, the walk-around labeling also gives rise to a TF code. Clearly, there can be many others.

12.   Unrooted Trees Again

We now return briefly to the problem of coding ordinary trees, that is, not planted, plane, or even rooted, for which, so far, only the binary code of Section 9 has been defined. Since it is often not convenient to work with binary strings, and since the trick of using the integer codes is not easily applicable if the trees are large, we need to have available some other kind of coding procedure.

Now, although the valency codes were defined in the last section specifically for planted plane trees, we can easily extend them to trees in general. The decoding algorithm (Algorithm 1) associates with any tree a unique planted plane tree, from which we can derive some canonical labeling. Once we have this labeling, then any of the codes of Section 11 are available to us. Thus, to construct a valency code for an unlabeled tree, we could first find its binary code, then decode this to get a canonical labeling, and then construct a valency code. This idea has two very obvious disadvantages.

(1)   Although we will have avoided binary strings in the code itself, they are still required in finding the binary code.

(2)   The procedure of coding, decoding, and then recoding is, to say the least, a bit devious. One could wish for a more direct method.

What we would like is a coding procedure which works with integer sequences throughout, and which, like Algorithm 2, does the whole job in just one pass through the tree, from the outside to the root. There seems to be little hope of achieving this ideal for a BUV code. The bottom-up labeling depends critically on the position of the root, and this would not be known until the end of an algorithm of the type desired. It is quite otherwise with the WAV code, however. The portions making up this code relate to the branches at certain nodes, and these can be recognized, when the node becomes ripe, even though it is not yet known which node will end up as the root. We now describe a direct algorithm for finding the WAV code of an unrooted tree. This algorithm is similar to Algorithm 2. The differences are

(1) The tags are strings of integers,

(2) The rule for the preferred ordering of tags will be different. It does not matter much what it is, and for definiteness we can suppose that tags are ordered first by increasing length, and that tags of the same length are ordered lexicographically.

(3) The rule for the formation of the new tags is different.

ALGORITHM 5.   DIRECT INTEGER ALGORITHM FOR WAV CODES.   At each stage of this algorithm the nodes will be partitioned into two sets: A, the set of tagged nodes, and B, the set of untagged nodes. The tags are strings of integers. Initially A consists of the end nodes of the tree, each having the tag 0.

Step 1:   Construct the set R of those nodes of B that are adjacent to, at most, one other node of B.

Step 2:   For each node $X \in R$, consider the tags of the adjacent nodes that are in A. Catenate these tags in the preferred order, and prefix to the result the number of tags that were catenated to produce it.

Step 3:   (1)   If B has only one element X, then the tag of X is the code for the tree, and the algorithm terminates.

(2)   If B has exactly two elements, determine which of the tags comes first in the preferred ordering. Insert this tag after the first term of the other tag. Then increase by 1 the first term in the resulting sequence. This last move is necessary since, once one of the bicentral nodes has been chosen to be the root, the bicentral edge will then go upwards from that node.

(3)   Otherwise put $A = A \cup R$, $B = B - R$, and go back to Step 1.

This algorithm is illustrated in Fig. 11.

The fact that the binary code of a tree determines a canonical labeling of the tree implies that we can use on an unlabeled tree any of the several algorithms that are known for coding labeled trees. All we need do is to give the tree its canonical labeling before using the coding algorithm in question. Algorithms that can be used in this way include the Prüfer sequence, see [8], and the

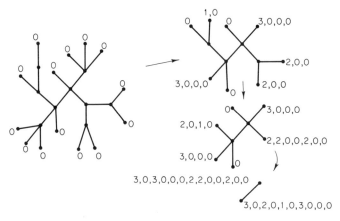

Fig. 11. Direct integer algorithm for WAV codes: the code is
4,3,0,2,0,1,0,30,0,0,0,3,0,0,0,2,2,0,0,2,0,0.

several variations on this as given by Neville [10]. It is doubtful whether there is any advantage in using codes of this type for our present purpose. The Prüfer sequence, for example, has $p-2$ terms. Hence, it gives no saving in length over the WAV code, abbreviated by omission of its first term. Moreover, the process of obtaining the Prüfer sequence would be more complex, since the canonical labeling would have to be found first.

A coding procedure for labeled trees which, at first sight, seems to offer possible advantages is that described by Smolenskii [15]. If a tree has k end nodes, then there are $k(k-1)/2$ distances between these nodes. Smolenskii showed that if a set of integers is the set of distances between end nodes of some tree, then this tree is unique. He did not give a method of telling whether such a set belonged to a tree in this way, or of finding the tree if it did. Both these questions were answered later by Zaretskii [18]. Thus, if the end nodes are labeled $1, 2, 3, \ldots, k$ in some order, and $d(i, j)$ denotes the distance from end node i to end node j, then the Smolenskii code is the sequence

$$d(1, 2), \quad d(1, 3), \ldots, d(1, k), \quad d(2, 3), \ldots \quad \ldots, d(k-1, k)$$

This sequence of integers could be quite short. For a path it consists of just one term. It will be shorter than the WAV code if $k(k-1)/2 < p-2$.

We naturally ask how likely this is to happen. In order to answer this question, assuming that all trees on p nodes are equally likely, we would need to know the number of trees that have at most K end nodes, where K is the largest value of k for which $k(k-1) < (p-2)$. The enumeration of unlabeled trees by the number of end nodes is, I believe, still an open problem, though the solution of the analogous problem for labeled trees is well known (see Moon

TABLE V

The Proportional Numbers of Trees for Which the Smolenskii Code Gives a Saving in Length Compared with the Abbreviated WAV Code

p:	6	7	8	9	10	11	12	13	14	15	16	17
Total:	6	11	23	47	106	235	551	1301	3159	7741	19320	48629
$k=2$	1	1	1	1	1	1	1	1	1	1	1	1
$k=3$	2	3	4	5	7	8	10	12	14	16	19	21
$k=4$	2	4	8	14	23	36	52	76	108	148	199	262
$k=5$	1	2	6	14	32	64	123	219	377	616	978	1496
K	3	3	3	4	4	4	4	5	5	5	5	5
$k \leqslant K$	3	4	5	20	31	45	63	308	500	781	1196	1780
Ratio	0.50	0.36	0.22	0.43	0.29	0.19	0.11	0.24	0.16	0.10	0.062	0.037

[8]). Thus, the exact answer cannot be given. However, the rough answer seems to be, "not very likely."

Table V gives the total numbers of unlabeled trees on 4 to 17 nodes. These numbers were taken from Riordan [14]. It also gives the subtotals of these having 2 to 5 end nodes. These were calculated by using Pólya's Hauptsatz [11], to add notes of valency 2 to the homeomorphically irreducible trees listed at the end of [4]. From these numbers the fraction r_p of those trees whose Smolenskii code is shorter than the WAV code can be found, and is given in Table V. It will be seen that r_p is small, and seems generally to decrease as p gets larger. It is reasonable to conjecture that $r_p \to 0$ as $p \to \infty$.

Thus, the Smolenskii code does not seem to have any advantages in the present context, except possibly for applications in which the trees that are being handled tend to have few end nodes. One can remark also that, in any case, the fact that trees with the same number of nodes may have Smolenskii codes of different lengths is liable to be, for the programmer, an embarrassment rather than an asset.

References

1. Busacker, R., and Saaty, T., "Finite Graphs and Networks." McGraw-Hill, New York, 1965.
2. Cayley, A., On the analytical forms called trees, with applications to the theory of chemical combinations, *Rep. Brit. Assoc. Advan. Sci.* 257–305 (1875).
3. de Bruijn, N. G., and Morselt, B. J. M., A note on plane trees, *J. Combinatorial Theory* **2**, 27–34 (1967).
4. Harary, F., and Prins, G., The number of homeomorphically irreducible trees and other species, *Acta Math.* **101**, 141–162 (1959).

5. Harary, F., Prins, G., and Tutte, W. T., The number of plane trees, *Nederl. Akad. Wetensch. Proc. Ser. A* **67** No. 3 (*Nederl. Akad. Wetensch. Indag. Math.* **26** No. 3), 319–329 (1964).

6. Klarner, D. A., Correspondences between plane trees and binary sequences, *J. Combinatorial Theory* **9**, 401–411 (1970).

7. Knuth, D. E., "The Art of Computer Programming," Vol. 1. Addison-Wesley, Reading, Massachusetts, 1968.

8. Moon, J. W., Counting labelled trees: A survey of methods and results, *Publ. Dep. of Math., Univ. of Alberta*, undated.

9. Morris, P. A., A catalogue of trees on *n* nodes, *n* < 14, *Mathematical Observations, Research and Other Notes*, No. 1 StA, Department of Mathematics, University of the West Indies (unpublished mimeograph).

10. Neville, E., The codifying of tree-structure, *Proc. Cambridge Philos. Soc.* **49**, 381–385 (1953).

11. Pólya, G., Kombinatorische Anzahlbestimmungen für Gruppen, Graphen und chemische Verbindugen, *Acta Math.* **68**, 145–254 (1938).

12. Parris, R., The coding problem for graphs, M.Sc. thesis. University of the West Indies, 1968.

13. Read, R. C., How to grow trees, *in* "Combinatorial Structures and their Applications," pp. 343–347. Gordon & Breach, New York, 1970.

14. Riordan, J., "An Introduction to Combinatorial Analysis." Wiley, New York, 1958.

15. Smolenskii, Ye, A., A method for the linear recording of graphs, *USSR Comput. Math and Math. Phys.* **2**, 396–397 (1963).

16. Thalwitzer, K., Charakterisierung eines Labyrinthes durch ein Wort über einem zweistellige Alphabet, *in* "Beiträge zur Graphentheorie" (H. Sachs, H.-J. Voss, and H. Walther, eds.), pp. 157–161. Teubner, Leipzig, 1968.

17. Tutte, W. T., The number of planted plane trees with a given partition, *Amer. Math. Monthly* **71**, 272–277 (1964).

18. Zaretskii, K., Constructing a tree on the basis of a set of distances between the hanging vertices, *Uspehi Mat. Nauk* **20**, 90–92 (1965).

A GRAPH-THEORETIC STUDY OF THE
NUMERICAL SOLUTION OF SPARSE
POSITIVE DEFINITE SYSTEMS
OF LINEAR EQUATIONS

Donald J. Rose[†]

Department of Mathematics
University of Denver
Denver, Colorado

† Present address: Aiken Computation Laboratory, Harvard University, Cambridge, Massachusetts.

1. Introduction

The necessity to solve linear systems

$$(1) \qquad\qquad Mx = b,$$

where M is an $n \times n$ sparse[†] symmetric positive-definite matrix, arises frequently in physical applications. These include classical electrical network analysis, analysis of structural systems, and nonlinear hydraulic problems. In such problems understanding and controlling sparsity is essential for efficient computer solution, since, in general, many systems with the same zero–nonzero structure will be solved.

To solve systems like (1) by elimination, it is standard procedure (Forsythe and Moler [12] and Westlake [27]) to decompose, or factor, M as

$$(2) \qquad\qquad M = GG^{\mathrm{T}} \quad \text{or} \quad M = LDL^{\mathrm{T}},$$

where G and L are lower triangular, $L = (l_{ij})$ with $l_{ii} = 1$, and $D = (d_{ij})$ is diagonal with $d_{ii} > 0$. Since M is symmetric and positive definite, we may decompose M for any *a priori* ordering of the linear system, that is,

$$A = PMP^{\mathrm{T}},$$

where P is an $n \times n$ permutation matrix. Experience and simple examples show that the choice of ordering is an important consideration in obtaining an efficient elimination scheme. Such considerations lead to the study of the following basic question: what is the effect of the order of elimination upon sparse positive-definite systems?

We have restricted our class of matrices to those M which are symmetric positive definite, because this allows us to examine only the equivalence class PMP^{T} rather than the class PMQ, P and Q being permutation matrices. Equally important is the fact that the decompositions of (1), especially the Cholesky $M = GG^{\mathrm{T}}$ decomposition, are stable with respect to rounding error, see Wilkinson [28, pp. 220, 231–232, 244], for any *a priori* ordering P. Formally, our analysis extends to the more general class of matrices M, such that PMP^{T} can be decomposed as

$$PMP^{\mathrm{T}} = LDU = LU'$$

for any permutation matrix P, and such that $m_{ij} \neq 0 \leftrightarrow m_{ji} \neq 0$. Here, of course, both L and U must be stored.

We use the word "formally" above because the decomposition must be computed in finite precision arithmetic. In the more general case of sparsity

---

† We leave *sparse* undefined formally, but informally we think of matrices M, many of whose entries m_{ij} are zero.

symmetry above, and in the case where M is simply a nonsingular matrix row, or a column, interchanges are usually effected (see Wilkinson [28, p. 205] or Forsythe and Moler [12, p. 34]), to avoid zero pivot elements and to maintain stability with respect to round-off error. Whereas the decomposition of PMP^T for any P is stable when M is positive definite, it is clear that in these more general cases pivoting to control stability and pivoting to control sparsity are not *a priori* compatible.

With M as in (1) we associate an undirected graph G. In Section 2 we formulate the elimination process of decomposition (2) as vertex elimination on this graph. We call this formulation the *combinatorial elimination process*. In addition to formulating the elimination process as vertex elimination on a graph, we seek a graph-theoretic description of those matrices M, such that in the decomposition L or G has exactly the same zero–nonzero structure as the lower triangular part of M. We call such graphs *monotone transitive graphs*.

The suggestion that graph theory might be a convenient way to study elimination is due to Parter [17], although he does not pursue a detailed graph-theoretical study. He analyzes the special case when the matrix M has a graph-theoretic representation as a tree, and he shows that trees can be ordered so that they are monotone transitive. A well-known example of the simple elimination scheme which results from choosing a monotone transitive ordering for such a graph, or matrix, is the case when M is tridiagonal.

There are two interesting implications of Parter's study which were not pursued in the literature. First, since trees are without cycles and can be ordered to be monotone transitive, we are led to investigate monotone transitivity in more general graphs by studying their cycle structure. Second, although a tridiagonal band matrix is represented by a special tree, Parter's elimination scheme applies to any matrix represented by a tree. The elimination process involved has absolutely nothing to do with the bandwidth of the matrix M. This is significant given the recent activity in minimizing bandwidth for sparse matrix calculations [1, 2, 9, 19].

Following the formulation of elimination as a combinatorial process, in Section 3 we gain considerable insight into the elimination process by studying the evolution of the cycle structure and the vertex-separator, or cut-set, structure of a graph under elimination. We show that monotone transitive graphs are triangulated graphs, and conversely, as defined by Berge [3]. This is a cycle characterization. In addition, we characterize monotone transitive graphs by a property of their separators.

In Section 4 we study criteria from which we may define best or good orderings. By counting the arithmetic operations necessary to effect the decompositions, we relate these criteria for optimization to the computational complexity of calculations involving the elimination process. We note that in the literature the study of optimal ordering contains many subjective decisions and implicit assumptions which are not always clearly presented. One reason for some of the confusion which exists is that for practical applications there is an implicit constraint that any ordering algorithm to be used be reasonably efficient. Otherwise, such an algorithm may become too time- or storage-consuming to be feasible. Several authors (Tinney and Walker [26], Tinney [29, p. 25], and Tewarson [29, p. 35]) have developed algorithms which have been partially successful in producing good orderings but which do not, in general, produce optimal orderings for the criterion they choose. Most of this work has been experimental. In Section 4 we discuss these algorithms in view of the results developed in Section 3 and the results of our generalized study of criterion functions.

Another interesting graph-theoretic approach for dealing with sparse systems with respect to Gaussian elimination is to attempt to find permutation matrices P, Q such that

$$(3) \qquad\qquad A = PMQ$$

is block lower triangular, since in this case it is necessary only to decompose the diagonal blocks of PMQ. Naturally such a transformation does not preserve symmetry. Harary [13, 14] solves this problem algorithmically with the restriction that $Q = P^{\mathrm{T}}$. His results have application to the algebraic eigenvalue problem. Steward [21] and Dulmage and Mendelsohn [10, 11] have solved the more general problem and have algorithms for producing P and Q. These results are not applicable when M is symmetric positive definite and irreducible, since the algorithm would then produce only one diagonal block, M itself. Even when applicable, this theory does not differentiate between reorderings of the system within the diagonal blocks.

Our interest in sparse linear systems was motivated initially by its application to the potential flow network problem [18]. Several examples of sparse linear systems arising in applications can also be found in [29]. Other theoretical considerations on sparse linear systems and numerical linear algebra are reported in Brayton et al. [6]. Finally, we wish to emphasize the importance of new approaches to data handling and efficient use of memory hierarchy which are required for the successful machine implementation of sparse matrix methods. While we do not discuss computer implementation here, considerable progress on this aspect of sparse matrix research is reported in Gustavson et al. [15].

2. The Elimination Process

In this section, we study the combinatorial nature of the elimination process upon sparse symmetric positive-definite matrices. We will see that it is useful to regard elimination as vertex elimination on a graph. We first review the well-known LDL^T decomposition theorem for positive-definite matrices where L is a lower triangular matrix with unit diagonal elements, and D is a diagonal matrix with positive nonzero entries.

Unfortunately, sparse matrices tend to *fill in* during elimination. That is to say, in general, the number of nonzeros in L of the decomposition $M = LDL^T$ is greater than the number of nonzeros in the lower triangular part of M. In Section 2.2 we ask for the class of matrices so that we can find an ordering (permutation P) such that no zeros are lost in the decomposition of PMP^T. Of course, this class of matrices is special, but we will see that, after elimination, any matrix is transformed into a matrix of this special class. More precisely, we will see that L^T has this property. This question leads us to our notions of elimination graph, monotone transitivity, and perfect elimination processes.

2.1. Decompositions

We begin by stating a well-known (Forsythe and Moler [12, pp. 27–29]), theorem of numerical linear algebra. Let M be a real $n \times n$ matrix and let M_k, $k = 1,...,n-1$ denote the principal submatrices of M consisting of the first k rows and columns of M.

THEOREM 1. Let M and M_k be as above, and assume $\det(M_k) \neq 0$, $k = 1, 2,...,n-1$. Then there exist unique matrices L, D, U, such that

(4) $$M = LDU,$$

where $L = (l_{ij})$ and $U = (u_{ij})$ are real $n \times n$ unit lower ($l_{ii} = 1$) and unit upper ($u_{ii} = 1$) triangular matrices, respectively; and D is an $n \times n$ real diagonal matrix.

If M is a real symmetric positive-definite matrix, PMP^T satisfies the hypothesis of the theorem for any permutation matrix P. Furthermore, by uniqueness it follows that

(5) $$M = LDL^T,$$

In this case, D has positive diagonal entries. Also,

(6) $$M = GG^T,$$

where $G = LD^{1/2}$. The factorization (6) is due to Cholesky [12, p. 114].

For sparse matrices M, it is significant that L, D, and G of (5) and (6) are unique. This means that the zero–nonzero structure of M uniquely determines the zero–nonzero structure of L or G, independent of the method used to compute L, D, or G. Note also that since M is symmetric, the decompositions (5) or (6) are more efficient than the $M = L'U'$ decomposition, where only

L' or U' is unit triangular, because for the decompositions (5) and (6) we need only store the upper triangular part of M and either L and D of (5) or G of (6).

Since the symmetric Gaussian elimination scheme and Cholesky's method are the two most generally accepted methods for obtaining (5) and (6) respectively, we state them now in algorithmic form which we will need in Section 2.2 and Section 4.

SYMMETRIC GAUSSIAN ELIMINATION

This method is also known as the method of congruent transformations (see Westlake [27, p. 21]). To explain the algorithm, it suffices to exhibit the first major step since the algorithm then proceeds recursively on lower order principal submatrices. Let the $n \times n$ positive definite matrix

$$M^{(1)} = \begin{bmatrix} a & r^T \\ r & \overline{M} \end{bmatrix},$$

where a is 1×1, r is $(n-1) \times 1$, and $\overline{M}$ is $(n-1) \times (n-1)$. Then

(7) $$M^{(1)} = \begin{bmatrix} 1 & 0 \\ r/a & I \end{bmatrix} \begin{bmatrix} a & 0 \\ 0 & \overline{M}-rr^T/a \end{bmatrix} \begin{bmatrix} 1 & r^T/a \\ 0 & I \end{bmatrix}$$

$$= L_1 \begin{bmatrix} a & 0 \\ 0 & M^{(2)} \end{bmatrix} L_1^T,$$

where $M^{(2)} = \overline{M} - rr^T/a$ and I is the $(n-1) \times (n-1)$ identity matrix. Note that $M^{(2)}$ is positive definite because

$$(-y^T r/a \,|\, y^T) M^{(1)} \left(\frac{-y^T r/a}{y} \right) > 0$$

for any $(n-1)$ vector y. We may compute L^T and D which replace the strictly upper triangular and diagonal parts of $M^{(1)}$ respectively, by the algorithm

(8)

for i: $= 1$ step 1 until $n-1$ do;

for j: $= i+1$ step 1 until n do;

begin;

s: $= M[i,j]/M[i,i]$;

for l: $= j$ step 1 until n do;

$M[j,l]$: $= M[j,l] - s \times M[i,l]$;

$M[i,j]$: $= s$;

end.

To compute the Cholesky decomposition, note that (7) can be rewritten as

$$M = \begin{bmatrix} \sqrt{a} & 0 \\ r/\sqrt{a} & \tilde{G} \end{bmatrix} \begin{bmatrix} \sqrt{a} & r^{\mathrm{T}}/\sqrt{a} \\ 0 & \tilde{G}^{\mathrm{T}} \end{bmatrix},$$

where $\tilde{G}\tilde{G}^{\mathrm{T}}$ is the Cholesky factorization of $\bar{M} - rr^{\mathrm{T}}/a$ and the algorithm (8) can be changed appropriately. Usually, however, the elements of G are computed column by column, which requires exactly the same operations executed in a different order, as in the following algorithm from Forsythe and Moler [12, p. 114]

for $j = 1$ step 1 until n do;

begin $G[j,j] := \mathrm{sqrt}\left(M[j,j] - \sum_{k=1}^{j-1} G[j,k]^2\right);$

(9)   for $i = j + 1$ step 1 until n do;

$G[i,j] := \left(M[i,j] - \sum_{k=1}^{j-1} G[i,k] \times G[j,k]\right)\Big/G[j,j];$

end.

Finally, to solve $Mx = b$ by symmetric Gaussian elimination, we solve

(10) $$Lz = b,$$

(11) $$Dy = z,$$

(12) $$L^{\mathrm{T}}x = y.$$

Since (10) and (11) involve triangular systems, this is merely back-solving. Similarly for Cholesky's method, we compute

(13) $$Gy = b,$$

(14) $$G^{\mathrm{T}}x = y.$$

Let $M = (m_{ij})$ be an $n \times n$ symmetric positive-definite matrix with decomposition $M = LDL^{\mathrm{T}}$. It is clear, because the unique decomposition can be generated by (7), that the set of pairs $\{i,j\}$ with $l_{ij} = 0$ is, in general, a subset of the pairs with $m_{ij} = 0$, that is, the triangular factor L cannot, in general, be more sparse than the lower triangular part of M.

Given M, if there exists a permutation matrix P such that

$$A = PMP^{\mathrm{T}} = LDL^{\mathrm{T}}$$

and

(15) $$a_{ij} = 0 \Rightarrow l_{ij} = 0, \qquad i > j,$$

then we say M is a *perfect elimination matrix*. It is straightforward from (7) (see also Parter [17, Theorem 1]) that A has this property (property P) if, and only if, for all $1 \leqslant i < j < k \leqslant n$

(16) $\qquad a_{ij} \neq 0 \quad$ and $\quad a_{ik} \neq 0 \Rightarrow a_{jk} \neq 0.$

We give a graph theoretic interpretation of property P in the next section where we introduce monotone transitive graphs and perfect elimination processes.

2.2. The Combinatorial Elimination Process

We now temporarily abandon the arithmetic aspects of the elimination process in order to study its combinatorial nature. We begin by associating with each symmetric positive definite matrix an ordered and an unordered graph. First, some graph-theoretic terminology.

For our purposes, a *graph* will be a pair, $G = (X, E)$, where X is a finite set of $|X|$ elements called vertices, and

$$E \subseteq \{\{x, y\} \mid x, y \in X, \quad x \neq y\}$$

is a set of $|E|$ vertex pairs called *edges*. Given $x \in X$, the set

$$\mathrm{adj}(x) = \{y \in X \mid \{x, y\} \in E\}$$

is the set of vertices *adjacent* to X. For distinct vertices $x, y \in X$ a *chain* from x to y of length $l = n$ is an ordered set of distinct vertices

$$\mu = [p_1, p_2, \ldots, p_{n+1}], \qquad p_1 = x, \quad p_{n+1} = y$$

such that $p_{i+1} \in \mathrm{adj}(p_i)$, $i = 1, \ldots, n$. Similarly, a *cycle* of length $l = n$ is an ordered set of n distinct vertices

$$\mu = [p_1, p_2, \ldots, p_n, p_1]$$

such that $p_{i+1} \in \mathrm{adj}(p_i)$, $i = 1, \ldots, n-1$ and $p_1 \in \mathrm{adj}(p_n)$. We will always assume that the graph G is *connected*, that is, for each pair of distinct vertices $x, y \in X$, there is a chain from x to y.

For a graph $G = (X, E)$ with $|X| = n$ an *ordering* of $|X|$ is a bijection

$$\alpha \colon \{1, 2, \ldots, n\} \leftrightarrow X.$$

We sometimes indicate an ordering by the shorthand $X = \{x_i\}_{i=1}^n$. If $G = (X, E)$ and X is ordered by α, then $G_\alpha = (X, E, \alpha)$ is an ordered graph associated with G. Given an ordering α of X, the set of vertices *monotonely adjacent* to a vertex x is denoted by $M \operatorname{adj}(x)$ and defined by

$$M \operatorname{adj}(x) = \operatorname{adj}(x) \cap \{z \in X \,|\, \alpha^{-1}(z) > \alpha^{-1}(x)\}.$$

We associate with each $n \times n$ symmetric matrix $M = (m_{ij})$ an ordered graph $G_\alpha = (X, E, \alpha)$ such that vertex x_i corresponds to row i and $\{x_i, x_j\} \in E$, if and only if $m_{ij} \neq 0$ and $i < j$. The unordered graph $G = (X, E)$ then represents the equivalence class of matrices PMP^{T}, where P is any permutation matrix. For convenience we assume that for no P can PMP^{T} be represented as a direct sum of lower-order matrices, that is, M is irreducible, so that G is connected.

Consider again (7) which represents the first major step of elimination, the elimination of x_1. We proceed to interpret this step graph theoretically. Let $G = (X, E)$ be a graph and α be an ordering of X. The *deficiency*, $D(x)$, is the set of all distinct pairs of $\operatorname{adj}(x)$ which are not themselves adjacent, that is,

$$D(x) = \{\{y, z\} \,|\, y, z \in \operatorname{adj}(x),\ y \neq z,\ y \notin \operatorname{adj}(z)\}.$$

Similarly, the *monotone deficiency*, $MD(x)$, is the set

$$MD(x) = \{\{y, z\} \,|\, y, z \in M \operatorname{adj}(x),\ y \neq z,\ y \notin \operatorname{adj}(z)\}.$$

Finally, for a graph $G = (X, E)$ and subset $A \subseteq X$, the *section graph* $G(A)$ is the subgraph

$$G(A) = (A, E(A)),$$

where $E(A) = \{\{x, y\} \in E \,|\, x, y \in A\}$.

Given a vertex y of a graph G, the graph G_y obtained from G by

(1) deleting y and its incident edges;
(2) adding edges such that all vertices in the set $\operatorname{adj}(y)$ are pairwise adjacent

is the *y-elimination graph* of G (compare Parter [17, p. 120]). Thus

$$G_y = (X - \{y\}, E(X - \{y\}) \cup D(y)).$$

For an ordered graph $G = (X, E, \alpha)$, the *order sequence* of elimination graphs $G_1, \ldots, G_{n-1}$ is defined recursively by $G_1 = G_{x_1}$ and $G_i = (G_{i-1})_{x_i}$, $i = 2, \ldots, n-1$.

Since the graphs G_i determine the evolution of the process of vertex elimination, we formally define the *elimination process* on a graph $G = (X, E)$ with ordering α as the ordered set

$$P(G; \alpha) = [G = G_0, G_1, \ldots, G_{n-1}].$$

An elimination process $P(G; \alpha)$ is perfect if

$$G_i = G\left(X - \bigcup_{j=1}^{i} \{x_j\}\right).$$

DEFINITION.[†]   The ordered graph $G = (X, E, \alpha)$ is *monotone transitive* when, for all $x \in X$, we have

$$y \in \text{M adj}(x) \quad \text{and} \quad z \in \text{M adj}(x) \Rightarrow y \in \text{adj}(z).$$

The significance of monotone transitivity is given in the following lemma which merely summarizes our definitions and relates them to perfect elimination matrices. It is immediate that monotone transitivity is the graph-theoretic interpretation of the perfect elimination matrix condition of (16).

LEMMA 1.   Let M be a symmetric positive definite matrix with unordered graph $G = (X, E)$. Then the following are equivalent:

(1) M is a perfect elimination matrix;
(2)   there exists an ordering α such that $G_\alpha = (X, E, \alpha)$ is monotone transitive;
(3)   in G_α, $MD(x) = \varnothing$ for all $x \in X$;
(4) $P(G; \alpha)$ is a perfect elimination process.

Thus, in a monotone transitive graph, vertex elimination adds no edges. Suppose, however, that $G_\alpha = (X, E, \alpha)$ represents a matrix M which is not a perfect elimination matrix. If elimination is carried out on M, vertex elimination of G_α, then for each $1 = i < j < k$ such that $m_{1j} \neq 0$ and $m_{1k} \neq 0$ but $m_{jk} = 0$, a new nonzero element will be created in the (j, k) position of $M^{(2)} = \overline{M} - rr^{\text{T}}/a$, see (7). Clearly, the graph of $M^{(2)}$ is the elimination graph G_1. Continuing inductively, we see that the study of monotone transitive graphs is interesting even if G_α is not monotone transitive, because the elimination process may be regarded as transforming the graph G_α, matrix M, into its *monotone transitive extension* $\text{M TE}(G; \alpha)$, where

$$\text{M TE}(G; \alpha) = \left(X, E \bigcup_{i=1}^{n-1} \tau_i \right), \quad \tau_i = D(x_i) \quad \text{in} \quad G_{i-1}.$$

and $\text{M TE}(G; \alpha)$ is the graph of L^{T}.

3.   Triangulated Graphs

3.1.   Preliminaries

In Section 2 we studied the role of ordered monotone transitive graphs in the elimination process. Here we shall characterize monotone transitive

[†] By way of motivation, a graph is *transitive* [8, p. 31] if $y \in \text{adj}(x)$ and $x \in \text{adj}(z)$ implies $y \in \text{adj}(z)$. Since we are dealing with undirected graphs, the adjacency relation is symmetric, that is, $x \in \text{adj}(z) \Leftrightarrow z \in \text{adj}(x)$. It is easy to see that any connected transitive graph on n vertices is the complete graph on n vertices, because then between any two vertices x and y

graphs by their cycle structure and their separating sets of vertices. Monotone transitive graphs are shown to be triangulated graphs as defined by Berge [3]. The theory developed in this section shows very clearly why sparse matrices must fill in during elimination.

Recall that we are dealing only with connected graphs, and for a graph $G = (X, E)$ and subset $A \subseteq X$, the *section graph* $G(A)$ is the subgraph

$$G(A) = (A, E(A)), \qquad E(A) = \{\{x, y\} \in E \mid x, y \in A\}.$$

A *separator* of a graph $G = (X, E)$ is a subset $S \subset X$ such that the section graph $G(X - S)$ consists of two or more *connected components*, say $C_i = (V_i, E_i)$. The section graphs $G(S \cup V_i)$ are then the *leaves* of G with respect to S. A *minimal separator* is a separator no subset of which is also a separator. Similarly, given $a, b \in X$ with $a \notin \mathrm{adj}(b)$, an a, b separator is a separator such that a and b are in distinct components, say C_a and C_b, respectively. Note that (see the Example) a minimal separator is a minimal a, b separator for some $a, b \in X$, but a minimal a, b separator is not, in general, a minimal separator. A *clique* C of a graph is a subset of vertices which are pairwise adjacent. A *separation clique* is a separator which is also a clique.

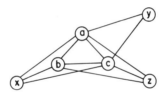

Fig. 1. Graph $G = (X, E)$.

Example: Consider the graph $G = (X, E)$ shown in Fig. 1. The set $S = \{a, b, c\}$ is a separator and $G(X - S)$ consists of the three components $(\widehat{x}, \phi)$, $(\widehat{y}, \phi)$, and $(\widehat{z}, \phi)$. The leaf containing $\widehat{x}$ is the clique on $\{\widehat{x}, \widehat{a}, \widehat{b}, \widehat{c}\}$. Note that S is not minimal because $S' = \{\widehat{a}, \widehat{c}\}$ is also a separator. S' is, however, minimal and it is a minimal x, y separator. On the other hand, S is a minimal x, z separator. In addition, both S and S' are separation cliques.

The following definition is due to Berge [3, p. 158].

DEFINITION.   A graph G is *triangulated*, if for every cycle $\mu = [p_1, \ldots, p_n, p_1]$ of length $n > 3$, there is an edge of G joining two nonconsecutive vertices of μ. Such edges are called *chords* of the cycle.

there exists a chain from x to y. Hence, a matrix represented by a transitive graph is a full matrix. In Section 3 we will see how monotone transitive graphs are built up of smaller intersecting complete graphs.

Remark: Note that any section graph of a triangulated graph is triangulated because any cycle in $G = (X - A)$ is a cycle in G itself, and the chord of this cycle in G must be an edge in $G(X - A)$.

3.2. Main Results

THEOREM 2.[†]   For a graph $G = (X, E)$ the following statements are equivalent:

(1) G is triangulated;
(2)   every minimal a, b separator is a clique;
(3)   there exists an ordering α of X such that $G_\alpha = (X, E, \alpha)$ is monotone transitive.

Theorem 2 and Lemma 1 (Section 2) characterize monotone transitive graphs and thus perfect elimination matrices. Statements (1) and (2) give the structure of the unordered graph while statement (3) is a property of a corresponding ordered graph. Of the three equivalent properties above, statement (3) is clearly the most algorithmic in the sense that its verification is straightforward. In fact Lemma 1, Section 2 shows how to test for monotone transitivity, because in each successive elimination graph there must always exist a vertex with empty deficiency.

THEOREM 3.   Let $G = (X, F)$ be triangulated with subgraph $\hat{G} = (X, E)$, $E \subseteq F$. Then $\hat{G}$ is triangulated, if and only if for each $e = \{x, y\} \in F - E$ there exists an x, y separation clique, S_e, of $\hat{G}$.

If $G = (X, F)$ is triangulated, an arbitrary subgraph of G obtained by removing a subset of edges need not remain triangulated. Theorem 3 gives a necessary and sufficient condition that the subgraph be triangulated. It has an important corollary which requires anticipating a notion of Section 4.

Suppose a graph $G = (X, E)$ is not triangulated. Then for any ordering α of X the set $T(\alpha)$ of M TE$(G; \alpha) = (X, E \cup T(\alpha))$ is a *triangulation* of G generated by the ordering α. A *minimum triangulation* would be a triangulation, $T(\hat{\alpha})$, such that

$$|T(\hat{\alpha})| = \min_{\alpha} |T(\alpha)|$$

COROLLARY 1.   Let $G = (X, E)$ be a graph with separation clique S with components C_i and leaves L_i. Then any minimum triangulation T of G contains only edges $e = \{x, y\} \in T$ with x and y in the same component C_j, or edges $e = \{x, y\} \in T$ with $x \in C_j$ and $y \in S$.

---

† For statements of parts of this theorem see Boland and Lekkerkerker [5a], Dirac [9a], and Fulkerson and Gross [12a].

Proof: If the triangulation T contains a nonempty subset of edges with incident edges in C_j and C_k, $j \neq k$, these edges may be deleted, and by Theorem 3 the resulting set $\hat{T}$ is still a triangulation.

Thus, in a graph with a separation clique, the problem of finding a minimum triangulation reduces to finding a minimum triangulation for each leaf.

THEOREM 4. Let $G = (X, E)$ be triangulated, and α be a monotone transitive ordering. If S is a minimal a, b separation clique of G, then $S = \mathrm{M\,adj}(x_j)$ for some $x_j \in X$. Conversely, for any $x_i \in X$, such that the vertices of the elimination graph G_{i-1} are not a clique, $\mathrm{M\,adj}(x_i)$ is a separation clique of G.

Hence, in a triangulated graph, all minimal a, b separators are generated in the elimination process. Note that although the sets $\mathrm{M\,adj}(x_i)$ are separation cliques, if G_{i-1} is not a clique, they need not be minimal a, b separation cliques. For example, in G of Fig. 2 below, $\{③, ④\}$ is the only minimal a, b separator.

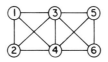

Fig. 2

Remark: One interesting application of Theorem 4 is that in a non-triangulated graph $G = (X, E)$ with $|X| = n$, at most, $n - \mu$ minimal a, b separators of G remain minimal a, b separators of $\hat{G} = (X, E \cup T)$, where T is any triangulation of G and $\mu = \max\{|C|, C$ a clique of $G\}$. This follows because any clique in G is also a clique in $\hat{G}$, and because, as we shall see in the proof of Theorem 2 in the next section, a monotone transitive ordering for $\hat{G}$ can be found which orders vertices in any clique last.

3.3. Proofs and Corollaries

We begin the proof of Theorem 2 by generalizing slightly Theorem 3 of Berge [3, p. 160].

LEMMA 2. In a triangulated graph $G = (X, E)$ every minimal a, b separator is a clique.

Proof: Let S be a minimal a, b separator and C_a and C_b be the components of $G(X - S)$ containing a and b, respectively. Since S is minimal, each $s \in S$

is adjacent to some vertex in C_a and some vertex in C_b. Let $x, y \in S$, and let μ_i be the shortest chains of the type

$$[x, c_{i,1}, c_{i,2}, ..., c_{i,p_i}, y], \qquad i = 1, 2, \quad c_{1,j} \in C_a, \quad c_{2,j} \in C_b.$$

The cycle containing x and y formed by μ_1 and μ_2 has length $l \geqslant 4$, and the only possible chord is $\{x, y\}$.

LEMMA 3. Let $G = (X, E)$ be a graph with separation clique S and leaves L_i, $i = 1, ..., n$. If S_0 is a separator of some L_i, then S_0 is a separator of G. Furthermore, if S_0 is a minimal a, b separator of L_i, then S_0 is a minimal a, b separator of G.

Proof: Let D_j, $j = 1, ..., m$ be the components of L_i with respect to S_0. Since S is a clique, vertices in S can be in only one component, say D_k. Thus, S_0 is a separator of G, because any chain from a vertex $x \in (L_j - S_0)$ with $j \neq i$ to a vertex $y \in D_l$, $l \neq k$ must contain a vertex of S_0. This proves the first statement.

For the second statement, note that S_0 is a separator of G as we have just shown. It must be an a, b separator, for the same a, b, because $D_i \cap S \neq \phi$ for at most one i. Finally, S_0 must be minimal in G since any a, b separator $\hat{S}_0 \subset S_0$ in G must be an a, b separator in L_i.

LEMMA 4. Let $G = (X, E)$ satisfy statement (2) of Theorem 2. Then either X is a clique, or given any clique $C \subset X$, there exists a vertex $x \notin C$ such that $D(X) = \varnothing$.

Proof: The proof is by induction on $|X|$ and the case $|X| = 1$ is clear. Assuming any case with $|X| \leqslant k$, let $G = (X, E)$ be such a graph with $|X| = k + 1$ and C be any clique. Either X is a clique or there exists by Lemma 2 some a, b separation clique of G, say C_1. Let D_a, D_b and L_a, L_b be the corresponding components and leaves of G containing a and b respectively. Clearly, the vertices in $C - C_1$ can be in, at most, one component. Suppose such vertices are in D_a. Consider the leaf L_b. By Lemma 3 it inherits statement (2) of Theorem 2. Writing $L_b = (W, F)$, we have $|W| \leqslant k$ and, hence, by induction, either W is a clique or there exists a vertex $x \notin C_1$ such that $D(x) = \varnothing$ in L_b. In either case then, since W must contain at least one vertex not in C_1, there exists an $x \notin C_1$ with $D(x) = \varnothing$ in L_b. Finally, $D(x) = \varnothing$ in G because x is not adjacent to a vertex in any component other than D_b. Clearly, $x \notin C$, the original clique.

Lemmas 3 and 4 yield the following two corollaries concerning the existence of vertices with $D(x) = \varnothing$. The first corollary will imply that the ordering α guaranteed by statement (3) of Theorem 2 is not unique.

COROLLARY 2. Let G be as in Lemma 4, and S be any separation clique of G with components C_i and leaves L_i. Then for each component C_i, there exists a vertex $c_i \in C_i$ with $D(c_i) = \varnothing$ in G.

Proof: By Lemma 3, each L_i has Property (2) of Theorem 2. Thus by Lemma 4, for each L_i there exists a vertex $c_i \notin S$ of L_i with $D(c_i) = \varnothing$ in L_i and, therefore, in G.

COROLLARY 3. Let G be as in Lemma 4. Then, for any $x \in X$, one, and only one, of the following statements is true:

(1) $D(x) = \varnothing$;
(2) $x \in S$, where S is a minimal a, b separation clique.

Proof: If (2) is true, clearly (1) must be false. We show by induction on $|X|$ that (1) or (2) must be true. The case $|X| = 1$ is clear, and we suppose the case $|X| \leqslant k$. Note that if X is a clique, the result is immediate. Assuming otherwise, let S be a minimal a, b separation clique of G. Let $x \in X$. If $x \in S$, the proof ends, so let $x \in (L_a - S)$. By the induction hypothesis and Lemma 3, either $D(x) = \varnothing$ in L_a, and hence in G, or $x \in \hat{S}$, where $\hat{S}$ is a minimal c, d separation clique of L_a, and hence of G.

LEMMA 5. Let $G = (X, E)$ be as in Lemma 4. Then there exists an ordering α of X such that for all $x \in X$, $M D(x) = \varnothing$.

Proof: The proof is by induction on $|X|$. The case $|X| = 1$ is clear, and we suppose the case $|X| = k$. If G is such a graph with $k + 1$ vertices, then, by Corollary 2 above, there exists a vertex x_1 such that $D(x_1) = \varnothing$. Let $G_1 = (X_1, E_1)$ be the x_1-elimination graph. Since $\text{adj}(x_1)$ is a separation clique, if X itself is not a clique, G_1 satisfies the hypothesis of the lemma by Lemma 3, and G_1 has $|X_1| = k$. By induction there exists an ordering α_1 of the vertices of G_1 such that

$$\alpha_1(i) = x_{i+1}, \qquad i = 1, \ldots, k, \qquad \text{defining } x_i$$

with $M D(x_i) = \varnothing$. Finally, in G, choose the ordering

$$\alpha(i) = x_i, \quad i = 1, \ldots, k + 1.$$

Then $M D(x_i) = \varnothing$ with this ordering in G.

Note that the ordering α assured by Lemma 5 is not unique in view of Corollary 2. This means that if $G = (X, E)$ is not triangulated, any triangulation $T(\alpha)$ generated by an ordering α will also be generated by other orderings α'. Also, note that another way of stating Lemma 5 is that there exists an ordering α such that the order sequence of elimination graphs of G, that is, $G = G_0, G_1, \ldots, G_{n-1}$, has $D(x_i) = \varnothing$ in G_{i-1}. Finally, we shall call any ordering guaranteed by Lemma 5 a *monotone transitive ordering*.

LEMMA 6.   A monotone transitive graph is a triangulated graph.

Proof: Let α be the ordering and μ be any cycle with $l > 3$. Let $p^* \in \mu$ be the vertex such that

$$\alpha^{-1}(p^*) = \min_{p \in \mu} \alpha^{-1}(p).$$

Since p^* is adjacent to two nonconsecutive vertices by monotone transitivity, μ has a chord.

Proof of Theorem 2:   Statement (1) $\Rightarrow$ statement (2) by Lemma 2, statement (2) $\Rightarrow$ statement (3) by Lemma 5, and statement (3) $\Rightarrow$ statement (1) by Lemma 6 and Lemma 1.

The following corollary shows that in a triangulated graph a monotone transitive ordering can be found such that any given clique is ordered last.

COROLLARY 4.   Let $G = (X, E)$ be triangulated with clique $C \subseteq X$. Then there exists a monotone transitive ordering α such that $\alpha(j) \in C$ for $j = k+1, k+2, \ldots, |X|$, where $k = |X| - |C|$.

Proof: The proof follows from Lemma 4 and the induction argument of Lemma 5.

Corollary 4 has the following interesting interpretation. Suppose $G = (X, E)$ is not triangulated, and we wish to find an ordering which generates a triangulation $T(\alpha)$ with a specific property, for example, a minimum triangulation. Since any clique in G remains a clique in the triangulated graph $\hat{G} = (X, E \cup T(\alpha))$, the corollary implies the existence of other orderings α' such that $T(\alpha) = T(\alpha')$, and such that α' orders the clique last. We will see in Section 4 that if only the unknowns represented by the vertices in the clique are desired, ordering the clique last will reduce the number of backsolving operations [see (12) and (14)].

We begin the proof of Theorem 3 with

LEMMA 7.   Let $G = (X, F)$ be triangulated with a subgraph $\hat{G} = (X, E)$, $E \subset F$. Suppose S is a separation clique of $\hat{G}$ such that for each edge $e = \{x, y\} \in F - E$, x and y are in different components. Then $\hat{G}$ is triangulated.

Proof: Let μ be any cycle in $\hat{G}$ with $l \geqslant 4$. If μ is entirely with some leaf of $\hat{G}$, then μ contains a chord, because μ is also a cycle in G. If μ has vertices in more than one component, then μ must contain at least two distinct vertices of S. These vertices are adjacent; hence, μ has a chord.

Proof of Theorem 3: The "if" part of the theorem follows by successive applications of Lemma 7. Given some S_e, discard all edges in $F - E$ with incident vertices in different components. S_e is then a separation clique of this new graph $\bar{G}$. By Lemma 7, $\bar{G}$ is triangulated. Continue for each edge in $F - E$ not already discarded. The converse is clear by Lemma 2, because for each $e = \{a, b\} \in F - E$, there exists a minimal a, b separator S_e in $\hat{G}$, and S_e is a clique.

Proof of Theorem 4: To prove the first assertion, let $C_1 = (V_1, E_1)$ and $C_2 = (V_2, E_2)$ be the components of G with respect to S containing a and b respectively. For each V_i let v_i^* be the vertex such that

$$\alpha^{-1}(v^*) = \max_{v \in V_i} \alpha^{-1}(v).$$

Choose $\hat{v} \in \{v_1^*, v_2^*\}$ such that

$$\alpha^{-1}(\hat{v}) = \min(\alpha^{-1}(v_1^*), \alpha^{-1}(v_2^*)).$$

Because S is minimal, each $s \in S$ is adjacent to some vertex in C_a and C_b. Hence, if $j = \alpha^{-1}(\hat{v})$ by monotone transitivity and the connectivity of C_a and C_b, we have $S = \mathrm{M\,adj}(x_j)$.

To prove the second assertion, note first that $\mathrm{M\,adj}(x_1)$ is a separation clique of G, unless X is a clique. Also, the elimination graph G_1 is a leaf of G, which is triangulated, with respect to $\mathrm{M\,adj}(x_1)$, and $G_1 = (X_1, E_1)$ has $|X_1| = |X| - 1$. The assertion then follows by induction on $|X|$ and Lemma 3.

3.4. Examples

As our first example, we will discuss in detail the ladder graph (see Fig. 3), since it illustrates the notions of Section 2 and Corollary 1, as well as anticipating some of the developments in the next section. The remaining examples are classes of graphs which illustrate our theoretical results.

3.4.1. LADDER GRAPH

Figure 3a shows the ladder graph on $2n$ vertices with two ordering α_1 and α_2, written in the form $(\alpha_1(x), \alpha_2(x))$ at each vertex x. Figure 3b shows the zero-nonzero structure of the matrices corresponding to the two ordered graphs. Finally, Fig. 3c shows the upper triangular factor L^T [see (5)] for each matrix. Clearly M_1 and M_2 are not perfect elimination matrices, because the graph has nonchorded cycles.

Note that the decomposition using α_1 requires $O(n^2)$ cells of storage, while the decomposition using α_2 requires only $O(n)$ cells. We will see in Section 4,

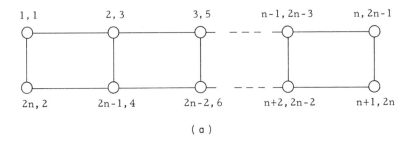

(a)

$$
M_1 = \begin{bmatrix} * & * & & & & & & & * \\ * & * & * & & & & & * & \\ & * & * & * & & & * & & \\ & & * & * & * & & & & \\ & & & * & * & * & & & \\ & * & & & * & * & * & & \\ & * & & & & * & * & * & \\ * & & & & & & * & * & \end{bmatrix} \quad ; \quad M_2 = \begin{bmatrix} * & * & * & & & & & & \\ * & * & 0 & * & & & & & \\ * & 0 & * & * & * & & & & \\ & * & * & * & 0 & * & & & \\ & & * & 0 & * & * & * & & \\ & & & * & * & * & 0 & * & \\ & & & & * & 0 & * & * & \\ & & & & & * & * & * & \end{bmatrix}
$$

(b)

$$
L_1^T = \begin{bmatrix} * & * & 0 & 0 & 0 & 0 & 0 & * \\ & * & * & 0 & 0 & 0 & * & * \\ & & * & * & 0 & * & * & * \\ & & & * & * & * & * & * \\ & & & & * & * & * & * \\ & & & & & * & * & * \\ & & & & & & * & * \\ & & & & & & & * \end{bmatrix} \quad ; \quad L_2^T = \begin{bmatrix} * & * & * & & & & & \\ & * & * & * & & & & \\ & & * & * & * & & & \\ & & & * & * & * & & \\ & & & & * & * & * & \\ & & & & & * & * & * \\ & & & & & & * & \end{bmatrix}
$$

(c)

Fig. 3. (a) Ladder graph with two orderings. (b) Matrices corresponding to the two orderings, where * indicates nonzero elements. (c) Upper triangular factors.

Theorem 5 that $O(n^3)$ arithmetic operations are required to effect the decomposition with α_1, while only $O(n)$ operations are needed with α_2. The difference is significant.

It follows from Corollary 1 that α_2 generates a minimum triangulation of the ladder graph. Since the pairs of vertices connected by each of the $n-2$ inner vertical rungs form separation cliques, the problem of finding a minimum triangulation for the ladder reduces to finding a minimum triangulation of the cycle on four vertices which requires only one edge. Thus, a minimum triangulation for the $2n$ vertex ladder requires $n-1$ edges, and one such triangulation is generated by α_2.

3.4.2. TREES AND k TREES

A *tree* is a connected graph which has no cycles. Equivalently a tree with $|X| = n > 1$ vertices is a connected graph with $n-1$ edges. Apparently Parter [17] was the first to realize that the matrix M represented by a tree was a perfect elimination matrix, although he does not use this term. Parter gives a specialized algorithm [17] for Gaussian elimination on such a matrix.

Trees are clearly triangulated graphs, and any tree must have at least two *pendent* vertices, that is, vertices adjacent to only one edge. Pendent vertices x are the only vertices in a tree with $D(x) = \varnothing$, otherwise the tree would have a cycle. A generalization of a tree is a k tree defined recursively as follows: A k tree on k vertices is a clique on k vertices. Given any k tree $T_k(n)$ on n vertices, a k tree on $n+1$ vertices is obtained when the $(n+1)$st vertex is adjacent to the vertices of a clique on k vertices in $T_k(n)$.

If we order the vertices x_i, $i = 1, 2, \ldots, n$ in the construction of a k tree on n vertices as defined above, then clearly this graph is monotone transitive with ordering $\alpha(i) = x_{n+1-i}$, $i = 1, \ldots, n$. Then, k trees are triangulated graphs. They also have the following property.

PROPOSITION 1. Every minimal separator S of a k tree $T_k(n)$ has $|S| = k$.

Proof: Since $T_k(n)$ has a monotone transitive ordering such as α above, $|\text{M adj}\,(\alpha(i))| = k$ for $i = 1, 2, \ldots, n-k$. By Theorem 4, $S = \text{M adj}(x_i)$ for some such i, since neither the set $W = \{x_i\}_{i=1}^{k-1}$ nor any subset of W is a separator.

3.4.3. THE CYCLE

Let $C = (X, E)$ be the cycle on $|X| = n$ vertices. With respect to triangulating C, we have

PROPOSITION 2. Let $C = (X, E)$ be a cycle with $|X| \geqslant 3$ vertices. Then a minimum triangulation $\hat{T}$ of C has $|\hat{T}| = |X| - 3$. Furthermore, if α is any ordering and $\text{M TE}(C; \alpha) = (X, E \cup T(\alpha))$, then $T(\alpha)$ is a minimum triangulation.

Proof: Both conclusions are proved easily by induction on $|X|$, and the case $|X| = 3$ is immediate. Let $C = (X, E)$ with $|X| = k+1$ assuming these assertions for such graphs with $|X| \leqslant k$. Let $e \in \hat{T}$, where $\hat{T}$ is a minimum triangulation of G. Clearly the vertices incident on e form a separation clique S in $\hat{C} = (X, E \cup \hat{T})$, by Theorem 3. Hence, by the corollary $\hat{T} = T_1 \cup T_2 \cup \{e\}$, where T_1 and T_2 are minimum triangulations of the leaves of $\hat{C}$ with respect to S, say $L_1 = (V_1, E_1)$ and $L_2 = (V_2, E_2)$. L_1 and L_2 are cycles with $|V_i| \leqslant k$,

$i = 1, 2$, and $|V_1| + |V_2| = |X| + 2$. By induction, $|T_i| = |V_i| - 3$, $i = 1, 2$, imply-
ing $|T| = |X| - 3$. For the second statement, note that $|D(x)| = 1$ for any
$x \in X$, and that the elimination graph $C_x = (X_1, E_1)$ is a cycle with $|X| = k$
vertices. By induction, any ordering $\hat{\alpha}$ on X_1 gives a minimum triangulation
of C_x. The assertion now follows.

3.4.4. COMPLETE BIPARTITE GRAPHS

A graph $G = (X, E)$ is bipartite if $X = R \cup B$ with $R \cap B = \varnothing$, and for each
$e = \{x, y\} \in E$ either $x \in R$, $y \in B$ or $y \in R$, $x \in B$. Equivalently, G is bipartite
if every cycle has even length [8, p. 86]. Because of the second condition,
trees are the only bipartite graphs which are triangulated.

Let $G = (X, E)$ be a bipartite graph with $X = B \cup R$ and $|R| \leqslant |B|$. If each
vertex $x \in R (x \in B)$ is adjacent to each vertex $y \in B (y \in R)$, the resulting
graph is a *complete bipartite graph*, denoted by $C_{n,m}$ ($n = |R|$, $m = |B|$).

By Theorem 2, in any triangulation of $C_{n,m}$, there must exist a vertex with
$D(x) = \varnothing$. Hence, to triangulate $C_{n,m}$ at least $n(n-1)/2$ edges are necessary.
However, this number of edges is clearly sufficient by taking the MTE
generated by the ordering $\alpha(i) = b_i$, $i = 1, \ldots, m$, $B = \{b_i\}_{i=1}^m$ and $\alpha(i) = r_i$,
$i = n+1, \ldots, n+m$, $R = \{r_i\}_{i=m+1}^{n+m}$.

4. Optimal Ordering and Algorithms

In this section we examine carefully several criteria by which we may
evaluate "optimal," and we relate these criteria to the computational complexity of the elimination process on sparse matrices. We give, first, a count of
the number of operations needed to effect the decompositions and backsolving
operations associated with solving symmetric sparse linear systems $Mx = b$.
In Section 4.2 we discuss criterion functions in a general setting, and in
Section 4.3 we present some results which give bounds for triangulations T
of a nontriangulated graph. Finally in Section 4.4 we discuss ordering
algorithms.

4.1. Operation Counts and Practical Criteria

Let M be an $n \times n$ symmetric positive definite matrix with ordered graph
$G = (X, E, \alpha)$. Denote by $d(\alpha(i))$ the degree of the vertex $\alpha(i)$ in the elimination
graph G_{i-1}, that is $d(\alpha(i)) = |\mathrm{adj}(\alpha(i))|$ in G_{i-1}. Where it causes no con-
fusion, $d(\alpha(i))$ will be written d_i. Using this notation we present the following:

THEOREM 5. Let M and G be as above. Counting multiplications and divisions as multiplications and operations, $a+0$, $a \neq 0$, which occur whenever $D(x_i) \neq \varnothing$ in G_{i-1} as additions, we have

(a) the LDL^T decomposition [see (8)] requires

(17) $$\sum_{i=1}^{n-1} d_i(d_i+3)/2 \quad \text{multiplications}$$

and

(18) $$\sum_{i=1}^{n-1} d_i(d_i+1)/2 \quad \text{additions};$$

(b) the Cholesky decomposition $M = GG^T$ [see (9)] requires the same number of multiplications and additions as in (a) and also n square roots;

(c) for a general n-vector b, the back-solving operations

(1) $Lz = b$,
(2) $Dy = z$,
(3) $L^T x = y$

require

(19) $$2\sum_{i=1}^{n-1} d_i + n \quad \text{multiplications}$$

and

(20) $$2\sum_{i=1}^{n-1} d_i \quad \text{additions.}$$

(d) the back-solving operations $Gy = b$ and $G^T x = y$ require n more multiplications than (19) and the same number of additions as (20).

Proof: By the discussion in Section 2.1 [see (7)–(9)] we see that (b) follows easily from (a). The proof of (a) is by induction on n. The case $n = 2$ is immediate. Suppose the theorem is true for $2 < n = k-1$ and let $G = (X, E, \alpha)$ have $|X| = k$. Referring to (7) and (8), the first step of elimination requires that we compute $s = r/a$ and $M - sr^T$ for all $1 = i \leqslant j \leqslant n$. This requires d_1 multiplications and $d_1(d_1+1)/2$ multiplications and additions. Hence, in total the first step of elimination requires

(21) $\quad d_1(d_1+3)/2$ multiplications $\quad$ and $\quad d_1(d_1+1)/2$ additions.

Since the graph of $M^{(2)}$ is the elimination graph G_1, we have, by induction that the decomposition of $M^{(2)}$ requires

(22) $\sum_{i=2}^{n-1} d_i(d_i+3)/2$ multiplications $\quad$ and $\quad \sum_{i=2}^{n-1} d_i(d_i+1)/2$ additions.

Adding (21) and (22) gives (17) and (18).

To verify (c), recall that the graph of L^T is $MTE(G;\alpha) = (X, E \cup \tau_i)$ and note that one addition and one multiplication are required for each edge in $E \cup \tau_i$ in the operations (1) and (3). Since

$$\sum_{i=1}^{n-1} d_i = |E \cup \tau_i|,$$

the result follows.

The Cholesky backsolving operations (d) require n more multiplications than the total in (c) because G has, in general, a nonunit diagonal.

These counts show that for a sparse $n \times n$ matrix M as above, the importance of n as a measure of computational complexity is relatively minor. For example, for such an arbitrary irreducible matrix we know, *a priori*, only that the number of multiplicative operations θ for the decomposition A satisfies

$$2(n-1) \leqslant \theta \leqslant \tfrac{1}{6}n(n-1)(n+4).$$

We consider three practical criteria for optimal ordering of a symmetric matrix M for elimination. While the minimum arithmetic criterion is suggested naturally by the operation counts given above, the minimum "fill in" and minimum bandwidth criteria are the two most commonly used.

4.1.1. MINIMUM ARITHMETIC

Let M be a symmetric matrix with ordered graph $G = (X, E, \alpha)$, $|X| = n$. Define

$$L(\alpha) = \sum_{i=1}^{n-1} d(\alpha(i)),$$

$$Q(\alpha) = \sum_{i=1}^{n-1} d^2(\alpha(i)),$$

and

$$J(p, q; \alpha) = pL(\alpha) + qQ(\alpha), \qquad p > 0, \quad q > 0.$$

Then, criteria based on minimizing arithmetic operations counted by Theorem 5 can be formulated as attempting to find $\hat{\alpha}$ such that

$$J(p, q; \hat{\alpha}) = \min_a J(p, q; \alpha)$$

for specific p and q. For example, to solve $Mx = b$ using the LDL^T decomposition and backsolving requires

$$J(\tfrac{1}{2}, \tfrac{7}{2}; \alpha) + n \quad \text{multiplications} \quad \text{and} \quad J(\tfrac{1}{2}, \tfrac{5}{2}; \alpha) \quad \text{additions.}$$

To compute $\det(M)$, which is the product of the diagonal entries of (D) requires

$$J(\tfrac{1}{2},\tfrac{3}{2};\alpha) + (n-1) \quad \text{multiplications} \quad \text{and} \quad J(\tfrac{1}{2},\tfrac{1}{2};\alpha) \quad \text{additions.}$$

The operation counts for these two specific computations above suggest a difficulty with the minimum arithmetic criterion. To define "optimal" for either of these computations requires a decision about the relative cost of additions, multiplications, and storage. Furthermore, an optimal ordering for solving $Mx = b$ is not necessarily an optimal ordering for the problem of computing $\det(M)$. Since both computations involve the decomposition $M = LDL^T$, it may be unsatisfactory, from the viewpoint of having a general sparse matrix package, to consider four different criteria in order to define optimal ordering for these two very similar computations. Specifically, the difficulty arises because *a priori* we cannot be assured that there exists an ordering α which minimizes $L(\alpha)$ and $Q(\alpha)$ simultaneously. In practice, it is common to attempt to minimize the less stringent fill in criterion $L(\alpha)$. To relate the $L(\alpha)$ criterion to the more general minimum arithmetic criterion, the following bound is relevant.

PROPOSITION 3. Let $G = (X, E, \alpha)$ be monotone transitive and

$$\mu(\alpha) = \max_{1 \leqslant i \leqslant n-1} d(\alpha(i)).^{\dagger}$$

Then

$$Q(\alpha) \leqslant \mu L(\alpha) - (\mu-1)\mu(\mu+1)/6,$$

and there exist graphs for which this bound is sharp.

Proof: The elimination graph G_i must contain a clique with μ vertices if $d(\alpha(i)) = \mu$ is the degree of the vertex $\alpha(i)$ in G_{i-1}, since G is monotone transitive. Hence, for all integers $l = \mu-1, \mu-2, \ldots, 1$ there exists an integer $k_l > i$ with $d(\alpha(k_l)) = l$.
Let

$$p_i = \{j \in I \mid d(\alpha(j)) = i\}, \qquad I = \{1, 2, \ldots, n-1\}.$$

Then

$$Q(\alpha) = \sum_{i=1}^{n-1} d^2(\alpha(i)) = \sum_{i=1}^{\mu} i^2 + \sum_{i=1}^{\mu} (|p_i|-1)i^2$$

$$\leqslant \sum_{i=1}^{\mu} i^2 + \mu \sum_{i=1}^{\mu} (|p_i|-1)i$$

$$= \mu \sum_{i=1}^{n-1} d_i - \sum_{i=1}^{\mu} (\mu-i)i = \mu \sum_{i=1}^{n-1} d_i - (\mu-1)\mu(\mu+1)/6.$$

Finally, the monotone transitive graph of Fig. 4 shows that equality is possible.

† Note that $\mu+1$ is the number of vertices in the largest clique of G.

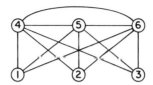

Fig. 4. Monotone transitive graph, where

$d(1) = d(2) = d(3) = 3, d(4) = 2, d(5) = 1.$

4.1.2. Minimum Fill-In

Let $G = (X, E)$ be a graph with monotone transitive extention $\hat{G} = (X, E \cup T(\alpha))$. Then, since $\hat{G}$ is monotone transitive,

(23) $$L(\alpha) = \sum_{i=1}^{n-1} d(\alpha(i)) = |E| + |T(\alpha)|$$

because each edge in $E \cup T(\alpha)$ is counted once, and only once, in some $d(\alpha(i))$. Thus, by minimizing $L(\alpha)$ over all orderings, we minimize the fill in $T(\alpha)$ caused by elimination. Then, $T(\alpha)$ is a minimum triangulation of the graph.

Various authors [20, 23, 25, 26, 29, p. 25] have taken the criterion of minimum fill in as the "appropriate" criterion for defining optimal orderings. However, the effect of minimizing $L(\alpha)$ upon the count of necessary arithmetic operations for certain computations seems to have been overlooked in the literature. It is certainly not the case, as is evident from the discussion above, that minimizing $L(\alpha)$ necessarily minimizes arithmetic. Note, however, that for any α we must store $L(\alpha) + |X|$ nonzero numbers for D and L in the decomposition of M corresponding to G. We call this *primary storage*, as opposed to the secondary storage necessary to determine which elements of L are nonzero.

We think the advantages of using $L(\alpha)$ as a criterion for optimal ordering are as follows:

(1) minimizing $L(\alpha)$ minimizes primary storage;
(2) minimizing $L(\alpha)$ minimizes the backsolving operations of Theorem 5(c);
(3) for a graph in which $\mu(\alpha)$ of Proposition 3 can be bounded independent of $|X| = n$, a satisfactory bound on arithmetic operations can be given which is minimized with $L(\alpha)$;
(4) if $L(\alpha)$ is minimum, the triangulation $T(\alpha)$ of G is a minimum triangulation, that is, the function $L(x)$ on a graph has graph-theoretic significance.

4.1.3. MINIMUM BANDWIDTH

For an $n \times n$ symmetric matrix M, it is natural to define M to have *bandwidth* $k \geq 0$ if

$$(24) \qquad k = \max_{\{i,j\} \in A} (j - i),$$

where

$$A = \{\{i,j\} \mid i < j \text{ and } m_{ij} \neq 0\}.$$

Thus, a symmetric tridiagonal matrix has bandwidth $k = 1$. Equation (24) is consistent with the recent paper of Cuthill and McKee [9], although other authors count the diagonal and subdiagonals in their definition of bandwidth [12, p. 15]. If $G = (X, E, \alpha)$ is the ordered graph of M, clearly M has bandwidth k, if and only if

$$k = \max_{1 \leq i \leq n-1} \ \max_{y \in M \, \text{adj}(x_i)} (\alpha^{-1}(y) - i).$$

Furthermore, since $|\text{adj}(x)| = |M \, \text{adj}(x)| + |\{y \in X \mid x \in M \, \text{adj}(y)\}|$, it follows easily that

$$(25) \qquad k \geq \max_{x \in X} \{[\tfrac{1}{2} |\text{adj}(x)|]\},$$

where $[p]$ is the least integer $l \geq p$.

Matrix bandwidth minimization has enjoyed considerable popularity in matrix methods of structural analysis, see, for example, Livesley [16], McCormick [29, p. 155], Cuthill and McKee [9], and Rosen [19]. By using bandwidth methods, these authors attempt to limit fill in and arithmetic to a level acceptable for their applications. The popularity of bandwidth methods is partially justified by the following two properties of bandwidth analysis. First, for a symmetric matrix M of bandwidth k and ordered graph $G = (X, E, \alpha)$, all the fill in of M due to elimination is constrained within the bandwidth. That is, the graph $M \, \text{TE}(G; \alpha)$ also has bandwidth k.[†] Second, the special elimination scheme for a symmetric matrix of bandwidth k is relatively easy to implement on a digital computer primarily because necessary data handling and indexing is simplified (see the discussions in McCormick [29, p. 155] and Cuthill and McKee [9, Section 1]).

Note, however, that the effectiveness of bandwidth implicitly presupposes that the band width k of a symmetric $n \times n$ matrix M will be small relative to n. Bandwidth analysis is crude, in general, because k need not be small relative to $|X| = n$, and because this analysis takes no account of the zero–nonzero structure within the band. To substantiate this claim, we appeal to Theorem 4

[†] This is clear since $k \geq |M \, \text{adj}(x)|$, and making $M \, \text{adj}(x)$ a clique does not increase k.

and the corollary to Theorem 3. Theorem 4 states that the sets $M\,adj\,(x_i)$ are separation cliques in the extension graph $M\,TE(G;\alpha)$. Suppose that the vertices are being ordered by some sequential scheme to attain a minimum, or approximate minimum, bandwidth.[†] If, in the elimination graph G_i, the set $S = M\,adj(x_i)$ is a separation clique breaking G into $c \geqslant 2$ components, Theorem 3 and Corollary 1 (Section 3) implies that in G_i the vertices in S should be ordered after those in all but one component. Bandwidth minimization, however, will tend to order the vertices in S immediately. This causes redundant edges in successive elimination graphs which may increase the bound (25) in subsequent elimination graphs.[‡] We illustrate this phenomenon in the following example.

Example: The snowflake graph shown in Fig. 5 provides an example where bandwidth ordering (Cuthill–McKee algorithm) orders separating sets too early. Note also that for this graph, (25) gives the overly optimistic bound $k \geqslant 3$. In fact, this ordering gives $k = 6$, where k is not small relative to $|X| = 18$.

4.2. Criterion Functions

Let $G = (X, E, \alpha)$ be a monotone transitive graph with $|X| = n$. Then

$$\sum_{i=1}^{n-1} d(\alpha(i)) = |E|,$$

that is, the $(n-1)$ integers $d(\alpha(i))$ form a *partition*, or degree partition, of $|E|$.

For two ordered monotone transitive graphs $G_\alpha = (X, E, \alpha)$ and $G_\beta = (X, F, \beta)$ with $|X| = n$, the partitions of $|E|$ and $|F|$ generated by the $d(\alpha(i))$ and the $d(\beta(i))$, respectively, will be called *equal*, if there exists a permutation π on the integers $1, 2, ..., n-1$ such that

$$d(\alpha(i)) = d(\beta(\pi(i))), \qquad i = 1, 2, ..., n-1.$$

Similarly, the partition generated by the $d(\alpha(i))$ *dominates* the partition generated by the $d(\beta(i))$, if

$$d(\alpha(i)) \geqslant d(\beta(\pi(i))), \qquad i = 1, 2, ..., n-1.$$

[†] As, for example, the algorithm presented by Cuthill and McKee [9], which is probably the best available for large order graphs ($|X| = 10^3$–10^5). It can be combined with the recent algorithm of Rosen [19] for further improvements (see Cuthill and McKee [9, p. 12]). See Akyuz and Utku [1] and Alway and Martin [2] for other bandwidth algorithms.

[‡] This analysis explains the results of Cuthill and McKee [9, p. 15] where it is reported that, for several sets of graphs generated randomly with $|X| = 50$ and $100 \leqslant |E| \leqslant 150$, the average bandwidth, after using the Cuthill–McKee algorithm or the Cuthill–McKee–Rosen modification, ranged from $k = 17$ to $k = 28.2$. k is not found to be small relative to $|X|$ in these experiments.

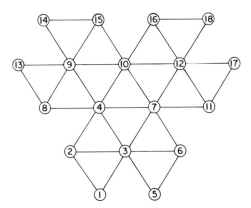

Fig. 5. Snowflake graph with bandwidth ordering given by the Cuthill–McKee algorithm [9], where $k = 6$. Vertices ③ and ④ are ordered too early.

We consider a class of functions defined on the quantities $d(\alpha(i))$ each of which may represent a cost of elimination, or if the graph $G = (X, E)$ is not triangulated, these functions can be considered as *criterion functions* for choosing an optimal ordering.

As criterion functions for the graph $G = (X, E)(|X| = n)$, we choose the class of symmetric isotone functions, that is, real valued functions

$$F(a_1, a_2, ..., a_{n-1}), \qquad a_i \quad \text{integer}$$

such that

(1) $F(a_1, a_2, ..., a_{n-1}) = F(a_{\sigma(1)}, a_{\sigma(2)}, ..., a_{\sigma(n-1)})$, where σ is any permutation on $\{1, 2, ..., n-1\}$;

(2) $F(a_1, a_2, ..., a_{n-1}) \geqslant F(b_1, b_2, ..., b_{n-1})$ when $a_i \geqslant b_i$, $i = 1, ..., n-1$.

We now show, see Theorem 6, that if F is a criterion function for a triangulated graph $G = (X, E)$ with distinct monotone transitive orderings α and β, then

$$F(d(\alpha(1)), d(\alpha(2)), ..., d(\alpha(n-1))) = F(d(\beta(1)), d(\beta(2)), ..., d(\beta(n-1))).$$

Furthermore, by Theorem 7, if γ is any nonmonotone transitive ordering of X, then $G_\alpha = (X, E, \alpha)$ is a subgraph of $\mathrm{MTE}(G_i; \gamma)$ and we show

$$F(d(\gamma(1)), d(\gamma(2)), ..., d(\gamma(n-1))) > F(d(\alpha(1)), d(\alpha(2)), ..., d(\alpha(n-1))).$$

Thus, with respect to criterion functions on triangulated graphs, monotone transitive orderings may be regarded as *optimal*.

THEOREM 6. Let $G = (X, E)$ be triangulated and let α and β be two distinct monotone transitive orderings of X. Then the two partitions of $|E|$ generated by the $d(\alpha(i))$ and the $d(\beta(i))$ are equal.

Proof: We use induction on $|X|$. The case $|X| = 2$ is clear. Suppose that, in the case $|X| = k-1$, we consider G with $|X| = k$. If $\alpha(1) = \beta(1) = x$, the result follows immediately from the induction hypothesis on the elimination graph G_x. Suppose, then, that $\alpha(1) = y$ and $\beta(1) = z$. Note that if $y \in \text{adj}(z)$ then $\text{adj}(y) - \{z\} = \text{adj}(z) - \{y\}$ by monotone transitivity and $|\text{adj}(y)| = |\text{adj}(z)|$. Consider the new monotone transitive orderings $\hat{\alpha}$ and $\hat{\beta}$ defined by

$$\hat{\alpha}(1) = y, \qquad \hat{\beta}(1) = z,$$

$$\hat{\alpha}(2) = z, \qquad \hat{\beta}(2) = y,$$

$$\hat{\alpha}(i) = \hat{\beta}(i) = \gamma(i-2), \qquad i = 3,\dots,n,$$

where γ is any monotone transitive ordering of the triangulated section graph $G(X - \{x, y\})$. By the first part of the proof, the partitions generated by $d(\hat{\alpha}(i))$ and $d(\alpha(i))$ are equal, as are those generated by $d(\hat{\beta}(i))$ and $d(\beta(i))$. It remains to show that the partitions generated by $d(\hat{\alpha}(i))$ and $d(\hat{\beta}(i))$ are equal. Now $\hat{\alpha}$ yields the partition $|\text{adj}(y)|$, $|\text{adj}(z)|$, $d(\gamma(i-2))$, $i = 3,\dots,n$, if $y \notin \text{adj}(z)$ and $|\text{adj}(y)|$, $|\text{adj}(z)| - 1$, $d(\gamma(i-2))$, $i = 3,\dots,n$, if $y \in \text{adj}(z)$. But β gives an equal partition in each case because $|\text{adj}(y)| = |\text{adj}(z)|$ if $y \in \text{adj}(z)$.

LEMMA 8.   Let $G = (X, F)$ be triangulated with triangulated subgraph $\hat{G} = (X, E)$, $E \subseteq F$. If α is any monotone transitive ordering for both G and $\hat{G}$, then the degree partition of F dominates the degree partition of E.

Proof: Clearly, $\hat{d}(x_1) \leqslant d(x_1)$, and the elimination graphs $\hat{G}_1 = (X - \{x_1\}, E_1)$ and $G_1 = (X - \{x_1\}, F_1)$ are triangulated with $E_1 \subseteq F_1$. The proof then follows by induction on $|X|$.

LEMMA 9.   Let $G = (X, E)$ be triangulated and $x \in X$. Then $\hat{G} = (X, E \cup D(x))$ is triangulated.

Proof: Assuming $D(x) \neq \varnothing$, we need only show that cycles in $\hat{G}$ of the form

$$\mu = [x_1, y_1, p_1, \dots, p_n, x_1], \qquad n \geqslant 2$$

with $\{x_1, y_1\} \in D(x)$ have a chord. These are two cases.

  (1)   if some $p_i \in \text{adj}(x)$, then there is a chord $\{x_1, p_i\}$ (or $\{y_1, p_n\}$, if $i = n$) in $E \cup D(x)$;
  (2)   if no $p_i \in \text{adj}(x)$, the cycle $\mu' = [x_1, x, y_1, p_1, \dots, p_n, x_1]$ in G has a chord $\{x_1, p_i\}$, $\{y_1, p_j\}$, or $\{p_i, p_j\}$ in E, which is also in chord in $\hat{G}$.

LEMMA 10.   Let $\hat{G} = (X, E)$ and $G = (X, F)$ be triangulated with strict inclusion $E \subset F$. Then there exists a monotone transitive ordering α for G such that in $\hat{G}$, $\text{M TE}(\hat{G}; \alpha) = (X, E \cup T(\alpha))$ with strict inclusion $(E \cup T(\alpha)) \subset F$.

Proof: If X is a clique in G, the lemma is true for any α which is a monotone transitive ordering for $\hat{G}$. Hence, we assume X is not a clique in G and prove the assertion by induction on $|X|$. One easily verifies the cases when $|X| = 4$, and, assuming the case $|X| = k - 1$, we consider such graphs G and $\hat{G}$ with $|X| = k$.

Let $S = \{y \in X \mid D(y) = \varnothing \text{ in } G\}$. We first dispense with two cases.

(1) If for some $y \in S$, $[\text{adj}(y)]_{\hat{G}} \subset [\text{adj}(y)]_G$, that is, there is an edge $e = \{y, x\} \in F - E$, then by choosing any monotone transitive ordering for G with $\alpha(1) = y$ we have $(E \cup T(\alpha)) \subset F$.

(2) If some $y \in S$ with $[\text{adj}(y)]_{\hat{G}} = [\text{adj}(y)]_G$ has $D(y) = \varnothing$ in $\hat{G}$ also, then by choosing $\alpha(1) = y$, the lemma follows by the induction hypothesis on the elimination graph G_y and $\hat{G}_y$.

These cases being dismissed, we may assume that for each $y \in S$, $[\text{adj}(y)]_{\hat{G}} = [\text{adj}(y)]_G$, and that the clique $\text{adj}(y)$ in G contains at least one pair of vertices $e_y = \{v_1, v_2\} \in F - E$. By Corollary 2 and Lemma 6 (Section 3), since X in G is not a clique, there exists $y, z \in S$ with $y \notin \text{adj}(z)$. For such vertices $e_y \neq e_z$, because if $e_y = e_z = \{v_1, v_2\}$, the cycle $\mu = [y, v_1, z, v_2, y]$ has no chord in E, so $\hat{G}$ could not be triangulated.

Hence, for some $y \in S$, choose $\alpha(1) = y$ and consider the y-elimination graphs $\hat{G}_y = (X - \{y\}, E_1)$ and $G_y = (X - \{y\}, F_1)$. It is clear from the above that strict inclusion $E_1 \subset F_1$ holds because there exists a $z \in S$ with $y \notin \text{adj}(z)$ and such that $e_z \in F_1$, but $e_z \notin E_1$. Also, by Lemma 9, $\hat{G}_y$ is triangulated, as is G_y. Hence, the lemma follows by using induction on the graphs G_y and $\hat{G}_y$.

These lemmas give us Theorem 7.

THEOREM 7. Let $\hat{G} = (X, E)$ and $G = (X, F)$ be triangulated with $E \subseteq F$. Let α and β be monotone transitive orderings of $\hat{G}$ and G, respectively. Then the degree partition of $|F|$ dominates the degree partition of $|E|$.

Proof: We use induction of $|F|$. If $|F| = |X| - 1$, that is, G is a tree, then $E = F$, because both G and $\hat{G}$ are assumed connected, and the conclusion follows from Theorem 6. Suppose the theorem is true whenever $|X| - 1 < |F| \leqslant k - 1$, and let G and $\hat{G}$ be as above with $|F| = k$. If the subgraph $\hat{G} = (X, E)$ has $E = F$, then again the conclusion follows from Theorem 6. Assume then $E \subset F$ (strict). By Lemma 10, there exists a monotone transitive ordering $\hat{\alpha}$ for G such that $\text{MTE}(\hat{G}; \hat{\alpha}) = (X, E \cup T(\hat{\alpha}))$ and $(E \cup T(\hat{\alpha})) \subset F$ (strict). By the induction hypothesis, the degree partition of $|E \cup T(\hat{\alpha})|$, generated by $\hat{\alpha}$ in the triangulated graph $\text{MTE}(\hat{G}; \hat{\alpha})$, dominates the degree partition of $|E|$, and by Lemma 8 the degree partition of $|F|$ generated by $\hat{\alpha}$ dominates the degree partition of $|E \cup T(\hat{\alpha})|$. By Theorem 6, the degree partitions of $|F|$ generated by β and $\hat{\alpha}$ are equal.

Theorem 7 has the following important implication: If T is a triangulation of a graph $G = (X, E)$, the T is *minimal* if no $\hat{T} \subset T$ is also a triangulation of G. Clearly, a minimum triangulation is minimal, but a minimal triangulation need not be minimum. Theorem 7 implies that if T is a nonminimal triangulation of G, and $\hat{T} \subset T$ is also a triangulation, then the cost of elimination with a monotone transitive ordering of $G_1 = (X, E \cup T)$ is greater than the cost of elimination with a monotone transitive ordering of $G_2 = (X, E \cup \hat{T})$ for any criterion function.

4.3. Bounds for Triangulations

Since the size of a triangulation T of a nontriangulated graph G is one indication of the computational complexity of G, that is, M, with respect to elimination, we seek bounds on $|T|$ which are related to the structure of G. Corollary 5 below relates the size of a minimal triangulation to the size of minimal a, b separators in G. Theorem 9 shows that if k edges of G can be deleted to yield a triangulated graph, then G itself can be triangulated with $|T| \leqslant kn$.

THEOREM 8. Let $G = (X, E)$ be a graph with minimal triangulation T. Then every minimal a, b separator of $\hat{G} = (X, E \cup T)$ is a minimal a, b separator of G.

Proof: If S is an a, b separator of $\hat{G}$, clearly it is an a, b separator of G. Suppose S is minimal in $\hat{G}$ but not in G, that is, $S' \subset S$ is also an a, b separator in G. Let C_i be the components of G with respect to S'. Since some vertices in S are in the C_i, returning to $\hat{G}$ where S is minimal implies there must be edges $T_0 \subset T$ with vertices in different components C_i. S' is a clique in $\hat{G}$, and by removing edges in T_0 the graph $\tilde{G} = (X, E \cup T - T_0)$ is triangulated by Theorem 3. Thus, T is not minimal, which contradicts our hypothesis.

COROLLARY 5. Let $G = (X, E)$ be a graph, $|X| = n$, such that every minimal a, b separator S of G satisfies $|S| \leqslant k$. If T is a minimal triangulation of G, then

$$|T| \leqslant (n - \mu)(k(k+1)/2)$$

where

$$\mu = \max\{|C|, C \text{ a clique of } G\}.$$

Proof: By the remark following the statement of Theorem 4, there are, at most, $n - \mu$ minimal a, b separators in $\hat{G} = (X, E \cup T)$. The proof then follows by the above theorem and the hypothesis $|S| \leqslant k$.

THEOREM 9.[†] Let $T = (X, E)$ be triangulated and $G = (X, E \cup F)$. Let $V \subseteq X$ be a set of vertices covering F, that is, if $f = \{x, y\} \in F$, then $x \in V$ or $y \in V$. If $|X| = n$ and $|V| = m$, G can be triangulated with, at most,

$$nm - (m(m+1)/2) \quad \text{edges.}$$

Proof: Let $H = (X, E \cup F \cup T)$, where $T = \{\{v, x\} \mid v \in V, x \in X, v \neq x\}$. Note that $|T| = nm - (m(m+1)/2)$. We show H is triangulated. If μ is a cycle of H with length $l \geqslant 4$ and

(1) μ contains no vertex in V, then μ is a cycle of T and, hence, has a chord;

(2) μ contains a vertex $v \in V$, then H contains the chord $\{vx\}$ for any $x \in \mu$ not adjacent to v.

Note that the bound in Theorem 9 can be improved, if G has some separation cliques which are also separation cliques of T.

4.4. Ordering Algorithms

4.4.1. DYNAMIC PROGRAMMING

Given any criterion function F as defined in Section 4.2 and a graph G, it is possible to find an ordering $\hat{\alpha}$ which minimizes F by using the dynamic programming technique of Bertele, Brioschi, and Even [4, 5, 7], who consider the specific criterion function

$$y(\alpha) = \max_{1 \leqslant i \leqslant n-1} d(\alpha(i)).$$

However, for a graph with n vertices, the complexity of this algorithm and the storage requirements increase as 2^n. Hence, this algorithm is not feasible for large graphs, and no other general algorithm ensuring optimality is known.

In practice, it is tacitly agreed that a near optimal ordering is acceptable if the ordering algorithm is efficient. For example, the complexity grows only as n^p, p small. In the literature[‡] it is assumed that the next two algorithms to be discussed produce near optimal orderings. While some experimental results reported by Tinney [29, p. 25], confirm this assumption, no detailed study of these algorithms has been reported.

---

† From a private communication from A. Hoffman, IBM research, Yorktown Heights, New York.

‡ See, for example, Sato and Tinney [20], Tinney and Walker [26], and the summary paper by Tinney [29, p. 25], who use these algorithms for ordering sparse symmetric matrices. For similar algorithms applied to the nonsymmetric case, see Tewarson [29, p. 35].

4.4.2. MINIMUM DEGREE ALGORITHM

Let $G_0 = G = (X, E)$. The minimum degree algorithm orders X as follows:

(1) set $i = 1$;

(2) in the elimination graph G_{i-1}, choose x_i to be any vertex such that

$$|\mathrm{adj}(x_i)| = \min_{y \in X_{i-1}} |\mathrm{adj}(y)|$$

where

$$G_{i-1} = (X_{i-1}, E_{i-1});$$

(3) set $i = i + 1$;

(4) if $i > |X|$, stop;

(5) go to Step (2).

The advantage of this algorithm is its speed. $n(n+1)/2$ vertices are tested, and each test simply counts adjacent vertices. The disadvantages of the algorithm are

(1) the algorithm does not, in general, produce a monotone transitive ordering when the graph is triangulated (see Fig. 6)[†];

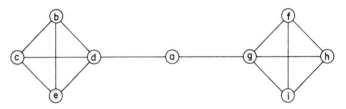

Fig. 6. Vertex ⓐ has minimum degree in the triangulated graph G. Since {a} is a separation clique, ordering ⓐ first leads to a nominimal triangulation.

(2) the algorithm does not, in general, produce a minimal triangulation (again see Fig. 6);

(3) there exist examples when the triangulation produced by this ordering is arbitrarily greater than a minimum triangulation (see following Example and Fig. 7).

Example: Let $n < m$ and C_{m-1} be a clique on $m-1$ vertices. Each of the vertices a_i is adjacent to each vertex of the clique C_{m-1}. Vertex x is adjacent to each a_i. Vertex x has minimum degree, $|\mathrm{adj}(x)| = n$, and the elimination graph G_x is the clique C_{n+m-1}. This triangulation obtained by ordering x first requires $n(n-1)/2$ edges. However, the triangulation obtained by ordering the a_i first (note that $|\mathrm{adj}(a_i)| = m-1$) requires only $m-1$ edges.

[†] It is easy to see that the algorithm will produce a monotone transitive ordering when G is a k tree.

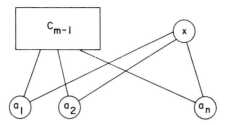

Fig. 7

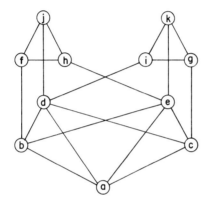

Fig. 8. Vertex $\textcircled{a}$ has $|D(a)| = 2$ and is a minimum deficiency vertex. However, in the elimination graph G_a, the edge $\{b,c\}$ is redundant in a triangulation given edge $\{d, e\}$ since $S = \{a, d, e\}$ is then a,b,c separation clique of G_a. Thus, given a deficiency set $D(x)$, only some subset of $D(x)$ may be necessary in a minimal triangulation.

4.4.3. MINIMUM DEFICIENCY ALGORITHM

Letting $G_0 = G = (X, E)$, the minimum deficiency algorithm orders X as follows:

(1) set $i = 1$;
(2) in the elimination graph G_{i-1} choose x_i to be any vertex such that

$$|D(x_i)| = \min_{y \in X_{i-1}} |D(y)|,$$

where

$$G_{i-1} = (X_{i-1}, E_i);$$

(3) set $i = i+1$;
(4) if $i > |X|$, stop;
(5) go to Step (2).

The advantages of the minimum deficiency algorithm are

(1) only $n(n+1)/2$ deficiency counts are needed to compute the ordering;
(2) the algorithm produces a monotone transitive ordering when the graph is triangulated; also, in this case, ordering a vertex as soon as the $D(x) = \emptyset$ condition is recognized leads to *fewer* than $n(n+1)/2$ deficiency counts.

The disadvantages are

(1) the algorithm is slower than the minimum degree algorithm, because in addition counting, or listing, vertices in $\mathrm{adj}(x)$, pairs of vertices in $\mathrm{adj}(x)$ must be edge tested;
(2) the algorithm does not, in general, produce a minimal triangulation (see Fig. 6, with edges $\{c,d\}$, $\{b,e\}$, $\{f,i\}$, $\{g,h\}$ deleted; see Fig. 8).

References

1. Akyuz, F. A. and Utku, S. An automatic relabeling scheme for bandwidth minimization of stiffness matrices, *AIAA J.* 6, 728–730 (1968).
2. Alway, G. and Martin, D., An algorithm for reducing the band width of a matrix of symmetrical configuration, *Comput. J.* **8**, 264–272 (1965).
3. Berge, C., Some classes of perfect graphs, "Graph Theory and Theoretical Physics" (F. Harary, ed.), pp. 155–166. Academic Press, New York, 1967.
4. Bertele, U. and Brioschi, F., A new algorithm for the solution of the secondary optimization problem in nonserial dynamic programming, *J. Math. Anal. Appl.* **27**, 565–574 (1969).
5. Bertele, U., and Brioschi, F., Contributions to nonserial dynamic programming, *J. Math. Anal. Appl.* **28**, 313–325 (1969).
5a. Boland J. and Lekkerkerker C., Representation of a finite graph by a set of intervals on the real line, *Fundam. Math.* **51**, 45–64 (1962).
6. Brayton, R. K., Gustavson, F. G., and Willoughby, R. A., Some results on sparse matrices, *Math. Comp.* (on press).
7. Brioschi, E., and Even, S., Minimizing the number of operations in certain discreet variable optimization problems, *Tech. Rep. 567, Harvard University, Div. Eng. and Applied Physics.*
8. Busacker, R. G., and Saaty, T. L., "Finite Graphs and Networks." McGraw-Hill, New York, 1965.
9. Cuthill, E., and McKee, J., Reducing the bandwidth of sparse symmetric matrices, *Proc. ACM 23rd National Conf.* (1969).
9a. Dirac, G. A., On rigid circuit graphs, *Abh. Math. Sem. Univ. Hamburg* **25**, 71–76 (1961).
10. Dulmage, A. L., and Mendelsohn, N. S., On the inversion of sparse matrices, *Math. Comp.* **16**, 494–496 (1962).
11. Dulmage, A. L., and Mendelsohn, N. S.,Two algorithms for biparite graphs, *SIAM J. Appl. Math.* **11**, 183–194 (1963).

12. Forsythe, G. E., and Moler, C. B., "Computer solution of Linear Algebraic Systems." Prentice-Hall, Englewood Cliffs, New Jersey, 1967.

12a. Fulkerson, D., and Gross, O., Incidence matrices and interval graphs, *Pacific J. Math.* **15**, 835–855 (1965).

13. Harary, F., A graph theoretic method for the complete reduction of a matrix with a view toward finding its eigenvalues, *J. Math. and Phys.* **38**, 104–111 (1959).

14. Harary, F., A graph theoretic approach to matrix inversion by partitioning, *Numer. Math.* **4**, 128–135 (1962).

15. Gustavson, F. G., Liniger, W., and Willoughby, R., Symbolic generation of an optimal Crout algorithm for sparse systems of linear equations, *J. Assoc. Comput. Mach.* **17**, 87–109 (1970).

16. Livesley, R. K., The analysis of large structural systems, *Comput. J.* **3**, 34–39 (1960).

17. Parter, S., The use of linear graphs in Gauss elimination, *SIAM Rev.* **3**, 119–130 (1961).

18. Rose, D. J., Symmetric elimination on sparse positive definite systems and the potential flow network problem, Ph.D. Thesis, Harvard University (1970).

19. Rosen, R., Matrix bandwidth minimization, *Proc. ACM 23rd National Conf.* 585–595 (1968).

20. Sato, N., and Tinney, W. F., Techniques for exploiting the sparsity of the network admittance matrix, *IEEE, PAS*, 944–950 (1963).

21. Steward, D. V., Partitioning and tearing systems of equations, *SIAM J. Numer. Anal.* **2**, 345–365 (1965).

22. Tewarson, R. P., On the product form of inverses of sparse matrices, *SIAM Rev.* **8**, 336–342 (1966).

23. Tewarson, R. P., The product form of inverses of sparse matrices and graph theory *SIAM Rev.* **9**, 91–99 (1967).

24. Tewarson, R. P., Row–column permutation of sparse matrices, *Comput. J.* **10**, 300–305 (1967).

25. Tewarson, R. P., Solution of a system of simultaneous linear equations with a sparse coefficient matrix by elimination methods, *BIT* **7**, 226–239 (1967).

26. Tinney, W. F., and Walker, J. W., Direct solutions of sparse network equations by optimally ordered triangular factorization, *Proc. IEEE* **55**, 1801–1809 (1967).

27. Westlake, J. R., "A Handbook of Numerical Matrix Inversion and Solution of Linear Equations." Wiley, New York, 1968.

28. Wilkinson, J. H., "The algebraic Eigenvalue Problem." Oxford Univ. Press (Clarendon), London and New York, 1965.

29. Willoughby, R. A. (ed.), IBM sparse matrix proceedings, *IBM Report RAI*, No. 11707 (1969).

INTELLIGENT GRAPHS: NETWORKS OF FINITE AUTOMATA CAPABLE OF SOLVING GRAPH PROBLEMS

P. Rosenstiehl
École Pratique des Hautes Études
Paris, France

J. R. Fiksel
Institut de Programmation
Paris, France

A. Holliger
École Polytechnique
Paris, France

1. Introduction to Myopic Algorithms

The value of myopic algorithms is well known, but their exact nature is poorly defined. Here we give an abstract formalization of the myopic property

219

in terms of finite automata. Intuitively, we say that an algorithm is *myopic* if, for each of the elementary activities into which the computation is divided, reference is made only to a restricted and well defined subset of the data on which the algorithm operates.

By *set of data* we mean, on the one hand, a mathematical structure on which computations are performed—in this chapter, a graph and various modules and structures of order defined on the graph—and, on the other hand, the partial results already obtained from previous computations. The myopic property relies on the storage of the given data within the graph itself, at the very places where those data were originally defined, or have been generated during the process of the algorithm.

At each vertex of the graph we install an automaton which is the local representative of the algorithm. It computes locally on data relating to that vertex and to part of the data relating to those other vertices to which it is directly connected. The graph itself is thus provided with means of computation, the vertex automata, for solving problems concerning itself. Hence, we obtain our title, *intelligent graphs*. The myopic property is often favorable for parallelism in the computation. Many local elementary computations whose order of execution matters little are conducted in parallel, each upon a restricted subset of the total data.

The automata used here are *state-output automata*, abstractly described by a 3-tuple $M = (E, L, \phi)$, where E is a set of states, L is a set of input letters, $\phi : E \times L \to E$ is the state-transition function, and the output of M is its state. The automaton is said to be finite if E is finite. We physically interpret such an automaton as a discrete-time system which, if at time t it is in state e and receives input letter l, then at time $t+1$ it will be in state $\phi(e, l)$.

A myopic algorithm for graphs is formalized by a network of identical finite automata. The input letter of automaton x consists of elements taken from the state of each of the automata to which x is directly connected. This formalization allows us to introduce such terms as computability by finite automata (fa) for a given graph problem, the computation time necessary for the automata to locally display a desired solution, and the number of states of each identical automaton, which is a measure of the complexity of the computation. The idea for this formalization of the myopic property in graphs came to us [15]

(1) from the study of some admirable myopic algorithms, the labyrinth algorithms of Tarry [19, 20];

(2) from the work of Moore in "The Shortest Path Through a Maze," where the number of states necessary per vertex is minimized [13];

(3) finally from our generalization, to finite connected networks, of the

well-known "firing squad problem" [12], introduced into switching and automata theory in its linear form by Myhill and Moore.

The class of fa-computable problems has proved to be larger than we first anticipated. For example, an intelligent graph can not only decompose itself into blocks, but also display its minimum tree for a given total ordering of the edges. Berstel [5] has shown, in the same vein, that an intelligent graph can solve certain graph problems such as the construction of a hamiltonian cycle. Here we prove that this can be done with five colors.

Although our abstract formalization of myopic algorithms requires as many automata as there are vertices of a graph, our results are interesting for any type of programming methods using a single machine. In particular, this is true of the following algorithms, which are dealt with in later sections, and are new in the literature on theory of graphs and computer science:

(1)   recoil algorithms for mazes;
(2)   direct construction of an Eulerian path;
(3)   edges-ordering compatible with 1-adjacency;
(4)   vertices-ordering compatible with 2-adjacency;
(5)   block decomposition;
(6)   Hamiltonian cycle.

The algorithms listed above could be adapted efficiently to the push-down methods for graphs developed by Dermiane and Pair [8].

The study of graph problems has led us to define three fundamental problems specific to networks of automata, which are

(1)   a self-synchronization process, which uses a three-position counter, making the network completely autonomous;
(2)   various generalizations of the famous firing squad problem to any network of automata;
(3)   a problem which appears to be just as important and basic as the firing squad problem, which we call the "early bird problem," and which opens the door to optimization computations.

We feel that it is not the purpose of this book to emphasize those aspects of our subject related to the algebraic theory of automata. Thus, we have stressed the algorithmic and computational aspects. Nevertheless, from the algebraic viewpoint, the model of a cellular automaton which constructs its input from the states of its neighbors is a rich and promising one, especially in the light of our stacking form of composition in which elementary automata may be combined to form more powerful ones.

In summary, the theory of networks of finite automata permits a measurement of the complexity of myopic computation, and, furthermore, offers an attractive model for acentric systems, that is, organizations where all elements are identical and none are indispensable. They are commonly observed in the biological, sociological, and technological domains. In the art of computing science, networks of automata are distinctly different from such structures as iterative arrays and tessellation automata [24], are more flexible and less vulnerable, and could represent a model for microprogramming rather than for the actual computing hardware.

2. Finite Graphs and Finite Automata

2.1. Networks of Finite Automata

We shall use the term *finite graph*[†] to designate a triplet (X, U, α), where X and U are two finite disjoint sets and α is a function with domain U, taking values in the set of unordered pairs of elements of X not necessarily distinct.

$x \in X$ is called a *vertex* of the graph, $u \in U$ is called an *edge* of the graph, $\alpha(u) = (a, b)$ consists of two vertices a and b, which are not necessarily distinct, called the *extremities* of the edge u. The *degree* $v(x)$ of a vertex x is the number of times that x appears in the elements of Im α, the image of α. We shall say that the *degree* d of the graph (X, U, α) is the maximum value of $v(x)$ for all $x \in X$, $d = \max_{x \in X} v(x)$.

We shall use the term *finite automaton* to designate a triplet (E, L, ϕ) where E and L are two finite sets, ϕ is a function from $E \times L$ into E

$$\phi : E \times L \rightarrow E,$$

$e \in E$ is called a *state* of the automaton, and $l \in L$ is called an *input letter* of the automaton. ϕ is called the *state-transition function*. $\phi(e, l)$ is the next state taken by an automaton which finds itself in state e and reads the input letter l. We shall agree that ϕ allows the automaton to change state from time t to time $t+1$, for all $t \in N$. We shall index e and l with the subscript t, and write

$$e_{t+1} = \phi(e_t, l_t).$$

In our network of automata, all the automata will be identical, and the input letter of each one will consist of a function of the state of each of the automata to which it is directly connected.

We shall use the term *network of finite automata* to designate a 5-tuple (X, d, ρ, E, ϕ). Here X is a finite set whose elements may be called vertices or automata, interchangeably. $d \in N$ is called the valence of the network, or the number of *limbs* of the automata of the network, where a limb of X is a pair (x, r) with $x \in X$ and $r \in [d] = \{1, 2, ..., d\}$. $\rho : X \times [d] \rightarrow X \times [d]$ is an *involution*. If $\rho(x, r) = (x, r)$, the limb (x, r) is said to be dead. If $\rho(x, r) = (y, s)$ with

---

† In the terminology of Berge [2, 3] we should say, to be more exact, *multigraph*.

$(x, r) \neq (y, s)$, the two limbs (x, r) and (y, s) are joined (x may be equal to y); we have by definition $\rho(y, s) = (x, r)$. E is a finite nonempty set disjoint from X, called the set of states. It is convenient to write $E \subset F^d$, where F is a finite set called the set of limb states. The element $e_{x,t}$ of E, the state of automaton x at time t, thus becomes a d-tuple the rth component of which is written

$$e^r_{x,t}, \quad r \in [d], \quad x \in X, \quad t \in N.$$

If (x, r) is a dead limb we shall say, by convention, that $e^r_{x,t} = \omega$ for all $t \in N$. Hence, $\omega \in F$. ϕ is the transition function of the network, $\phi : F^d \times F^d \to F^d$. It associates with each state-input letter pair (e, l) a new state. For automaton $x \in X$, this is written

$$e_{x,t+1} = \phi(e_{x,t}, l_{x,t}),$$

where $e_{x,t} \in E$, and the input letter $l_{x,t} \in F^d$, like $e_{x,t}$. Its components are written $l^r_{x,t}$ with $r \in [d]$, $x \in X$, $t \in N$, and simply defined as follows:

$$l^r_{x,t} = e^s_{y,t} \quad \text{if} \quad \rho(x, r) = (y, s)$$

Let us denote by $M(t)$ the state of the network at time t, which we interpret as an $X \times [d]$ matrix whose elements are the $e^r_{x,t}$, the time t being the same for all the automata.

If we fix the values of $e_{x,0}$ for all $x \in X$, that is $M(0)$, we define a sequence $M(t)$ which necessarily becomes periodic, since X is finite, and is stationary if the period is 1. In this case, we say that the network has operated from an initial state to a stationary state, with all the automata operating synchronously at each instant t.

IDLENESS CONVENTION. A special value of the components $e^r_{x,t}$, denoting idleness, is written $I, I \in F$. A state vector, all of whose components are either I or ω, is called an idle, or *quiescent state*. The function ϕ, by convention, is such that if A and B are quiescent states, we have

$$\phi(A, B) = A.$$

NETWORK, GRAPH, AUTOMATON. In a network of automata, $R = (X, d, \rho, E, \phi)$, we distinguish a graph and an automaton. Let us first define the graph $G(R)$ of the network R, by the triplet

$$G(R) = (X, U, \alpha),$$

where U is the set of couples of distinct elements of the matching ρ, and for the edge $u \in U$ corresponding to the couple $((x, r), (y, s))$,

$$\alpha(u) = (x, y),$$

which does not exclude the case $x = y$.

It is important to point out the following:

(a)   The dead limbs of automata of R do not generate any elements of $G(R)$.

(b)   The degree of the graph $G(R)$ is bounded by d, and is actually equal to d if an automaton of the network has no dead limbs.

(c)   In the case where $\alpha(u) = (x, x)$, u is a self-loop at the vertex x of the graph $G(R)$, The vertex x can have as many loops as its degree permits.

Now let us define the automaton $\mathcal{C}(R)$ of the network R, simply as

$$\mathcal{C}(R) = (E, L, \phi),$$

where $L = F^d$.

It is important to point out the following:

(a)   One copy of the automaton $\mathcal{C}(R)$ is associated with each element $x \in X$, and is also called x for convenience of notation. It takes state $e_{x, t}$ at time t, and reads the letter $l_{x, t}$ at time t.

(b)   For a fixed degree d, the independence of (x, ρ) and (E, ϕ) permits us to:

change the graph of a network of automata without changing the automaton: if $R = (X, d, \rho, E, \phi)$ and $R' = (X', d, \rho', E, \phi)$, then $G(R') \neq G(R)$ and $\mathcal{C}(R') = \mathcal{C}(R)$, where $G(R')$ may be any graph whose degree does not exceed the given d;

change the automaton of a network of automata without changing the graph: if $R = (X, d, \rho, E, \phi)$ and $R' = (X, d, \rho, E', \phi)$, then $G(R') = G(R)$ and $\mathcal{C}(R') \neq \mathcal{C}(R)$.

Finally, then, we see that a network of finite automata R and its evolution, that is, its sequence $M(t)$ for $t \in N$, are defined by the following:

(1)   a graph G of degree at most d, in which for each vertex x we define a one-to-one function from the edges incident to x into $[d]$;

(2)   an automaton $\mathcal{C}$ which takes its states and input letters from the same set F^d;

(3)   an initial state for each vertex automaton x.

2.2.   Graphical Problems Solvable by Networks of Finite Automata

Let P be a simple or multiple property of graphs, for example, connectivity, regularity, absence of cycles, bicoloration, etc. Let $d \in N$. Let $\mathcal{G}(P, d)$ be the family of finite graphs (X, U, α) with degree d or less, having the property P. Let K be a problem of the type that makes network R exhibit a stationary state $M(\theta)$, where stationary means that for all $t > \theta$, $M(t) = M(\theta)$, meaningful in terms of the elements of the graph $G(R)$. For example

(1) $M(\theta)$ displays a configuration such as a hamiltonian cycle of the graph, or the nonexistence of that configuration;
(2) $M(\theta)$ displays all vertices of the graph in a special state for the first time.

We say that a problem K is fa-computable by $\mathcal{A}$ in $\mathcal{G}(P, d)$ for a class of initial states $\mathcal{M}(\mathcal{G})$ if there exists a finite automaton $\mathcal{A}$ such that, for any $G \in \mathcal{G}(P, d)$, the network R which has $G(R) = G$ and $\mathcal{A}(R) = \mathcal{A}$, and its initial state $M(0) \in \mathcal{M}(\mathcal{G})$ will, at a finite time θ, exhibit a stationary state $M(\theta)$ associated with K. The minimal value of θ satisfying this definition is called the computation time of $\mathcal{A}$ for the problem K.

Notice that the automaton $\mathcal{A}$ which solves the problem K is independent of the choice of $G \in \mathcal{G}(P, d)$. In particular, we have card E independent of card X. Thus, though card X may be unbounded for $G \in \mathcal{G}(P, d)$, card E is bounded by a function of d.

Remarks: (1) In what follows, we will usually abbreviate a reference to the above definition as follows: the problem K is fa-computable by $\mathcal{A}$. Unless we explicitly state the property P, then this is understood to mean that K is fa-computable in $\mathcal{G}(\cdot, d)$, that is, for any finite graph of degree d or less. The class of initial states $\mathcal{M}(\mathcal{G})$ will be defined by the description of the initial state $M(0)$ of the network.

(2) The state vector $e_{x,\theta}$ gives the local solution for vertex x, and $M(\theta)$ gives the global solution to problem K.

The problem involved here is to show whether or not a problem K is fa-computable, and, if so, to minimize the computation time θ, which in general is a function not only of d, but also of card X, and finally for a given θ to minimize the number of states of the finite automaton $\mathcal{A}$ which computes K, that is, card E. For the latter problem, at our present stage of development, it is more often a matter of reduction than of minimization.

We would point out that the vector notation used for the states $e \in E$ $(E \subset F^d)$ is convenient to work with, but its redundancy does not at all affect the exact number of states required for the operation of $\mathcal{A}$. We may have card $E <$ (card $F)^d$, and in particular, if $e^r = e^s$, for all $r, s \in [d]$, then card $E =$ card F.

Remark: One might think that the synchronization of the automata to which we have referred until now considerably reduces the autonomy of the network. This is not at all true. We shall see in Section 2.3 that by associating with each automaton a 3-position counter, to count the transitions, modulo 3, we can achieve an autosynchronization of the network. Hence, the only external action upon the network's computation is the choice of $M(0)$.

As an illustration for our notation, we shall define the *boundary automaton*, which solves the rather easy problem of computing any boundary in a graph. In a graph (X, U, α), the linear operator "boundary," written as ∂, associates each edge with the sum of its two extremities

$$\partial u = a + b \quad \text{if} \quad \alpha(u) = (a, b).$$

The addition operation is commutative, and has the property $z + z = 0$. As for the boundary of a sum of edges, it is equal to the sum of their extremities. In other words, a vertex belongs to the boundary of a sum of edges if it has an odd number of limbs in common with these edges. We shall attempt to compute the boundary of a given sum L_0 of edges, namely $S_0 = \partial L_0$.

Let us describe the boundary automaton.

(a) *The states:* The set F of possible limb states will be written $F = \{\omega, I, A\}$, with ω for a dead limb, I for a quiescent limb, and A for an activated limb.

(b) *The initial state of the network:* Let the set of edges L_0 be defined by $\Lambda_0 \subset X \times [d]$ with

$$(x, r) \in \Lambda_0 \Rightarrow \rho(x, r) \in \Lambda_0.$$

At time $t = 0$, we have

$$e_{x,0}^{r} = \omega \quad \text{or} \quad I \quad \text{for all} \quad (x, r) \in X \times [d]$$

except

$$e_{x_0,0}^{r_0} = A \qquad \text{for all} \quad (x_0, r_0) \in \Lambda_0.$$

(c) *The possible cases for state transitions:*

(1) $\Leftrightarrow \operatorname{card} \{r \in [d] \mid e_{x,t}^r = A\}$ is even for given $x \in X$ and, given $t \in N$,

(1) $\Rightarrow e_x^r : A \leftarrow I.$

We define only the possible cases for change of state. In other cases, the function ϕ leaves the state of the automaton limbs unchanged. The possible cases are defined by propositions, such as (1) above and its complement $(\bar{1})$. Several propositions may be composed together (see the more complex example).

$((1) \Rightarrow)$ is to be read as "if proposition (1) is true at time t, then ..." In the transition instructions, (e_x^r) is to be read as "the change of state of limb e_x^r from time t to time $t+1$ is ..." $(P \leftarrow Q)$ is to be read as "if the state is P, then it is replaced by the state Q." $(\cdot \leftarrow Q)$ is to be read as "any element of F is replaced by Q."

(d) *Results:* S_0 is the sum of the elements of Σ_0, where

$$\Sigma_0 = \{x \in X \mid \exists r \in [d] : e_x^r = A\},$$

since the computation reaches a stationary state at time 1, we have $\theta = 1$. Card $F = 3$, card $E = 3^d$, and this is the minimum E.

THEOREM 1.   The boundary of any set of edges of a graph is fa-computable.

Let us now define the coboundary automaton. In a graph (X, U, α) the linear operator coboundary, written δ, associates with each vertex the sum of those edges having this vertex exactly once as an extremity

$$\delta x = \sum_{u \in C(x)} u,$$

where $C(x) = \{u \in U \,|\, \alpha(u) = (x, b), b \neq x\}$.

As for the coboundary of a sum of vertices, it is equal to the sum of their coboundaries, a commutative sum for which $z + z = 0$. In other words, an edge belongs to the coboundary of a sum of vertices if exactly one extremity of this edge is a member of the sum. We shall attempt to compute the coboundary of a given sum S_1 of vertices, namely $L_1 = \delta S_1$. Let us describe the coboundary automaton.

(a)   *The states:* $F = \{\omega, I, A, C\}$, with ω for a dead limb, I for a quiescent limb, A for an activated limb, and C for a limb, not dead, of an edge belonging to the coboundary.

(b)   *The initial state of the network:*   Let $S_1 \subset X$. $e^r_{x,0} = \omega$ or I, except for $e^{r_1}_{x_1,0} = A$ if $x_1 \in S_1$ and (x_1, r_1) is not dead.

(c)   *The possible cases for state transitions:*

$$(1) \Leftrightarrow e^r_{x,t} = I \quad \text{and} \quad e^s_{y,t} = A$$

or

$$e^r_{x,t} = A \quad \text{and} \quad e^s_{y,t} = I, \quad \rho(x,r) = (y,s),$$

$$(2) \Leftrightarrow e^r_{x,t} = A \quad \text{and} \quad e^s_{y,t} = A, \quad \rho(x,r) = (y,s),$$

$$(1) \Rightarrow e^r_x : \cdot \leftarrow C,$$

$$(2) \Rightarrow e^r_x : \cdot : I.$$

(d)   *Results:* L_1 is the sum of the elements of the set Λ_1, where

$$\Lambda_1 = \{((x,r),(y,s)) \in U \,|\, e^r_x = C\}.$$

$\theta = 1$ and card $F = 4$. Card $E = 3.2^d - 2$ by inspecting the possible states e_x, and this is the minimum E.

THEOREM 2.   The coboundary of any set of vertices of a graph is fa-computable.

Let us finally define the marking automaton which associates with $x_0 \in X$ and $U_0 \subset U$ the set X_0 of vertices of the graph (X, U, α) that are connected to x_0 in the partial graph $(X, U_0, \alpha|_{U_0})$.

(a) *The states:* $F = \{\omega, I, A, \bar{A}\}$, where I is the state of a limb which is not the extremity of an edge of U_0, A is the state of a limb which is the extremity of an edge of U_0, and $\bar{A}$ is the state of a limb (x, r), which is the extremity of an edge of U_0, with x connected to x_0 in the partial graph.

At time $t = 0$ we have

$$e^r_{x_0,0} = \bar{A} \qquad \text{for all} \quad r \in [d]_{x_0},$$

$$e^r_{x,0} = A \qquad \text{if} \quad u(x, r) \in U_0, \quad x \neq x_0,$$

$$e^r_{x,0} = \omega \qquad \text{or} \quad I \qquad \text{otherwise.}$$

(b) *The possible changes of state:*

$$(1) \Leftrightarrow l^r_{x,t} = \bar{A}$$

for at least one $r \in [d]$.

$$(1) \Rightarrow e^s_x : A \leftarrow \bar{A},$$

which means every limb of x in state A changes to state $\bar{A}$ if a component of the input letter is $\bar{A}$.

(c) *Results:* the set X_0 is defined by

$$X_0 = \{x \in X \mid e^r_{x,\theta} = \bar{A} \text{ for at least one } r \in [d]\}.$$

We have θ less than the length in the subgraph of the longest path from x_0 that does not pass through the same vertex twice.

2.3. Self-Synchronizing Networks

Let us recall what is meant by the synchronization of the automata of a network $R = (X, d, \rho, E, \phi)$. $M(t)$, the state matrix of R at time $t \in N$, consists of elements all of which have the same associated time: $e^r_{x,t}$ with $r \in [d]$ and $x \in X$ is the rth component of the state of automaton x after t transitions.

For each automaton x, the input letter at time t is composed of the states of neighboring limbs also at time t, which we write

$$l^r_{x,t} = e^s_{y,t},$$

where

$$\rho(x, r) = (y, s).$$

Abstractly, the sequence $M(t)$ for $t \in N$ is defined without reference to time, duration of state changes, or spacing of transitions.

Intuitively now, to allow us to speak in temporal terms, which is certainly legitimate if we wish to consider a physical realization of parallel computation, we can imagine that the synchronization is performed by an external clock, which causes a change in the states of all the network automata simultaneously, and does this at discrete instants spaced conveniently according to the duration of the state transitions and the readings of input letters. In fact, a connected network of the type described above may perform its own self-synchronization, as we shall demonstrate, by having a 3-position counter affixed to each automaton. The network with a given autonomous clock network, thus freed from any external clock, completely autonomous, is called a self-synchronized network.

Let us define a clock network $H = (X, d, \rho, T, \psi)$ for an arbitrary, finite, connected graph $G(H)$, and where $T = \{\hat{0}, \hat{1}, \hat{2}\}$, a set on which addition is taken to be mod 3. The state of x, written $\hat{e}_{x,t}$ is a constant vector whose transitions are written simply

$$\hat{e}_{x,t+1} = \hat{e}_{x,t} + \hat{1} \,(\text{mod } 3)$$

If the initial state of the network is

$$\hat{e}_{x,0} = \hat{0} \qquad \text{for all} \quad x \in X,$$

then

$$\hat{e}_{x,t} = t \,(\text{mod } 3) \qquad \text{for} \quad x \in X, \quad t \in N.$$

The initial states are

$$\hat{e}_{x_0,1} = \hat{1},$$

$$\hat{e}_{x,1} = \hat{0}, \qquad \text{for} \quad x \neq x_0.$$

We write $|x_0 x|$ for the length of the shortest path between x_0 and x in $G(H)$. Let us note that during the first $|x_0 x| - 1$ units of time, x undergoes transitions which accomplish nothing. It remains in an inactive state. Here, let τ_x be the total number of effective transitions, undergone by x. We define the age of x as

$$t_x = |x_0 x| + \tau_x.$$

For the autonomous clock network the state of x at age t_x is written e_{x,t_x}, which we shall also write as e_x. Now we define an admissible state of the network as a list of the states of the automata x satisfying the following rule of compatibility: two neighboring automata, and two automata having the same neighbor, which are in the same state, that is, reading the same number on the 3-position clock, have the same age.

The autonomous clock network thus approximates the rule of synchronization for externally synchronized networks within one unit of time between two neighboring automata.

The automata, though unable to store their age, can, nevertheless, maintain this compatibility in the following manner. X undergoes a transition, that is, $\hat{e}_x \leftarrow \hat{e}_x + \hat{1}$, if and only if for each neighboring y, $\hat{e}_y = (\hat{e}_x$ or $\hat{e}_x + \hat{1})$ and for $x \neq x_0$ there is at least one neighbor z of x for which $\hat{e}_z = \hat{e}_x + \hat{1}$ In effect, if the network is in an admissible state, then after x makes the above transition, it is still in an admissible state.

We have exactly the following situation. x has gained a unit of age, to equal the ages of his neighbors, z at the very least, who were one step ahead of him. The ages of any two neighbors, with the same state, x and z for example, will never differ by a multiple of 3, since $|x_0 x|$ and $|x_0 z|$ cannot differ by more than one unit, and τ_x and τ_z under the rule for compatibility cannot differ by more than one unit. Incidentally, we have proved here that 3 is the minimum number of states for the clock automaton $\mathcal{C}(H)$.

Let $R = (X, d, \rho, E, \phi)$ be a network, the graph $G(R)$ of which is connected, and let the autonomous clock network H described above be designed so that $G(H) = G(R)$. We see, without need of further formalism, that R can become autonomous if it is superposed onto H, with the automaton $\mathcal{C}(R)$ at x undergoing transitions whenever the corresponding $\mathcal{C}(H)$ does.

$\mathcal{C}(R)$ reads its input letter component by component, reading the component associated with its neighbor y when y's clock is at the same position as x's. This introduces no contradictions.

It is easy to show that the state of x at its tth transition in the externally synchronized network is equal for all $x \in X$ and all $t \in N$ such that $t > |x_0 x|$, to the state of automaton x with age t in the corresponding self-synchronized network.

To properly understand the weaker type of synchronization that we have introduced in R, consider two arbitrary vertices x and y. For a given state of the self-synchronizing network, the age of y cannot exceed the age of x by more than $|xy|$. In the externally synchronized network, the tth transition of x is independent of the $|xy| - 1$ states which y took prior to t. We thus see that the lead which y has on x in the self-synchronizing network does not hamper the proper functioning of y.

2.4. Undecidable Connectivity

We shall define a type of symmetry in a network of finite automata which permits the formulation of two fundamental problems of connectivity, unsolvable by networks of automata.

DEFINITION. A *reflection* of a network $R = (X, d, \rho, E, \phi)$ is defined as any matching

$$\pi = X \times [d] \to X \times [d]$$

having the following properties for all $p, q \in X \times [d]$:

(1) $\pi p \neq p$, no limb is its own image under π;
(2) $(P^1 p = P^1 q) \Rightarrow (P^1 \pi p = P^1 \pi q),$† compatibility with vertex equivalence;
(3) $P^2 p = P^2 \pi p$, invariance of limb number;
(4) $\rho(\pi p) = \pi \rho(p)$, compatible with the matching ρ.

Remarks: (i) $P^1 p \neq P^1 \pi p$ since $\pi p \neq p$ by (1) and $P^2 p = P^2 \pi p$ by (3);
 (ii) if p is a dead limb then so is πp by (4);
 (iii) $\pi(x, r) = (y, r)$ is always true for all $r \in [d]$, by (3), which justifies the notation $\pi x = y$.

If R admits a reflection π, R and $G(R)$ are said to be symmetric according to π. If a state $M(t)$ of R is such that

$$e^r_{x,t} = e^r_{y,t} \qquad \text{for all} \quad (x, r) \in X \times [d], \qquad \pi(x, r) = (y, r),$$

then $M(t)$ is said to be symmetric according to π.

PROPOSITION 1: *Proposition of Symmetric Evolution.* Let R be a symmetric network according to the reflection π, and $M(0)$ an initial state of R, symmetric according to π. Then, for all $t \in N$, $M(t)$ is symmetric according to π.

In effect, let us suppose that $M(t-1)$ is symmetric according to π, then

$$e_{x,t-1} = e_{\pi x, t-1} \qquad \text{for all} \quad x \in X$$

and hence

$$e^r_{px, t-1} = e^r_{\pi px, t-1} \qquad \text{for all} \quad x \in X.$$

This yields

$$e_{x,t} = e_{\pi x, t} \qquad \text{for all} \quad x \in X.$$

The proposition now may be proved by induction on $t \in N$.

An obvious example of a symmetric network is that of the union R of two identical disjoint networks S_1 and S_2, written $R = S_1 + S_2$. R admits a natural reflection from S_1 onto S_2.

We now define a transformation upon symmetric networks which leaves their reflection π invariant. Let the network $R = (X, d, \rho, E, \phi)$ be symmetric according to the reflection π. Let p_1 be a nondead limb of R, meaning that $\rho p_1 \neq p_1$, such that $\pi \rho p_1 \neq p_1$. We can associate with p_1 three other distinct

---

† P^1 signifies first projection of the pair of elements.

limbs: ρp_1, πp_1, and $\pi \rho p_1 = \rho \pi p_1$. Let us speak then of the quadruple of four distinct limbs associated with p_1,

$$(p_1, p_1', p_2, p_2')$$

where

$$\rho p_1 = p_1' \quad \text{and} \quad \rho p_2 = p_2',$$
$$\pi p_1 = p_2 \quad \text{and} \quad \pi p_1' = p_2'.$$

DEFINITION. The exchange on the quadruple of p_1 is defined as the operation upon R which generates the network

$$(p_1 \nabla R) = (X, d, \dot{\rho}, E, \phi)$$

different from R only for the quadruple associated with p_1, which becomes

$$(p_1, p_2', p_2, p_1')$$

which means by our conventions that we now have

$$\dot{\rho} p_1 = p_2' \quad \text{and} \quad \dot{\rho} p_2 = p_1'$$
$$\pi p_1 = p_2 \quad \text{and} \quad \pi p_1' = p_2'.$$

We shall write $p_1 \nabla G$ to denote the graph $G(p_1 \nabla R)$.

From the above definition, we immediately deduce the proposition of identical evolution for R and $p_1 \nabla R$.

PROPOSITION 2: *Proposition of Invariance under Exchange.* Let R be sym-
(2) If S_1 is a network that remains connected when the edge (p_1, p_1') is symmetric according to π, and has the same sequence of states $M(t)$ as R, provided that they both have the same initial state $M(0)$ symmetric according to π.

Remarks:

(1) $p_1 \nabla (p_1 \nabla R) = R$ and $p_1 \nabla R = p_1' \nabla R = p_2 \nabla R = p_2' \nabla R$.

(2) If S_1 is a network that remains connected when the edge (p_1, p_1') is removed, then the symmetric network $R = S_1 + S_2$, where S_2 is identical to S_1, is not connected, but $p_1 \nabla R$ is connected.

We now give a criterion for problems which are not fa-computable.

PROPOSITION 3: *Proposition of Indiscernability.* Let $G, G' \in \mathcal{G}$ be two finite graphs having the same set of vertices X. Let K be the problem of displaying a graph configuration (for example, a Hamiltonian cycle) present in G but not in G'. If for any pair of networks R and R' such that

$$G(R) = G \quad \text{and} \quad G(R') = G',$$

$$\alpha(R) = \alpha(R'),$$

$$M(0) = M'(0), \quad \text{where} \quad M(0) \in \mathcal{M}(\mathcal{G}),$$

we have $M(t) = M'(t)$ for all $t \in N$, then the problem K is not fa-computable in $\mathcal{G}$ for $\mathcal{M}(\mathcal{G})$.

Proof: There is no finite θ for which $M(\theta)$ will display the configuration while $M'(\theta)$ displays its nonexistence, since $M(\theta) = M'(\theta)$ for all $\theta \in N$.

Our first statement concerning undecidability is relative to an automaton which cannot tell whether two of its own limbs are joined, or whether they are in fact joined with two other limbs of a symmetric automaton.

THEOREM 1. Let K be the problem of displaying all the self-loops of a graph. If for some $\hat{G} \in \mathcal{G}(\cdot, d)$ symmetric according to a reflection π, there exists an $\hat{M}(0) \in \mathcal{M}(\mathcal{G})$, compatible with $\hat{G}$ and also symmetric according to the reflection π, then the problem K is not f.a.-computable in $\mathcal{G}(\cdot, d)$ for $\mathcal{M}(\mathcal{G})$.

Proof: Let $\hat{G}$ contain at least one nondead limb p_1 such that $p_1 \neq \pi \rho p_1$ and $P^1 p_1 = P^1 \rho p_1$, that is, p_1 forms self-loop with ρp_1. Consider the networks R and R', where

$$G(R) = \hat{G} \quad \text{and} \quad G(R') = p_1 \nabla \hat{G},$$

$$\alpha(R) = \alpha(R'),$$

$$M(0) = M'(0) = \hat{M}(0).$$

By Proposition 2, we have $M(t) = M'(t)$ for all $t \in N$. But notice that $p_1 \nabla \hat{G}$ does not have a self-loop between p_1 and ρp_1, since

$$P^1 \dot{\rho} p_1 = P^1 \pi \rho p_1 = P^1 \pi p_1 \neq P^1 p_1.$$

Thus, by Proposition 3, the problem K is not fa-computable in $\mathcal{G}(\cdot, d)$ for $\mathcal{M}(\mathcal{G})$.

Our second statement of undecidability is relative to an automaton which cannot determine whether one of its neighbors is connected to it by edges which belong to a given subset of the edges of the graph.

Let us denote by U_0 a set of nondead limbs such that

$$p \in U_0 \Leftrightarrow \rho p \in U_0.$$

The notation U_0 reminds us that U_0 is identifiable with a subset of U, and any subset of U is identifiable with a U_0.

U_0 is said to be symmetric according to π if $p \in U_0 \Leftrightarrow \pi p \in U_0$. Let us define an equivalence relation of connectivity in X associated with U_0. For $p \in U_0$, let us denote the adjacency of $P^1 p$ and $P^1 \rho p$ by

$$P^1 p \underset{o}{-} P^1 \rho p \qquad \text{and} \qquad P^1 \rho p \underset{o}{-} P^1 p.$$

The transitive and reflexive closure of the relation $\underset{o}{-}$ is denoted by $\underset{o}{\leftrightarrow}$, and $x \underset{o}{\leftrightarrow} y$ is read "x is connected to y by U_0."

THEOREM 2.   Let K be the following problem: For a given graph, with a given subset of limbs $U_0 \subset X \times [d]$ such that $p \in U_0 \Leftrightarrow \rho p \in U_0$, and a given limb $p_1 \notin U_0$ such that $P^1 p_1 \neq P^1 \rho p_1$, to determine whether or not the relation $P^1 p_1 \underset{o}{\leftrightarrow} P^1 \rho p_1$ holds.

If for some $\hat{G} \in \mathcal{G}(\cdot, d)$ which is symmetric according to a reflection π, there exists an $\hat{M}(0) \in \mathcal{M}(\mathcal{G})$, compatible with $\hat{G}$ and also symmetric according to the reflection π, then the problem K is not fa-computable in $\mathcal{G}(\cdot, d)$ for $\mathcal{M}(\mathcal{G})$.

Proof: Let us construct the graph $\hat{G}$ as follows: $\hat{G} = S_1 + S_2$, where S_1 and S_2 are two identical, disjoint, connected graphs. Let π be the natural reflection from S_1 onto S_2. Let p_1 and U_0 satisfy the above hypotheses, with U_0 symmetric according to π, and with p_1 in S_1 such that $P^1 p_1 \underset{o}{\leftrightarrow} P^1 \rho p_1$. Clearly, $p_1 \neq \pi \rho p_1$ since $\pi \rho p_1$ is in S_2, so that $p_1 \nabla \hat{G}$ exists. Notice that in $p_1 \nabla \hat{G}$, $P^1 p_1 \underset{o}{\not\leftrightarrow} P^1 \rho p_1$, since $\dot{\rho} p_1 = \pi \rho p_1$, $p_1 \pi p_1 \notin U_0$, and p_1 and πp_1 are the only limbs of S_1 connected to S_2. Now consider the networks R and R', where

$$G(R) = \hat{G} \qquad \text{and} \qquad G(R') = p_1 \nabla \hat{G}$$

$$\mathcal{O}(R) = \mathcal{O}(R'),$$

$$M(0) = M'(0) = \hat{M}(0).$$

By Proposition 2, we have $M(t) = M'(t)$ for all $t \in N$. But in $\hat{G}$ the connectivity relation holds, whereas in $p_1 \nabla \hat{G}$ it does not. Thus, by Proposition 3, K is not fa-computable in $\mathcal{G}(\cdot, d)$ for $\mathcal{M}(\mathcal{G})$.

The two foregoing theorems show us that, to avoid problems of undecidable connectivity when we do not know whether or not a graph is symmetric, we must make certain that in our computation procedures we always choose initial states $M(0)$, being not symmetric according to any reflection π, for example, with all states identical except one.

3.   Elementary Problem-Solving Automata

3.1.   *Labyrinth Problems: The Tarry Automaton*

Let us first define the complete words of a connected graph. For this we use the convenient notation of limbs $(x, r) \in X \times [d]$ and the matching ρ in $X \times [d]$, introduced in Section 2.1. We define the grammar of words of the graph G as the set of words constructed from the alphabet

$$\mathscr{A} = \{l \in X \times [d] \,|\, \rho(l) \neq l\}$$

and belonging to one of the following classes:

(1) Λ_x, where $x \in X$, called *null words* of G;
(2) l, where $l \in \mathscr{A}$, called *one-letter words*;
(3) $\sigma = l_1 \ldots l_k l_{k+1} \ldots l_p$, where $l_k \in \mathscr{A}$, for $k = 1, \ldots, p$ and $P^1 \rho(l_k) = P^1(l_{k+1})$, for $k = 1, \ldots, p-1$, called *p-letter words* of G.

Intuitively we interpret the one-letter word $l = (x, r)$ as the traversal of the edge of G beginning with l, that is, a traversal from x to y of the edge $((x, r), (y, s))$, if $\rho(x, r) = (y, s)$; the p-letter word σ as the traversal of a sequence of p edges beginning with $l_1, \ldots, l_p$ respectively.

The word $\sigma' = \rho(l_p) \rho(l_{p-1}) \ldots \rho(l_1)$ is called the *inverse word* of σ.

A word $\sigma = l_1 \ldots l_p$ is called *cyclic* if $P^1(l_1) = P^1 \rho(l_p)$.

A word $\mu = l_1 \ldots l_{2m}$ of a graph G with m edges is called a *complete word* if it contains each letter of the alphabet $\mathscr{A}$ exactly once.

We showed [18] that every complete word is cyclic, and that for a graph to possess a complete word, it is necessary and sufficient that the graph be connected.

Let G be a connected graph. We define a tree of G rooted in a, where $a \in X$, as any subset V of $\mathscr{A}$ with the properties

(1)   the function: $(x, r) \to P^1 \rho(x, r)$ restricted to V is a one to one mapping of V onto $X - \{a\}$;
(2)   for all $x \in X$, if $x \neq a$, there exists a word σ of G written in the restricted alphabet V, say $\sigma = l_1 \ldots l_p$, such that $P^1(l_1) = a$ and $P^1 \rho(l_p) = x$.

Specifying

$$V' = \{l \in \mathscr{A} \,|\, \rho(l) \in V\}$$

is clearly equivalent to specifying V. V' is called the inverse tree of tree V.

We showed [18] that for every complete word $\mu = l_1 \ldots l_{2m}$ of a graph G, the set

$$V(\mu) = \{l_k \in \mu \,|\, P^1 \rho(l_k) \neq P^1 l_h \text{ for } h < k\}$$

is a tree of G rooted in $P^1(l_1)$, called the entrance tree of the complete word μ. The element $l_k \in \mu$ is called the entrance letter of μ at the vertex $P^1 \rho(l_k)$.

We shall compute, using three types of finite automata, three types of complete words, each having a specific application. In general, these automata serve to induce, from the order defined by d at each vertex, circular orders on the elements of the graph: limbs, edges, and vertices.

Let us describe the Tarry automaton $\mathscr{L}^T$, which solves a labyrinth by constructing in every finite graph a complete word beginning at a given point.

(a) *Principles:* Tarry's rule is extremely simple; construct a complete word μ such that $V(\mu) = V(\mu')$. In other words

(T1) never use the same limb twice;
(T2) if $l \in V, \rho(l)$ is the entrance limb at $x = P^1 \rho(l)$, then ρl is the exit limb. It is only used at the last resort.

THEOREM. The choice of a complete word of a finite-connected graph is fa-computable or, the choice of a circular permutation of the limbs of a finite-connected graph is fa-computable.

(b) *The states of* $\mathscr{L}^T$: $F = \{\omega, I, V, \hat{V}, 1, 2, \dots, d, \hat{1}, \hat{2}, \dots, \hat{d}\}$, where $\hat{\ }$ is the state of the last letter written in the complete word μ, that is, the position of the signal which traces out μ, V indicates the entrance limb at x, and $r \in [d]$ is a state of (x, s) which stores the directions for μ. The signal leaving x by (x, s) had arrived there by (x, r); in other words, $(x, r) = \rho(l_k)$ and $(x, s) = l_{k+1}$. At time $t = 0$, we have $e^r_{x,0} = w$ or I except for $e^{r_0}_{x_0,0} = \hat{r}_0$.

(c) *The possible changes of state:*

(1) $\Leftrightarrow l^{r_1}_{x,t} = \hat{\ }$ the signal arrives at (x, r);

(2) $\Leftrightarrow e^r_{x,t} = \omega$ or I
      for all $r \in [d]$ the signal has never passed through
                                x;

(3) $\Leftrightarrow e^{r_2}_{x,t} = \omega$
      for all $r_2 \in [d], r_2 \neq r_1$, x has no active limb other than
                                (x, r_1);

(4) $\Leftrightarrow e^{r_2}_{x,t} = I$
      with $r_2 \in [d]$, minimum, x has nondead but unused limbs of
                                which r_2 is the smallest;

(5) $\Leftrightarrow e^{r_3}_{x,t} = V$ (x, r_3) is the entrance limb at x;
                                therefore, $x \neq x_0$;

(6) $\Leftrightarrow e^r_{x,t} = \hat{\ }$ the signal arrives at $\rho(x, r)$;

(7) $\Leftrightarrow r_4 = e_{x,t}^{r_4} = e_{x,t}^{r_5}$,
with $r_4 \neq r_5$

we have in mind $r_4 = r_0$, at a time when the signal returning to (x_0, r_0) has left via (x_0, x_5); r_0 is now the state of (x_0, r_0) and (x_0, r_5).

(12) $\Rightarrow e_x^{r_1} : I \leftarrow V$

(13) $\Rightarrow e_x^{r_1} : I \leftarrow \hat{V}$

($1\bar{3}4$) $\Rightarrow e_x^{r_2} : I \leftarrow \hat{r}_1$

($1\bar{4}5$) $\Rightarrow e_x^{r_3} : V \leftarrow \hat{V}$

(6) $\Rightarrow e_x^{r} : \,^\wedge \leftarrow \cdot$

($1\bar{4}\bar{5}7$) $\Rightarrow e_x^{r_4} : r_4 \leftarrow r_1$

(d)   *Results:* $\theta = 2m$, where $m = \text{card } U$. In effect, the network is stationary from the time $2m - 1$ if $\rho(x_0, r_0)$ is the last letter of μ, and from the time $2m$ if not. The automaton x_0 knows that $t = \theta$ as soon as it receives a signal and finds that it has no more limbs in state I. x_0 has always remembered that it was the initial automaton by the absence of a V among its limb states.

The complete word μ is stored locally at each vertex. In effect, each limb (x, r), placed in the state s, is the successor of the limb $\rho(x, s)$ in the word μ. As for the set

$$V' = \{(x, r) \in X \times [d] \,|\, e_{x,\theta}^r = V\},$$

it represents the inverse of the entrance tree $V(\mu)$ of μ. Card $F = (4 + 2d)$ and card $E < (4 + 2d)^d$.

In the case where we simply want to pass a signal along once, it is unnecessary to store the directions, we set $1 = 2 = \cdots = d = U$ and $\hat{1} = \hat{2} = \cdots = \hat{d} = \hat{U}$ which reduces the number of limb states to 6, $F = \{\omega, I, U, \hat{U}, V, \hat{V}\}$. If we do not even wish to store the tree, we can set $\hat{V} = \hat{U}$, leaving 5 states, $F = \{\omega, I, U, \hat{U}, V\}$. Finally, in the case of the traveler who marks his passage by coloring the limbs, ω and I are superfluous. If U and V are two colors, they are necessary and sufficient for traversing μ.

3.2.   Labyrinth Problems: The Recoil Automata

Let us define the *neutral words* of a graph. The operation of *reduction* performed on a word is defined as the removal of two consecutive inverse letters l and l'. A word is called *neutral* if it can be reduced to a null word by a sequence of such reductions. We showed [18] that every finite connected graph possesses complete neutral words.

Let us describe the *minirecoil automaton* $\mathscr{L}^{\min}$ which solves a labyrinth, by constructing in any finite graph a complete neutral word starting at a given point.

(a) *Principles:* The minirecoil rule is even simpler than Tarry's rule.

(R1) Give priority to limbs with l and $\rho(l)$ unused.

(R2) When there are none, retract the existing word in the opposite direction, without using the same limb twice.

We showed [18] that the minirecoil rule generates a complete neutral word, having the property of Tarry's words.

THEOREM. The choice of a complete neutral word of a finite-connected graph is fa-computable.

(b) *The states of* $\mathscr{L}^{\min}$: $F = \{\omega, I, 1, 2, ..., d, \hat{1}, \hat{2}, ..., \hat{d}\}$ where $\hat{\ }$ is the state of the last letter written. When (x, s) alone is in state r, $\rho(x, r)$ and (x, s) are used consecutively in the word. When (x, r) and (x, s), with $s \neq r$, are both in state r, then $(x, r) = (x_0, r_0)$.

At time $t = 0$, $e^r_{x,0} = \omega$ or I except for $e^{r_0}_{x_0,0} = \hat{r}_0$.

(c) *Possible changes of state:*

(1) $\Leftrightarrow l^{r_1}_{x,t} = \hat{\ }$

(2) $\Leftrightarrow e^{r_2}_{x,t} = l^{r_2}_{x,t} = I$
   with r_2 mini $\in [d]$

(3) $\Leftrightarrow e^{r_1}_{x,t} = r_3$

(4) $\Leftrightarrow e^{r_3}_{x,t} = r_4 \neq r_1$

(5) $\Leftrightarrow r_5 = e^{r_5}_{x,t} = e^{r_6}_{x,t}$
   with $r_5 \neq r_6$

(12) $\Rightarrow e^{r_2}_x : I \leftarrow \hat{r}_1$ (x, r_2) is written, and is not a recoil letter,

$e^r_x : \hat{\ } \leftarrow \cdot$

($1\bar{2}3$) $\Rightarrow e^{r_3}_x : I \leftarrow \hat{r}_1$ (x, r_3) is written, and is a recoil letter;

$1\bar{2}\bar{3} \Rightarrow e^{r_1}_x : I \leftarrow \hat{r}_1$ (x, r_1) is written just after $\rho(x, r_1)$;

($1\bar{2}345$) $\Rightarrow e^{r_5}_x : r_5 \leftarrow r_1$ we have in mind $x = x_0$ and $r_5 = r_0$.

(d) *Results:* $\theta = 2m$. The complete neutral word of $\mathscr{L}^{\min}$ and the associated entrance tree V are in a stationary state, displayed in the same way as for $\mathscr{L}^{T}$.

In Section 3.4, we give two immediate applications of minirecoil words by associating with $\mathscr{L}^{\min}$ other elementary automata, which we call stacked automata.

Let us describe the *maxirecoil automaton* $\mathscr{L}^{\max}$ which solves a labyrinth, by constructing in any finite graph a complete neutral word starting at a given point.

(a) *Principles:* The maxirecoil algorithm differs from the minirecoil algorithm in that it recoils as soon as possible instead of recoiling as late as possible [18].

We mean below by "vertex x is used" that at least one limb of x is already written. The maxirecoil rule is the following:

(S1) if the letter l brings to a vertex which is used, then leave by $\rho(l)$ provided that $\rho(l)$ has not yet been used;

(S2) in all other cases, give priority to limbs l such that l and $\rho(l)$ are unused.

THEOREM. The choice of a complete neutral word of the maxirecoil type of a finite-connected graph is fa-computable.

(b) *The states of $\mathscr{L}^{\max}$:* $F = \{\omega, I, U, \hat{U}, V, \hat{V}\}$.

This automaton will not store the traced complete neutral word μ, but only its entrance tree. $\hat{\ }$ is the state of the last letter written in μ, that is, the position of the signal which traces μ. V indicates the entrance limb, and U indicates a used limb.

At time $t = 0$ we have $e_{x,0}^{r} = \omega$ or I except for $e_{x_0,0}^{r_0} = \hat{U}$.

(c) *The possible changes of state:*

$$(1) \Leftrightarrow l_{x,t}^{r_1} = \hat{\ } \qquad \text{the signal arrives at } (x, r_1);$$

$$(2) \Leftrightarrow e_{x,t}^{r} = \omega \text{ or } I$$
$$\text{for all } r \in [d] \qquad x \text{ is unused};$$

$$(3) \Leftrightarrow e_{x,t}^{r} = \omega$$
$$\text{for all } \quad r \in [d], \quad r \neq r_1, \qquad x \text{ has no active limb other than } (x, r_1);$$

$$(4) \Leftrightarrow e_{x,t}^{r_2} = I$$
$$\text{with } r_2 \in [d], \quad \text{minimum}, \qquad x \text{ has nondead but unused limbs of which } (x, r_2) \text{ is the smallest};$$

$(5) \Leftrightarrow e^{r_3}_{x,t} = V$ (x, r_3) is the entrance limb;
                                                        therefore, $x \neq x_0$;

$(6) \Leftrightarrow e^{r}_{x,t} = \,\hat{}\,$ the signal arrives at $\rho(x, r)$;

$(8) \Leftrightarrow e^{r_1}_{x,t} = I$ (x, r_1) is unused;

$(6) \Rightarrow e^{r}_{x} : \,\hat{}\, \leftarrow \cdot$

$(13) \Rightarrow e^{r_1}_{x} : I \leftarrow \hat{V}$

$(1\bar{3}2) \Rightarrow e^{r_1}_{x} : I \leftarrow V$

$(1\bar{3}24) \Rightarrow e^{r}_{x} : I \leftarrow \hat{U}$

$(1\bar{3}28) \Rightarrow e^{r_1}_{x} : I \leftarrow \hat{U}$

$(1\bar{3}\bar{2}84) \Rightarrow e^{r_2}_{x} : I \leftarrow \hat{U}$

$(1\bar{3}\bar{2}\bar{8}45) \Rightarrow e^{r_3}_{x} : V \leftarrow \hat{V}.$

If $(1\bar{3}\bar{2}\bar{8}4\bar{5})$, the word is finished.

(d)   *Results:* $\theta = 2m$. Card $F = 6$. Card $F = 5$ if we do not wish to store
the entrance tree $(\hat{V} = \hat{U})$. Card $E = 6^d$.

3.3.   *Stacked Automata*

We say that the automaton $\mathcal{B}$ is stacked upon the automaton $\mathcal{A}$ if they
occupy the same vertex and satisfy the following conditions:

$$\mathcal{A} = (E_A, \phi_A, d), \quad \text{with} \quad E_A \subset F_A^{\,d},$$

$$\mathcal{B} = (E_B, \phi_B, d), \quad \text{with} \quad E_B \subset F_B^{\,d},$$

where

$$\phi_A : F_A^{\,d} \times F_A^{\,d} \rightarrow F_A^{\,d}$$

and

$$\phi_B : F_B^{\,d} \times F_B^{\,d} \times F_A^{\,d} \rightarrow F_B^{\,d}$$

so that for $\mathcal{A}$

$$^A e_{x,t+1} = \phi_A(^A e_{x,t}, {}^A l_{x,t}),$$

while for $\mathcal{B}$

$$^B e_{x,t+1} = \phi_B(^B e_{x,t}, {}^B l_{x,t}, {}^A e_{x,t+1}).$$

Thus, for its $(t+1)$th transition, $\mathcal{B}$ takes into account the state of $\mathcal{A}$ at
time $t+1$, the result of $\mathcal{A}$'s $(t+1)$th transition. This is intuitively acceptable,
since the information for $\mathcal{A}$'s $(t+1)$th transition is already available from

vertex x at time t. The scheme defined above is a particular case of the unique transition function for $\mathcal{O}$ and $\mathcal{B}$ together, for which $\mathcal{O}$ is independent of $\mathcal{B}$.

If $\mathcal{B}$ is stacked upon $\mathcal{O}$, the composition of the two may be considered as one automaton, which is written $(\mathcal{O} * \mathcal{B})$.

3.4 Eulerian Path and Edge Ordering

We seek to compute for an Eulerian graph, that is, a connected graph with all the vertices of even degree, an Eulerian word ε, that is, a word such that if any letter is used, its inverse is not used, and vice versa. An Eulerian word is thus cyclic.

We showed [18] that the sequence of letters in a minirecoil word which appear after their inverse make up an Eulerian word of G, if and only if G is Eulerian.

We shall design an automaton $\mathscr{L}^E$ which marks the recoil letters of the word μ for concatenating them suitably.

The stacked automaton $(\mathscr{L}^{min} * \mathscr{L}^E)$ solves the Eulerian problem.

(a) *Principles:* The automaton $\mathscr{L}^E$ marks, for every recoil letter of μ at x, the letter that precedes it in ε.

THEOREM. The choice of an eulerian word of a finite Eulerian graph is fa-computable.

(b) *The states of $\mathscr{L}^E$:* $F = \{\omega, I, U, 1, 2, ..., d\}$. We use the notation $\dot{e}_x^r$ and l_x^r for the state and input letter of $\mathscr{L}^E$. The initial state is

$$\dot{e}_{x,0}^r = I \qquad \text{for} \quad x \in X, \quad r \in [d].$$

For (x, r) the state s indicates that (x, r) follows $\rho(x, s)$ in ε. If $\rho(x, r) = (y, q)$ and there is not yet a p such that $\dot{e}_y^p = q$, the state s indicates that (x, r) is the last letter written in ε.

(c) *Possible changes of state:*

(I) $\Leftrightarrow l_{x,t}^{r_5} \neq I$
     and $\dot{e}_{x,t}^r \neq r_5$
     for all $r \in [d]$ the word ε is interrupted at the letter $\rho(x, r_5)$;

(II) $\Leftrightarrow e_{x,t+1}^{r_6} = \hat{\ }$
     and $c_{x,t+1}^{r_7} = r_6$ or $\hat{r}_6$,
     with $r_6, r_7 \in [d]$ (x, r_6) is the letter written in μ at time $t+1$ and $\rho(x, r_6)$ is already used for μ; thus, (x, r_6) is a recoil letter, the letter of ε written at time $t+1$;

$$(\text{III}) \Leftrightarrow \dot{e}^{r_7}_{x,t} = r_7$$

we have in mind $x = x_0$; (x_0, r_7) being the first recoil letter of μ, then the first letter of ε;

$$(\text{I}, \text{II}) \Rightarrow \dot{e}^{r_6}_x: I \leftarrow r_5$$

(x, r_6) follows $\rho(x, r_5)$ in ε;

$$(\bar{\text{I}}, \text{II}) \Rightarrow \dot{e}^{r_6}_x: I \leftarrow r_6$$

we have in mind $x = x_0$; (x_0, r_6) being the first letter of ε;

$$(\text{I}, \text{II}, \text{III}) \Rightarrow \dot{e}^{r_7}_x: r_7 \leftarrow r_5$$

we have in mind $x = x_0$; (x_0, r_7), the first letter of ε, must follow $\rho(x, r_5)$, the last letter of ε.

(d) *Results:* The stacking $(\mathscr{L}^{\min} * \mathscr{L}^{\text{E}})$ is in a stationary state at time $\theta = 2m$, as in the case of $\mathscr{L}^{\min}$. The desired eulerian word ε is a cyclic word, locally stored by the stationary state of $\mathscr{L}^{\text{E}}$ at each x. $(\dot{e}^r_x = s) \Leftrightarrow (x, r)$ follows $\rho(x, s)$ in ε.

Another application of complete neutral words, such as minirecoil words, is the computation of a circular permutation of the edges of G, such that every edge and its image under the permutation are either adjacent or adjacent to the same third edge. We shall construct a new stacking upon $\mathscr{L}^{\min}$, namely $\mathscr{L}^{\min} * \mathscr{L}^U$.

(a) *Principles:* We have seen that a complete word defines a circular permutation $\bar{\alpha}$ of the limbs of the graph $G(R)$, such that for each limb l, l and $\bar{\alpha}(l)$ are adjacent. We showed [18] that the sequence of edges corresponding to the odd-numbered letters of a minirecoil word μ constitutes a circular permutation $\bar{\beta}$ of the set U of edges of G, such that u and $\bar{\beta}(u)$ are either adjacent or adjacent to a same edge v. The automaton $\mathscr{L}^U$ will copy all the limb states of $\mathscr{L}^{\min}$ of the form $\hat{e}(\hat{e} = \hat{r}$ or $\hat{V})$, changing them into e if the limbs are even-numbered in μ.

THEOREM.   The choice of a circular permutation of the edges of a finite connected graph, such that each edge and its image are either adjacent or adjacent to the same third edge, is fa-computable.

(b) *The states of* $\mathscr{L}^U$: $F = \{\omega, I, 1, 2, ..., d, \hat{1}, \hat{2}, ..., \hat{d}\}$ where the state $\hat{s}$, for (x, r), indicates that the edge associated with (x, r), $u(x, r)$, is the follower in $\bar{\beta}$ of the edge associated with the limb which is followed by $\rho(x, s)$ in $\bar{\alpha}$, i.e.,

$$u(x, r) = \bar{\beta}(u(\bar{\alpha}^{-1}\rho(x, s))).$$

The state s, for (x, r), allows x to compute $\bar{\alpha}^{-1}(x, r)$, and allows automaton y, where $\rho(x, r) = (y, \rho)$, to know that (x, r) is even-numbered in the word μ. We use the notation $\dot{e}^r_x$ and $\dot{l}^r_x$ for $\mathscr{L}^U$.

The initial state is $\dot{e}^{r_0}_{x_0} = \hat{r}_0$. As with $\mathscr{L}^{\min}$, this state is not final. $\dot{e}^r_x = \omega$ or I elsewhere.

(c) *The possible changes of state:*

$$I \Leftrightarrow e^r_{x,t+1} = \hat{r}_1$$

$$II \Leftrightarrow e^r_{x,t+1} = \hat{V} \quad \text{and} \quad e_x^q \neq r_1 \quad \text{for all} \quad q \in [d]$$

$$III \Leftrightarrow l^{r_1}_{x,t} = {}^{\wedge}$$

$$(I\ III) \quad \text{or} \quad (II\ III) \Rightarrow \dot{e}_x^r : I \leftarrow r_1$$

$$(I\ \overline{III}) \quad \text{or} \quad (II\ \overline{III}) \Rightarrow \dot{e}_x^r : I \leftarrow \hat{r}_1$$

(d) *Results:* $\theta = 2m$ for $\mathscr{L}^{\min} * \mathscr{L}^U$, the same as for $\mathscr{L}^{\min}$. $\bar{\beta}$ is defined by the following equivalence:

$$(\dot{e}_x^r = \hat{s}) \Leftrightarrow (\beta^{-1}(u(x,r)) = u(\alpha^{-1}(\rho(x,s)))).$$

3.5. Rooted Tree of Minimal Paths

A path from b to a, see Section 3.1, where $a, b \in X, b \neq a$, is a word $l_1 l_2 \cdots l_p$ such that $P^1(l_1) = b$ and $P^1 \rho(l_p) = a$. We call p the length of the path. All paths of minimal length from b to a, called minimal paths, have their length written as $p = |b\,a|$.

The problem considered here is to find a tree of G, V, rooted at a, which for any $b \in X$ contains a unique minimal path from b to a. Using this tree, we should then be able to pass a signal along the shortest path from b to a.

(a) *Principles:* The principle of the rooted tree of minimal paths automaton, called $\mathscr{T}$, is defined in Moore [13].

We simply fan out from a, labeling each vertex with a number which counts its distance from a, modulo 3. Thus, a is labeled 0, all unlabeled neighbors of a are labeled 1, etc. At the tth step, where $t = 3m + q$, $m \in N$, $q \in \{0, 1, 2\}$, we label all unlabeled neighbors of labeled vertices with q. When no more vertices can be labeled, the algorithm is terminated. It is easy to prove that, for any $b \in X$, labeled q, the first limb of a minimal path from b to a is found by choosing a neighbor y of b that is labeled $q - 1$, modulo 3, and letting (b, r) be such that $P^1 \rho(b, r) = y$, and r is minimum. Hence, a unique rooted tree of minimal paths is induced by the labeling of the vertices and the local order on $[d]$.

THEOREM. The choice of a rooted tree of minimal paths of a finite-connected graph is fa-computable.

(b) *The states of $\mathscr{T}$:* $F = \{\omega, I, 0, 1, 2\}$ where ω is a dead limb, I is a quiescent limb and 0, 1, 2 are counting states, mod 3.

If (x, r) is dead, then $e^r_{x,t} = \omega$, for all $t \in N$. For convenience, let

$$[d]_x = \{r \in [d] \mid e_x^r \neq \omega\}.$$

At time $t = 0$, we have $e^r_{a,0} = 0$ for all $r \in [d]_a$ and $e^r_{x,0} = I$ for all $x \in X$, $x \neq a$, for all $r \in [d]_x$.

(c)   *The possible changes of state:*

$$(1) \Leftrightarrow e^{r_0}_x = I \quad \text{and} \quad l^{r_0}_x = q,\ r_0 \in [d]_x;$$

$$(1) \Rightarrow e^r_x\!: I \leftarrow q + 1\ (\mathrm{mod}\ 3) \quad \text{for all} \quad r \in [d]_x.$$

(d)   *Results:*   The tree is now uniquely determined. Let

$$\Delta_x = \{r \in [d]_x|\ l^r_x = q - 1\ (\mathrm{mod}\ 3)\}.$$

Then

$$V = \{(x, r_x)|\ x \in X,\ e^r_{x,\theta} = q,\ r_x = \min \Delta_x\}.$$

$\theta = \max_{x \in X} |xa|$, card $F = 5$, and card $E = 4.2^d$.

$\mathscr{T}$ has the property that all its active (nondead) limbs are always in the same state q or I. Thus, for an automaton x with no possibility of dead limbs, that is, with $[d]_x = [d]$, card E is reduced to 4.

Now we define the *minimal path automaton* $\mathscr{P}$, which traces out paths to a in the tree, when stacked upon $\mathscr{T}$.

(a)   *Principles:*   Suppose that we wish to find that specific path in the tree V which leads from b to a, where $b \in X$, $b \neq a$. The first limb of the path is that one with the minimum r which joins b to a vertex labeled one less than b, and so on.

(b)   *The states:* $F = \{\omega, I, 0, 1, 2, B, C\}$ where $\omega, I, 0, 1, 2$ have the same meanings here as the states of $\mathscr{T}$. B is the selected initial vertex of a minimal path, and C is the signal which traces the minimal path.

We shall denote the states of $\mathscr{P}$ as $\ddot{e}^r_{x,t}$ and the input letters as $\ddot{l}^r_{x,t}$. At time $t = 0$ we have

$$\ddot{e}^r_{b,0} = B \quad \text{for all} \quad r \in [d]_b$$

and

$$\ddot{e}^r_{x,0} = I \quad \text{for all} \quad x \in X\ |\ x \neq b, \quad \text{for all} \quad r \in [d]_x.$$

(c)   *The possible changes of state:*

$$(2) \Leftrightarrow \ddot{e}^r_{x,t} = B \qquad \text{for all} \quad r \in [d]_x$$

$$(3) \Leftrightarrow \ddot{e}^r_{x,t} = I \qquad \text{for all} \quad r \in [d]_x$$

$$(4) \Leftrightarrow e^r_{x,t+1} = q \qquad \text{for all} \quad r \in [d]_x$$

$$(5) \Leftrightarrow \dot{\Delta}_x = \{r \in [d]_x|\ \ddot{l}^r_x = q - 1\ (\mathrm{mod}\ 3)\} \neq \varnothing$$

$$(6) \Leftrightarrow \exists\, r_1 \in [d]_x|\ \ddot{l}^{r_1}_x = C$$

(34) $\Rightarrow \dot{e}_x^r: I \leftarrow q$

(245) $\Rightarrow \dot{e}_x^{\bar{r}}: B \leftarrow C, \qquad \bar{r} = \min \dot{\Lambda}_x$

(56) $\Rightarrow \dot{e}_x^{\bar{r}}: q \leftarrow C, \qquad \bar{r} = \min \dot{\Lambda}_x$

(d) *Results:* $\mathscr{P}$ follows the state changes of $\mathscr{T}$ until the vertex b is labeled. Then the minimal path from b to a is immediately traced back.

The path is $(x_1, r_1), \ldots, (x_k, r_k)$, where

(1) $x_1 = b$;
(2) (x_i, r_i) is such that $\dot{e}_{x_i,\theta}^{r_i} = C$, for $i = 1, 2, \ldots, k$;
(3) $P^1 \rho(x_k, r_k) = a$.

There is no more possible change of state when $\Delta_x = \Delta_a = \phi$. And so $\theta = 2|ba| - 1$, card $F = 10$, and card $E = 4.2^d + 3^d.2^{d-1} + 1$ or, with no dead limbs, card $E = 5 + 3d$. Furthermore, $(\mathscr{T} * \mathscr{P})$ may subsequently be used to find minimal paths to a from vertices other than b. The labeling of the vertices preserves the tree, and it is only necessary to reinitialize the $\mathscr{P}$ automata in order to begin another computation.

Let us define the *connection automaton* $\mathscr{K}$, which, when stacked upon $\mathscr{T}$ connects limbs with its neighbors in such a way that a complete neutral word is induced in the subgraph V of $G(R)$.

(a) *Principles:* The operation of $\mathscr{K}$ is extremely simple, and resembles that of $\mathscr{P}$, the minimal path automaton. As $\mathscr{T}$ labels the vertices, $\mathscr{K}$ merely marks those limbs which belong to the rooted tree V. The complete word v is induced by the circular order on $[d]$, as we shall see below.

(b) *The states:* $F = (\omega, I, T, \bar{T}, \hat{T}, V)$ where T, the limb, is a letter of v, $\hat{T}, \bar{T}$ are transient states used for making connections and V is the entrance limb for the rooted tree. We shall denote the states and input letters of $\mathscr{K}$ as $\dot{e}_{x,t}^r$ and $l_{x,t}^r$ respectively.

The root a of V is initialized as follows: $\dot{e}_{a,0}^r = \hat{T}$ for all $r \in [d]_a$ and $\dot{e}_{x,0}^r = I$ or ω otherwise.

(c) *The possible changes of state:*

(1) $\Leftrightarrow e_{x,t+1} = q, \dot{e}_{x,t}^r = I \qquad$ for all $\quad r \in [d]_x$
$\qquad$ and $\quad \dot{\Lambda}_x = \{r \in [d]_x : l_x^r = \hat{T}\} \neq \varnothing$

(2) $\Leftrightarrow \dot{e}_{x,t}^r = \bar{T}$

(3) $\Leftrightarrow l_{x,t}^r = V$
$\qquad$ and $\quad \dot{e}_x^r: \hat{T} \leftarrow \bar{T} \qquad$ under any conditions

(1) $\Rightarrow \dot{e}_x^r: I \leftarrow V, \qquad \bar{r} = \min \dot{\Lambda}_x,$
$\qquad$ and $\quad \dot{e}_x^r: I \leftarrow \hat{T} \qquad$ for all $\quad r \in [d]_x | r \notin \Lambda_x$

(23) $\Rightarrow \dot{e}_x^r$: $\overline{T} \leftarrow T$

($2\overline{3}$) $\Rightarrow \dot{e}_x^r$: $\overline{T} \leftarrow I$

(d)   *Results:*   By the above rules, $\dot{e}_{x,\theta}^r = T \Leftrightarrow l_{x,\theta}^r = V$, and all limbs of the rooted tree V are marked with a T or a V, V denoting the entrance limb of each vertex except a. $\theta = k(a) + 2$, where $k(a) = \max_{x \in X} |ax|$.

The extra two time units are lost at the beginning, since before making its connections $\mathcal{K}$ has to wait until $\mathcal{T}$ labels the successors of a. Card $F = 6$ and card $E < 6^d$. The complete word v may be found by the following rules, where $v = l_1, \ldots, l_{2n-2}$:

(1) l_1 is any limb of a marked with a T;
(2)   if $P^1 \rho(l_{i-1}) = x$, then l_i is the first limb of x, moving clockwise from $\rho(l_{i-1})$ on $[d]$, that is found marked with a T or a V.

Note that v', the inverse of v, may be found simply by beginning at l_{2n-2} and changing the term clockwise to counterclockwise; among the limbs of a marked with a T, l_1 and l_{2n-2} are adjacent; v and v' are cyclic words and v has $2n-2$ letters, that is, twice the number of edges in a tree of n vertices.

In this section we have separated the edges of the graph G into two parts, the tree V and its cotree W.

We may now construct, for each $u \in W$, the cycle which meets W only at u, by marking the edges of V that join the extremities of u. Such a computation is done by an automaton derived easily from $\mathcal{T} * \mathcal{P}$ of this section.

We may now construct, for any $u \in V$, the coboundary which meets V only at u. Such a computation is done by $\mathcal{M} * \mathcal{C}$, that is, by a marking automaton $\mathcal{M}$ (see Section 2.2) where x_0 is an extremity of u and $U_0 = V - \{u\}$, on which we stack a coboundary automaton.

3.6.   Vertex Ordering

Let us recall that we have constructed a circular permutation $\bar{\alpha}$ of the limbs of a connected graph, see the automaton $\mathcal{L}^T$ or $\mathcal{L}^{\min}$ (Section 3.1 and 3.2), and a circular permutation $\bar{\beta}$ of the edges of a connected graph, see the automaton $\mathcal{L}^U$ (Section 3.3). We shall now construct a circular permutation $\bar{\gamma}$ of X in a connected graph, such that for all $x \in X$, we have $|x, \bar{\gamma}(x)| \leqslant 3$.

We showed [18] that if we consider a complete neutral word v of a tree of G, beginning at a vertex x_0, and the sequence of vertices consisting of x_0 and, for each even-numbered letter l of v, of the vertex $P^1(l)$ if $\rho(l)$ comes before l in v and of $P^1 \rho(l)$ if $\rho(l)$ comes after l in v. Then this sequence is a circular permutation $\bar{\gamma}$ satisfying the above condition.

Now $\mathscr{K}$ (Section 3.5) defines the connections of a complete neutral word of a tree V of G rooted at x_0. $\bar{\gamma}$ then can be constructed by the stacking $\mathscr{T} * \mathscr{K} * \mathscr{L}^X$, with $\mathscr{L}^X$ defined as follows:

(a) *Principles:* $\mathscr{L}^X$ copies the complete neutral word v of $\mathscr{K}$, replacing the states T of odd-numbered limbs l_{2k-1} of v by S, then

(1) placing a hat $(S \leftarrow \hat{S})$ on the state of a limb l_{2k-1} if, according to the labeling of $\mathscr{T}$, l_{2k-1} joints its vertex to one of lesser value;
(2) placing a hat $(T \leftarrow \hat{T})$ on the state of limb l_{2k} if, according to the labeling of $\mathscr{T}$, l_{2k-1} joins its vertex to one of greater value.

As a result, at the end of the computation each automaton will have exactly one limb in the state $\hat{}$.

(b) *The states of $\mathscr{L}^X$:* $F = \{\omega, I, T, \hat{T}, S, \hat{S}\}$; T or $\hat{T}$ are even-numbered limbs of v, and S or $\hat{S}$ are odd-numbered limbs of v. If $\hat{}$ is the state of l, and if k is the first limb in state $\hat{}$ coming after l in v, then $P^1(k) = \bar{\gamma}(p^1(l))$.

We use the notation $e_x^r, \dot{e}_x^r, \ddot{e}_x^r$ for the states of $\mathscr{T}, \mathscr{K}$, and $\mathscr{L}^X$, respectively. $\bar{l}_x^r$ is the input letter of $\mathscr{L}^X$. Note that $\mathscr{L}^X$ does not depend directly upon e_x^r. Initially the network is idle, and $\ddot{e}_{x,0}^r = \omega$ or I for all r and x.

(c) *The possible changes of state:*

$(0) \Leftrightarrow \ddot{\Lambda}_x = \{r \in [d] \mid \ddot{e}_{x,t+1}^r = T\} \neq \varnothing$
  and $\ddot{e}_{x,t+1}^r \neq V$ for all $r \in [d]$ this condition occurs only once at vertex x_0 at time $t = 2$

$(1) \Leftrightarrow \bar{l}_{x,t}^{r_1} = T$ or $\hat{T}$

$(2) \Leftrightarrow \bar{l}_{x,t}^{r_1} = S$ or $\hat{S}$

$(3) \Leftrightarrow \ddot{e}_{x,t+1}^r = V$

$(4) \Leftrightarrow \ddot{e}_{x,t+1}^{r_1} = V$ r_2 is the first r after r_1 in the circular order such that $\ddot{e}_{x,t+1}^{r_2} = T$ or V;

$(0) \Rightarrow \ddot{e}_x^{r_2} : I \leftarrow \hat{S}$

$(14) \Rightarrow \ddot{e}_x^{r_2} : I \leftarrow \hat{S}$

$(1\bar{4}) \Rightarrow \ddot{e}_x^{r_2} : I \leftarrow S$

$(23) \Rightarrow \ddot{e}_x^{r_2} : I \leftarrow \hat{T}$

$(2\bar{3}) \Rightarrow \ddot{e}_x^{r_2} : I \leftarrow T$

(d)   *Results:*   Let $x \in X$ and let (x, r) be the limb of x such that $\ddot{e}^r_{x,\theta} = {}^\wedge$. Let (y, s) be the first letter after (x, r) in v such that $\ddot{e}^s_{y,\theta} = {}^\wedge$; then $\bar{\gamma}(x) = y$, $\theta = 2n$, card $F = 6$, and card $E \leqslant 6^d$.

THEOREM.   The choice of a circular permutation $\bar{\gamma}$ of the vertices of a finite-connected graph, such that for each vertex x and its image $\bar{\gamma}(x)$ we have $|x, \bar{\gamma}(x)| \leqslant 3$, is fa-computable.

3.7.   The Network Firing Squad Problem

The firing squad problem is one of the earliest problems dealt with in the literature on arrays of finite automata [1, 12, 22]. It may be stated as follows: Given a line of n automata, including a "general" at one end of the line who is activated at time $t = 0$, we must design the automata so that at some future date θ they will all, simultaneously and for the first time, enter a special *firing state*. A minimum-time 8-state solution for this linear problem, with $\theta = 2n - 2$ was given by Balzer [1].

We have formulated and solved a more general problem [15]. Given an arbitrary connected network of automata, with any vertex-automaton acting as general, we must have them all fire simultaneously at time θ. Here we give an alternate solution to this problem by stacking a generalized firing squad automaton $\mathscr{F}$ upon $\mathscr{T} * \mathscr{K}$ (see Section 3.5).

Surprisingly enough, we achieve a computation time of $\theta = 2n$, only two time units more than for the much simpler linear version.

Let us now define the generalized firing squad automaton, $\mathscr{F}$, which, when stacked upon $\mathscr{T} * \mathscr{K}$, uses the words v and v' to guide two linear firing squads around the tree.

(a)   *Principles:*   If the valence of R is d, then $\mathscr{F}$ has $2d$ limbs, circularly numbered $1_A, 1_B, \ldots, d_A, d_B$, in what we call the clockwise direction. Thus, $[d]$ is understood to represent the set $\{1_A, 1_B, \ldots, d_A, d_B\}$. Limbs r_A and r_B occupy the same edge of $G(R)$, so that the matching ρ becomes $\ddot{\rho}$, defined by

$$\rho(x, r) = (y, s) \Leftrightarrow \ddot{\rho}(x, r_A) = (y, s_B) \quad \text{and} \quad \ddot{\rho}(x, r_B) = (y, s_A).$$

In this way, a circular path of limbs corresponding to the word v may be found, simply by consulting $\mathscr{K}$ for the connections. A "double firing squad" process may then be launched, beginning at a, the general, who sends out signals in opposite directions which meet at the other side of the cyclic word v.

(b)   *The states:* F will be the same set of limb states used for the linear problem, with ω added.

Let us denote the states and input letters of $\mathscr{F}$ by $\ddot{e}^r_{x,t}$ and $\ddot{l}^r_{x,t}$ respectively and those of $\mathscr{K}$ by $\dot{e}^r_{x,t}$ and $\dot{l}^r_{x,t}$, respectively.

Initially the network is idle, and $\ddot{e}^r_{x,0} = I$ or ω for all r and x.

(c) *The possible changes of state:* Here it is best to use a verbal explanation rather than our conventional propositional notation. Now, in the linear firing squad problem, we have $d = 2$ and the solution involves a transition function $\Phi_2: F^2 \times F^2 \to F^2$. The automaton $\mathscr{F}$ uses that same function to operate on pairs of limbs which lie on the path of the complete word v. Those pairs are chosen as follows: let

$$L_x = \{r^1, r^2, \ldots, r^p\} \subset [d]_x$$

be the ordered set of limbs satisfying $\ddot{e}_{x,t+1}^{r^i} = T$ or V for $i = 1, \ldots, p$. We form a set of p pairs of elements of

$$L_x: P_x = \{(r_B^1, r_A^2), (r_B^2, r_A^3), \ldots, (r_B^p, r_A^1)\}.$$

Then, for every t such that $L_x \neq \phi$, we can say that Φ, the transition function for $\mathscr{F}$, is separable into p functions Φ_2, each operating on a pair from P_x. If $p = 1$, $P_x = \{(r_B^1, r_A^1)\}$. Thus, $\mathscr{F}$ may be thought of as an aggregation of p independent linear automata.

There are only two exceptions to the above transition rules.

(1)   When at $t = 2$ the connections of a have been decided, that is, $L_a \neq \phi$, the condition that $\ddot{e}_{a,1}^r \neq V$ for all $r \in [d]$ causes the following transition:

$$\ddot{e}_a^r: I \leftarrow M \qquad \text{for} \quad r = r^p, r_A^1.$$

Here M denotes the initial state of the general in the linear firing squad. This launches two signals which will meet at b, where $b = P^1 \rho(l_{n-1})$, the midpoint of v.

(2)   We include an instruction which will cause the transition

$$\ddot{e}_b^r: I \leftarrow M \qquad \text{for} \quad r = r_B^i, r_A^{i+1},$$

where

$$r_B^i = P^2(\rho(l_{n-1})), \qquad r_A^{i+1} = P^2(l_n).$$

This establishes the ends of the two firing squads. From that point on, their operation proceeds normally.

(d)   *Results:* Since both firing squads consist of n linear automata, they will fire simultaneously after $2n - 2$ time units. With the two-unit time delay caused by $\mathscr{K}$, we have $\theta = 2n$, card $F = 9$, and card $E < 9^d$.

We remark that $\mathscr{F}$ is dependent on $\mathscr{K}$, but $\mathscr{K}$ does not depend directly on $\mathscr{T}$, so that the original definition of stacking is preserved.

THEOREM.   The network firing squad problem is fa-computable by $\mathscr{T} * \mathscr{K} * \mathscr{F}$.

Note: By using the "generalized firing squad" algorithm of Moore and Langdon [14], where the general's position in the line is arbitrary, we may

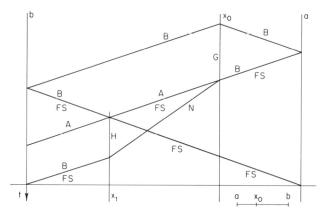

Fig. 1. Star firing squad, 2 rays.

achieve a further reduction of the computation time for a network firing squad, provided that $p > 1$. Any two pairs of a act as the generals for two firing squads, sending initial signals in both directions. For the linear case, $\theta = 2n - 2 - k$ where k is the general's distance from the nearer end of the line. Thus for the network problem we obtain $\frac{3}{2}n < \theta < 2n$.

We now solve the *star graph firing squad problem*. We shall consider a network R whose graph $G(R) = (X, U, \alpha)$ is a star graph with d rays, and with the general at the center. For the case $d = 2$, see Fig. 1.

If $d = 1$, this problem reduces to the original linear firing squad problem.

If $d = 2$, we have the firing squad problem of Moore and Langdon discussed above, in which the general is anywhere in the line. Let the general be x_0, and the ends of the two rays be a and b. Let $|x_0 a| = \alpha$, $|x_0 b| = \beta$, and assume $\alpha > \beta$. Moore and Langdon show without difficulty that the minimum solution time, which they achieve, is $\theta = 2\alpha + \beta$. (Note that $n = \alpha + \beta + 1$, where $n = \text{card } X$.) Here we give a solution that is also minimum time, but has a much simpler set of transition rules.

Furthermore, we have generalized our solution of the Star Graph Firing Squad Problem with 2 rays to the case with d rays where d is any finite number, while maintaining the same solution time $2\alpha + \beta$, α and β being the lengths of the longest and second longest rays of G.

Let us stack first the automaton $\mathscr{F}^2$, which resolves the ordinary linear firing squad problem, upon an automaton $\mathscr{S}^2$, which will coordinate the firing squad in the star graph G with 2 rays. We describe the automaton $\mathscr{S}^2$.

(a) *The states of $\mathscr{S}^2$*: $F = \{\omega, I, A, B, N_1, N_2, G, H\}$, where ω is a dead limb, I is a quiescent limb, A is a freezing signal (speed 1), B is an unfreezing signal (speed 1), N is a slow signal (speed $\frac{1}{2}$), G is a general, and H is a freezing wall. We denote the states of $\mathscr{F}^2$ and $\mathscr{S}^2$ by $\ddot{e}^r_{x,t}$ and $e^r_{x,t}$ respectively. Note that $[d] = \{1,2\}$.

At time $t = 0$, we have $e^r_{x_0,0} = B$, $r \in [d]$, and $\ddot{e}^r_{x,0} = e^r_{x,0} = I$ or ω otherwise.

(b) *The transition rules:* $\mathscr{F}^2$ functions normally unless we specify otherwise. See Section 3.8 for an explanation of signal propagation.

(1) x_0 immediately assumes the "general" state $(e^r_{x_0}: B \leftarrow G)$ as the two B signals propagate toward the ends a and b of the two rays.

(2) The B signals are reflected at a and b, but they cause the initiation of two firing squads $(\dot{e}^r_a, \dot{b}^r_b: I \leftarrow \hat{M})$.

(3) When the general x_0 transmits the first set of firing squad signals, indicated by the arrival from b of a B signal, he abandons the state G, changes the B signal to an A signal, and sends out an N signal in the same direction.

(4) When the two firing squads meet, one with an accompanying A signal and the other with a B signal, x_1, the automaton at which they meet assumes the state H and causes a wall to be formed for both firing squads

$$(e_{x_1,t+1} = H \Rightarrow \dot{e}_{x_1,t+1} = M).$$

The B signal is not transmitted, but the A signal is, and causes the shorter firing squad, emanating from b, to be frozen, whatever its state may be when the A signal makes contact with it.

(5) When the N signal arrives at x_1, in state H, x_1 sends a B signal in the same direction to unfreeze the firing squad, so that it may terminate normally.

(c) *Results:* It is easy to show that the shorter of the two firing squads is frozen for exactly $\alpha - \beta$ time units, so that both firing squads will fire at time $\theta = 2\alpha + \beta$. Card $F = 8$ for $\mathscr{S}^2$, card $E = 64$, and card $E = 16$ for $\mathscr{F}^2$ (Waksman's solution [22]).

STAR GRAPH: GENERAL CASE

(a) *Principles:* To operate the firing squad on a star graph having d rays, we basically use the same method as in the previous section for the two longest rays, having frozen the other firing squads and having determined the appropriate instants at which to unfreeze them. For the case $d = 5$, see Fig. 2.

(b) *The states of $\mathscr{S}^d$:* $F = \{\omega, I, N_1, N_2, A, R_{1p}, R_{2p}, R_{3p}, B_p, E_p, G, H\}$, where $p \in \{1, 2, ..., d\}$, R_{1p}, R_{2p}, R_{3p} are the slowest signals (speed $\frac{1}{3}$), and E_p is the storage of timing for firing squads. Initial conditions are the same as in the 2-rays star graph firing squad problem.

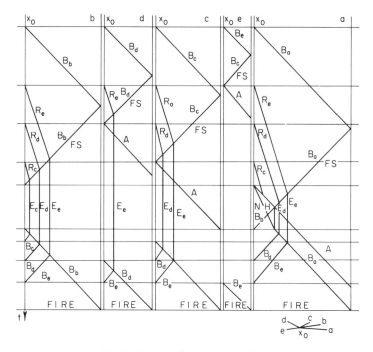

Fig. 2. Star graph firing squad, 5 rays.

(c) *The transition rules:*

(1) As above, a *B* signal is sent out on each ray, and x_0 takes state *G*.

(2) As above, the reflected *B* signals initiate firing squads.

(3) For the second-to-last *B* signal that arrives at x_0, the *A* and *N* signals are created as above.

(4) For other *B* signal arrivals, x_0 suppresses the *B* signal and sends out an R_q signal, where *q* is the number of the corresponding ray, onto any nonfrozen ray, and another *A* signal which freezes ray *q*.

(5) When an R_q signal meets a firing squad at an automaton x_2, $e^r_{x_2}: \cdot \leftarrow E_q$.

(6) The intersection of the firing squads with *A* and *B* signals follows the same rules as in the 2-rays star graph firing squad problem, with the *A* signal freezing the shorter ray, and the *N* signal prompting an unfreezing.

When R_q overtakes an *A* signal, R_q is destroyed.

(7) When a returning firing squad strikes an automaton in state E_q, a B_q signal is sent back toward x_0.

(8) When x_0 receives a B_p signal, it unfreezes the firing squad of ray *q* using a *B* signal.

(d)   *Results:*   It is easy to show that the point E_q on ray p is midway between the ends of rays p and q. Thus all the firing squads will fire simultaneously, again with $\theta = 2\alpha + \beta$, α and β being the longest and second longest lengths of rays. Notice that this time is considerably less than the computation time for a network firing squad as described in Section 3.7. If $d > 2$, $\alpha + \beta < n - 1$ so that $2\alpha + \beta < 2n$, card $F = 5d + 7$, and card $E \leqslant (5d + 7)^d$.

THEOREM.   The star graph firing squad problem with d rays is fa-computable by $\mathscr{S}^d * \mathscr{F}^d$.

3.8.   *The Early Bird Problem*

We now address ourselves to a problem which is not, strictly speaking, graph theoretical, but proves extremely important in the solution of certain graph problems, such as the minimum-tree problem, see Section 4.3. The early bird problem, as we call it [17] seems, in fact, to be a fundamental problem in the study of networks of finite automata, and is closely related to the problems of undecidability discussed in Section 2.4. It may be stated as

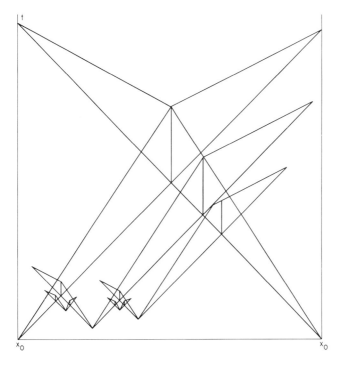

Fig. 3.   Early bird.

follows: given a network R with $d = 2$, whose graph $G(R)$ is an elementary cycle, and given that for some set $A \subset X$, each automaton $x, x \in A$, is excited by an external clock at a distinct time $t_x, t_x \geq 0$, we must design the automata in such a way that after a finite time θ, all the automata of X are in state I except for the single automaton x_0 who satisfies $t_{x_0} = \min_{x \in A} t_x$, and is thus the "early bird," signified by state E (see Fig. 3). Informally, we can say that the automata must determine which one of them was excited first.

Let us describe the early bird automaton, $\mathscr{E}^2$, which solves the problem described above.

(a) *Principles:* The easiest way to explain the operation of $\mathscr{E}^2$ is to speak of signals which are transmitted around the circle of automata at various speeds. We say that a signal is transmitted from x to its neighbor y if x assumes a certain state S at time t and y changes to state S at time $t + \delta$. We call $1/\delta$ the speed of the signal S. When one of the automata x is excited, it sends out signals in each direction. If $x = x_0$ then these signals will meet at the opposite end of the circle, and will be reflected back to x_0, informing it that it is the early bird. If $x \neq x_0$, its signals are destroyed before they meet, according to the transition rules for $\mathscr{E}^2$.

(b) *The states of $\mathscr{E}^2$:* $F = \{I, \hat{I}, E, \hat{E}, M, B, G, N_1, N_2, R_1, R_2, R_3\}$, where I is a quiescent limb, unchanged since $t = 0$, $\hat{I}$ is a quiescent limb which has transmitted a signal, E is an early bird, $\hat{E}$ is an excited automaton, M is a wall created by intersection of black signals, B is a blue signal (speed $= 1$), G is a green signal (speed $= 1$), N_1, N_2 are black signals (speed $= \frac{1}{2}$), and R_1, R_2, R_3 are red signals (speed $= \frac{1}{3}$).

To demonstrate the transmission of signals, suppose that at time t a left-traveling red signal is propagated through x and its left-hand neighbor y. Their changes of state are shown in Table I.

Initially, at time $t = 0$, the network is quiescent. Each $x \in A$ is excited at time t_x, but x responds to the excitation only if it is still quiescent. In other words, $e^r_{x,0} = I$ for all x, r such that $t_x \neq 0$; $e^r_{x,t_x} = \hat{E}$ for all $x \in A$, and for all $r \in [d]$, provided that when $t_x \neq 0$, and $l^r_{x,t_x-1} = e^r_{x,t_x-1} = I$. Notice that $[d] = \{1, 2\}$, and there are no dead limbs, since $G(R)$ is a cycle.

TABLE I

Time		t	$t+1$	$t+2$	$t+3$	$t+4$	$\cdots$
x	right limb	$\hat{I}$	$\hat{I}$	$\hat{I}$	$\hat{I}$	$\hat{I}$	$\cdots$
	left limb	R_1	R_2	R_3	$\hat{I}$	$\hat{I}$	$\cdots$
y	right limb	I	I	I	$\hat{I}$	$\hat{I}$	$\cdots$
	left limb	I	I	I	R_1	R_2	$\cdots$

(c) *The transition rules for $\mathscr{E}^2$:* For convenience, we list the rules verbally; unless otherwise stated signals are always transmitted, see Fig. 3.

(1) If $e_{x,t} = \hat{E}$, then x sends one black and one red signal in each direction, and then becomes quiescent.

(2) If x receives a black signal from both directions at once, then $e_x: \cdot \leftarrow M$ (x becomes a wall).

(3) If $e_{x,t} = M$ and x receives one red signal, then the wall disappears and the red signal is destroyed.

(4) If $e_{x,t} = M$ and x receives a blue signal then the wall disappears $(e_x^r: M \leftarrow \hat{I})$.

(5) If a black signal overtakes a red signal moving in the same direction, the red signal is destroyed.

(6) When a red signal is destroyed at x, x sends a blue signal in the direction that the red was following.

(7) If a blue signal overtakes a black signal moving in the same direction, both are destroyed.

(8) If x receives a red signal from both directions at once it sends one green signal in each direction. The red signals are destroyed.

(9) If a green signal overtakes a black signal moving in the same direction, the black is destroyed.

(10) If x receives a green signal at once from both directions, then the signals are destroyed, and x is the early bird x_0, $(e_x: \cdot \leftarrow E)$.

Note: Though we do not explicitly discuss the cases where signals meet at a pair of automata, rather than at just one, such cases are easily provided for in the transition function.

(d) *Results:* Under the above rules, the following assertions can be proved readily:

(1) If $t_{x_1} < t_{x_2}$, then the automaton x_3 which takes state M as a result of the black signals sent by x_1 and x_2 will receive the red signal of x_2 before that of x_1, provided that the reds are not destroyed beforehand. In any case, the red signal from x_2 will never be transmitted beyond the wall, x_3.

(2) If $t_{x_1} < t_{x_2}$, the blue signal resulting from the destruction of x_2's red signal will destroy x_2's black signal before it passes x_1.

(3) The red signals sent by x_0 cannot be destroyed.

The above statements imply that the only way green signals can be created is by the intersection of the two reds originating from the automaton x_0. We point out that the strict order on the set $\{t_x \mid x \in A\}$ is a crucial premise.

Thus, assuming that $t_{x_0} = 0$, we have $e_{x_0,\theta} = E$ and $e_{x,\theta} = \hat{I}$ for $x \neq x_0$, at $\theta = 2n$, where $n = \text{card } X$).

Card $F = 12$ and card $E = 4 + 8^2 = 68$.

THEOREM.   The early bird problem is fa-computable by $\mathcal{E}^2$.

Just as the firing squad problem was generalized from a line to a network, the early bird problem may be generalized from a circle to a network R with n vertices and degree d. The solution to the problem of finding the early bird in the network could be obtained by stacking a generalized early bird automaton $\mathcal{E}^d$, similar in principle to $\mathcal{F}$, upon the automaton $\mathcal{K}$, which induces a complete cyclic word in the tree defined by $\mathcal{T}$. The details of operation would parallel those given for the network firing squad computation (see Section 3.2) with $\theta = 4n - 2$.

THEOREM.   The network early bird problem is fa-computable by

$$\mathcal{T} * \mathcal{K} * \mathcal{E}^d.$$

Finally, we state and solve the flock of early birds problem. Given a connected network R, a complete cyclic word μ on a spanning tree of $G(R)$ which induces a vertex ordering for the vertices of R, and a forest of disjoint subtrees $\{T_i : i \in I\}$ of $G(R)$, we must find the early bird within each tree T_i, that is, the vertex $x_i \in T_i$ such that x_i is minimal in the vertex ordering.

For the solution, we use the automaton $\mathcal{L}^x$ operating on μ as the source of excitations for the vertices (see Section 3.6) rather than an external clock. Since the T_i are already defined, they each induce a complete cyclic word ν_i in the subtree, according to the rules given in Section 3.5. Thus, we may use $\mathcal{E}^d$ to perform the early bird computations on each cyclic word ν_i. Note that this problem could not be solved for a bounded number of states without the early bird algorithm.

THEOREM.   The flock of early birds problem is fa-computable by $\mathcal{L}^x * \mathcal{E}^d$.

4. More Complex Problems

4.1. Block Decomposition

It is well known since Whitney [23] that the blocks and the articulation points of a connected graph form the edges and the vertices, or the vertices and the edges, of a hypergraph [3] without a cycle.

DEFINITION.   We state the block decomposition problem as follows: For each vertex $x \in X$ of a connected graph $G = (X, U)$, determine a partition of the edges incident to x such that each class will consist of edges which are part of the same block, that is, which have an elementary cycle passing through any two of them. If x is not an articulation point, then this partition will contain only one class, which includes all the edges incident to x.

THEOREM. The block decomposition problem is fa-computable.

(a) *Principles:* Let us consider any vertex x_0 of the graph, sending a maxirecoil signal $M1$, as described in Section 3.2, through one of its limbs (x_0, r_0). When the signal returns via that limb, it will have traversed all limbs of x_0 in the block containing (x_0, r_0), and no other limb of x_0. In effect, the maxirecoil signal returns to x_0 via (x_0, r) before (x_0, r_0), if and only if there is an elementary cycle containing (x_0, r_0) and (x_0, r). This is easily proved by referring to the maxirecoil rules, see Section 3.2.

At the beginning, each vertex has only one class of nondead limbs.

A given vertex x_0 begins alone the identification of its partitioning, class after class, by sending out a maxirecoil signal $M1$. Each time a class of the vertex x, x_0 for example, is identified by the criteria given above, x sends out a signal in every direction within that class to erase the maxirecoil word states generated during the identification of that class. And at the same time x sends out a maxirecoil signal $M2$ through one limb of that class. The first automaton to receive signal $M2$ which has not yet completed its partitioning stops $M2$ and recommences the partitioning process. Parallel computation is then performed, with at most one partitioning in operation per block.

Notice that the erasing process cannot affect any active signal $M1$ or $M2$ since the hypergraph of blocks has no cycle.

(b) *The states of $\mathscr{L}_B^{\max}$:* To solve the block decomposition problem, we use a special automaton, $\mathscr{L}_B^{\max}$, of the same type as $\mathscr{L}^{\max}$, see Section 3.2, with some simple additional features stated below.

(1) $\mathscr{L}_B^{\max}$ can stop a maxirecoil word in process ($M1$ or $M2$) under the two conditions indicated above and send out at the same time a new maxirecoil word (respectively, $M2$ or $M1$);
(2) $\mathscr{L}_B^{\max}$ can store a partitioning of its limbs;
(3) $\mathscr{L}_B^{\max}$ can keep a maxirecoil signal in operation inside a given class of limbs, the class of the departure limb being the class of the arrival limb;
(4) $\mathscr{L}_B^{\max}$ is able to send and transmit erasing signals inside a class of limbs.

As far as the partitioning activity is concerned, d states suffice to class the limbs: $F_1 = \{I, 1, ..., d\}$. As far as the maxirecoil activity is concerned, we notice that there is no overlapping of signals $M1$ and $M2$ in the same limb since two such signals cannot exist in the same block at once. Furthermore, it is not necessary to record the word connections or the entrance tree. Then three active states suffice for each signal: $F_2 = \{I, U_1, \hat{U}_1, V_1, U_2, \hat{U}_2, V_2\}$.

We have $F = F_1 \times F_2 \cup \{(\omega, \omega)\}$. Initially, at time $t = 0$, the maxirecoil part of the state $e^{r_0}_{x_0, 0}$, is set to $\hat{U}$, representing the launch of the first $M1$ signal. Otherwise every limb is in the quiescent state.

(c)   *The transition rules for $\mathcal{L}^{max}_B$:*   The transition rules of $\mathcal{L}^{max}$ are listed in Section 3.2. It is unnecessary to explain any further the transition rules governing the additional features stated in (b).

(d)   *Results:*   For a given vertex, all limbs with the same class number $1, 2, \ldots, d-1$ or I, at time θ, belong to the same block.

Parallel computation in different blocks begins as soon as a first articulation point has completed its partitioning. Nevertheless, θ is difficult to estimate. We have roughly $\theta \leqslant \text{card } X \times \text{card } U$. Card $F = 7(d+1)+1$ and card $E \leqslant 7^d d!$.

Remark: We could easily supply $\mathcal{L}^{max}_B$ with an additional feature so that the block decomposition would include a labeling of all limbs in the same block.

4.2.   *Hamiltonian Cycle*

We seek to construct an automaton which tests any graph for the existence of a Hamiltonian cycle, and displays such a cycle if one exists.

THEOREM.   The Hamiltonian cycle problem is fa-computable.

Berstel [5] has already verified this theorem. Here we present an approach which is simpler and more rapid than his solution, by profiting from certain shortcuts in the enumeration of the cycles. Let us describe the automaton $\mathcal{H}$ which solves this problem with 5 colors per limb.

(a)   *Principles:*   Consider the class of words σ originating at a given vertex $x_0 \in X$,

$$\sigma = l_1 \ldots l_k l_{k+1} \ldots l_p,$$

such that

$$pr_1 l_1 = x_0$$

and

$$pr_1 l_h \neq pr_1 l_k \quad \text{for} \quad h < k, \ k = 2, \ldots, p.$$

Notice that no vertex is repeated and that the word must terminate if $pr_1 \rho(l_p) = pr_1 l_h$ for some $h < p$. It is clear that this class of words includes Hamiltonian words, which pass through every vertex once. The order on the set $[d]$ of automaton limbs induces a lexicographic order on this class of words. Thus, if we select any initial vertex $x_0 \in X$ and construct these words one by one, according to the lexicographic order, we must eventually discover a Hamiltonian cycle if one exists.

We follow a sequence of limbs from x_0 until we arrive at a vertex which has already been traversed. If this vertex is x_0, and if no traversed vertex has a neighbor that has not been traversed, a Hamiltonian cycle has been found. To do so, if the word reaches x_0, an 8 signal will retrace the word back until it reaches an automaton with an untraversed neighbor. Then a new word is tried by taking the first next limb met. It may be shown that none of the words skipped by this procedure can be a Hamiltonian cycle. Thus, we reduce the required number of computations significantly.

Note that to avoid confusion the vertices which have been retraced should not be marked "untraversed" until after the new limb has been selected at x_1. In fact, they are marked "untraversed" only when they see the forward passage of a new word through one of their neighbors.

(b)   *The states of $\mathcal{H}$*: $F = \{\omega, I, \bar{I}, \hat{I}, A, B, \bar{B}\}$, where ω is a dead limb, I is a quiescent limb of an untraversed vertex, $\bar{I}$ is a quiescent limb of a traversed vertex, $\hat{I}$ is a quiescent limb of a retraced vertex, A is a word construction signal, B is a retracing signal, and $\bar{B}$ is a retracing signal and also a transient indicator of forward passage of A. At time $t = 0$, we have $e_{x_0}^{r_0} = A$ where $r_0 = \min[d]_{x_0}$, and $e_x^r = I$ or ω otherwise.

(c)   *The transition rules for $\mathcal{H}$*: When x receives the A signal ($l_x^r = A$), then when $\bar{r}$ is the minimal limb such that $e_x^{\bar{r}} = I$, $\hat{I}$ or B, then $e_x^{\bar{r}} \leftarrow A$ if $l_x^{\bar{r}} = I$ or $\hat{I}$, and $e_x^{\bar{r}} \leftarrow B$ if $l_x^{\bar{r}} = B$ (not to be confused with a return of the word). Also, in both cases, x places its other limbs except for $r(e_x^r : \cdots \leftarrow \bar{I})$ in state $\bar{B}$ for one unit of time, and then in state $\bar{I}$ (traversed vertex). The $\bar{B}$ is a transient indicator of forward passage through x.

(1)   If x cannot transmit the A signal as above, since $\bar{r}$ does not exist, then it sends a $\bar{B}$ signal to retrace the word (if $l_x^r = A$ then $e_x^r : I \leftarrow \bar{B}$).

(2)   If x receives a B signal ($l_x^r = B$), and (a) $e_x^r = B$, then x treats the B signal as if it were an A signal. (b) $e_x^r = I$ or $\hat{I}$, then x transmits a $\bar{B}$ signal if it has an untraversed neighbor, and a B signal if not.

(3)   If a vertex is retraced (transmits a $\bar{B}$ or a B signal), then it places all its limbs in state $\hat{I}$, except for a limb transmitting a B signal, which remains unchanged.

(4)   If x has an untraversed neighbor, and receives a $\bar{B}$ signal, ($l_x^{r_1} = \bar{B}$) then it selects the first limb above r in the order if it exists, and transmits an A signal along that limb, or else transmits $\hat{B}$.

(5)   If x has its limbs in state $\hat{I}$, and sees a forward passage through one of its neighbors ($l_x^r = \bar{B}$), then x changes its limbs to state I.

(6)   If x_0 receives an A signal, x_0 being the only vertex whose limbs are in state I except for one in state A, then x_0 retraces the word .by

transmitting a B (if x_0 has no untraversed or retraced neighbors) or a $\bar{B}$ (if x_0 has a neighbor such that $l_x^r = I$ or $\hat{I}$). This constitutes a test for a Hamiltonian cycle. When x_0 receives the signal B that it transmitted, a Hamiltonian cycle has been found.

(7) If x is such that $l_x^{r_1} = A = l_x^{r_2}$, $r_1 \neq r_2$ and $e_x^{r_3} = A$, $e_x^{r_1} = \bar{I}$, then x does as if $l_x^{r_2} = \bar{I}$ and $e_x^r = \bar{I}$, for all $r \neq r_3$ (self-loop).

(d) *Results:* If a Hamiltonian cycle is found, it is marked by those limbs of the graph which are in state B. If no such cycle exists, the procedure will terminate when no new words can be formed. The computation time is extremely difficult to calculate, but the efficiency of the algorithm has been improved to a great extent by the omission of certain words which could not lead to a Hamiltonian cycle. Card $F = 7$, card $E \leqslant 7^d$, and card $E = 2d(d-1) + 6d = 2d^2 + 4d$ if there are no dead limbs. Then a traveler in the graph would solve the problem with 5 colors only.

4.3. The Minimum Tree Problem

The well-known minimum tree problem, as we showed [16], can be stated as follows.

Given a total order† upon the set of edges of a graph, determine the spanning tree which is first in the lexicographic order, when all the spanning trees are written as lists of their branches put in the increasing order.

The algorithm that best fits our goal here is the one known as Sollin's algorithm [3], which is a variation on the third algorithm of Kruskal [11]. It consists of

(1) choosing the minimal edge in the coboundary of each vertex, and making it a branch of the minimal tree V;

(2) choosing the minimal edge in the nonempty coboundaries of sets of vertices of each connected subtree of V already formed, and making it a branch of the minimal tree V.

Instruction (2) is repeated as many times as possible. Notice that the algorithm involves only local decisions and features parallel computation since all the coboundaries are considered simultaneously. This makes it an ideal application for computation by networks of finite automata.

Let us define the *total order* for the set of edges U in the framework of our theory. As we did for the early bird problem, we may suppose an external source of excitation acting on the edges in the prescribed order. More precisely, an edge enumeration clock functions as follows: the clock contacts each of

---

† In [16] we extended the problem to a weak order, which we do not introduce here for the sake of simplicity.

the edges to check whether they are all ready for an enumeration. This is done in the prescribed order, the checking process being performed by repeated circular inspections. As soon as the clock has contacted all the edges, and found that they are in the ready state, it runs through the order once, exciting each edge from the first to the last at a distinct instant of time, and then continues its normal contacting procedure. Though we speak casually about exciting or contacting edges, the clock actually interacts with vertex automata, each edge being associated with a specific limb of one of the vertex automata to which it is incident.

A second ordering is necessary due to an interesting undecidability problem involved here which generalizes Theorems 1 and 2 of Section 2.4, namely, that an automaton in a subtree V_1 of G cannot tell whether an edge incident to it and not belonging to V_1 is a chord of V_1 or an edge of the coboundary of the vertices of V_1, connecting V_1 to another subtree V_2. To resolve this problem we must escape the risks of undecidable connectivity due to symmetry. It can be shown that this is achieved by introducing a vertex enumeration clock, which functions continuously, exciting each of the vertices of X according to a circular order. We point out that no first vertex exists in the order. An excitation consists of a vertex being made to enter a special excited state, not accessible via its transition function.

It is understood, of course, that in addition to the two clocks mentioned above, the network is synchronized in the usual way by a synchronization clock, which prompts the state transitions of the automata at discrete intervals of time. Notice that all three clocks operate at the same speed. In effect, the two enumeration clocks can be regarded as synchronized by the synchronization clock.

Remark: As explained before, the three external clocks mentioned here may be incorporated into the network as follows:

(1) the synchronization clock can be replaced by the self-synchronization automaton of Section 2.3;

(2) the continuous vertex enumeration clock can be replaced by the vertex-ordering automaton of Section 3.6;

(3) the edge enumeration clock can be replaced by the edge-ordering automaton of Section 3.4 with an additional feature for recognizing the ready state.

On the basis of these definitions we can now prove the following theorem.

THEOREM. The minimal tree problem is fa-computable.

We define the minimum tree automaton $\mathscr{E}_M$, derived from the elementary early bird automaton $\mathscr{E}$ (see Section 3.8) for solving the minimum tree problem.

(a) *Principles of* $\mathcal{E}_M$: The algorithm will be performed using two types of iterations: the first type, which we call *labeling procedure*, for identification of boundaries, the second type, which we call *minimization procedure*, for selecting the minimal edge in a coboundary.

The labeling procedure involves, for a given vertex x belonging to a subtree V_1 of V already determined, labeling among the edges incident to x and not belonging to V_1 those belonging to the coboundary of the vertices of V_1.

To effect this algorithm, we use a modification of the early bird algorithm, see Section 3.8. When a vertex is excited by the vertex enumeration clock, it becomes a leader, causing itself and each of its neighbors along edges not belonging to its subtree V_1 to begin sending independent early bird signals within their own subtrees at the same time.

It is easy to show, due to the strict circular order of enumeration, that two automata will simultaneously enter the early bird state if and only if they began sending signals at the same instant along the same complete cyclic word of their common subtree.

Thus, at least one leader in each subtree will be able to label each of its edges as either chords or coboundary, according to whether the corresponding neighbor becomes an early bird at the same time as it does. Since it may have d neighbors, it must send out as many as d different sets of early bird signals.

We point out here that we have generalized the early bird problem to the case where there may be one or two early birds in the circle, the first ones excited. At the end of the computation, each early bird knows whether there were two or one. Here the roles of early birds are played by leader and neighbor.

Problems of symmetry leading to undecidability are avoided by the vertex-enumeration clock, which ensures that the state $M(t)$ of the network is asymmetric at each instant t.

Once a vertex has all its edges labeled, they are considered "ready" by the edge-enumeration clock, and that vertex ignores all further excitations by the vertex-enumeration clock. Thus, the labeling procedure continues until all edges are ready, at which time the minimization procedure begins.

The minimization procedure involves selecting the minimal edge in the coboundary of each subtree. This is done using the early bird algorithm in the standard version. As the edge-enumeration clock excites each coboundary edge, the corresponding vertex sends early bird signals along the complete cyclic word of its subtree. Excitations of other edges are simply ignored. Clearly, the early bird will be that vertex associated with the minimal edge, since it was excited first according to the strict order on U. Each subtree thus selects an edge which becomes a new branch of V, and the vertices of that subtree immediately begin accepting leader excitations from the vertex-enumeration clock, thus recommencing the labeling procedure.

We point out that, at this stage of the algorithm, there may be temporary gaps in the cyclic word of a subtree. For example, if two subtrees V_1 and V_2 are to be joined by the minimal edge u_1 of the coboundary of V_1, to form a new subtree V_{12}, V_1 may select u_1 before V_2 has terminated the minimization procedure. However, the vertices of V_1 will immediately commence the labeling procedure, as if the cyclic word of V_{12} already existed. In fact, there will be a gap in the word at the limbs corresponding to u_1 until V_2 is ready to establish its connections with neighboring subtrees. We have proved that, with very slight modifications to the early bird algorithm, the presence of such a gap does not affect the proper functioning of the early bird signals, although it may delay the labeling of edges in V_1.

(b) *The states of $\mathscr{E}_M$:* For the labeling procedure, we require $2d$ classes of early bird signals, $1_L, 2_L, \ldots, d_L$ and $1_N, 2_N, \ldots, d_N$, where k_L signals are sent out by leaders and k_N signals by their neighbors. Excluding the state ω, there are 11 states in the automaton $\mathscr{E}$ (see Section 3.8), so that we require 11^{2d} limb states in all. In addition, we require the states $\hat{X}, C, T, S$ where $\hat{X}$ is excited by the vertex enumeration clock, C is a limb corresponding to a chord of the subtree, T is a limb corresponding to an edge of the coboundary, and S is a limb belonging to the subtree.

For the minimization procedure we require only one set of early bird signals, as well as the limb states where $\hat{U}$ is excited by the edge enumeration clock and M is the minimal edge of a coboundary.

At time $t = 0$, the network is quiescent. The edge enumeration clock causes the algorithm to be initiated by the labeling procedure, labeling for the first step all the edges except the self-loops.

(c) *The transition rules for $\mathscr{E}_M$:* All early bird signals travel along cyclic words whose connections are determined according to the rules given in Section 3.5 with branches of the subtree being designated by the limb state S.
The labeling procedure is as follows:

(1) If an automaton x has no limbs in state S, then it places all its nondead limbs in state T.

(2) If x has limbs in state S, and has other limbs that are unlabeled in state I or $\hat{I}$, then it assumes state $\hat{X}$ when excited. Otherwise, it ignores the excitation of the vertex enumeration clock.

(3) If x is in state $\hat{X}$, and has an unlabeled limb k, then it sends out early bird signals of class k_L. At the same time the neighbor $pr_1 \rho(x, k)$ sends out early bird signals of class k_N. Both vertices may subsequently be reexcited, and will treat their still-quiescent limbs in the same way.

(4) If x is the early bird and receives green signals that are mixed, that is, of both class k_L and k_N, then limb (x, k) is placed in state C.

(5) If x is the early bird and receives green signals that are pure, that is, of either class k_L or k_N, then limb (x, k) is placed in state T.

(6) Edges that have both associated limbs in state T, S, or C are considered ready by the edge enumeration clock.

(d) *Minimization procedure:* We shall say that (x, k) is excited if that limb receives the excitation of the edge u with which it is associated.

(1) If all the nondead limbs of x are in state T, and (x, k) is excited by the edge enumeration clock, then (x, k) assumes state M and all other nondead limbs of x assume state I.

(2) If x has limbs in state S, and limb (x, k) is in state T when excited, and no other limbs of x have been excited during the current iteration, then (x, k) assumes state $\hat{U}$ and x sends out early bird signals. Otherwise the excitation of the edge enumeration clock is ignored.

(3) If x receives green signals indicating that it is the early bird, and has limb (x, k) in state $\hat{U}$, then (x, k) assumes state M, and all limbs of x in state T assume state I.

(4) If x transmits a single green signal, then all limbs of x formerly in state T are placed in state I.

(5) If a limb in state M finds itself connected to another limb in state M or state I, then both limbs assume state S.

(d) *Results:* The process terminates with all edges having both associated limbs in either state S or state C. Those in state S constitute the required minimal spanning tree.

Because of the parallel computation, with each subtree operating locally, the computation time is relatively small. The labeling procedure can last at most $2n^2$ time units, under very rare conditions, while the minimization procedure can last at most $2n$ time units. The maximum possible number of successive iterations of both procedures is $\log_2 n$. Thus we have the very rough upper bound

$$\theta < (\log_2 n)\, 2n(n+1).$$

Card $F = 11^{2d} + 3 + 11 + 2 + 1 = 11^{2d} + 17$ and card $E < 11^{2d} + 17^d$.

References

1. Balzer, R. M., An 8-state minimal time solution to the firing squad synchronization problem, *Information and Control* **10**, No. 1, 22–42 (1967).
2. Berge, C., and Ghouila-Houri, A., "Programming, Games and Transportation Networks" (M. Merrington and C. Ramanujacharyulu, transl.). Wiley, New York, 1965.
3. Berge, C., "Graphes et hypergraphes." Dunod, Paris, 1970.
4. Berstel, J., Résolution par un réseau d'automates du problème des arborescences dans un graphe, *C. R. Acad. Sci.* **264**, 388–390 (1967).

5. Berstel, J., Quelques applications des réseaux d'automates à des problèmes de la théorie des graphes (Thèse 3° cycle), Paris, 1967.

6. Burks, A. W., "Essays on Cellular Automata." University of Illinois Press, Urbana, 1970.

7. Cole, S. N., Real time computation by iterative arrays of finite-state machines, Ph.D. thesis. Harvard University, 1964.

8. Dermiane, J. C., and Pair, C., Problèmes de cheminements dans les graphes, *in* "Monographies d'Informatique," No. 8. Dunod, Paris, 1971.

9. Gabrielian, A., The theory of interacting local automata, *Information and Control* **16**, 360–377 (1970).

10. Hennie, F., "Iterative Arrays of Logical Circuits." M.I.T. Press, Cambridge, Massachusetts, 1962, and Wiley, New York, 1962.

11. Kruskal, J. B., Jr., On the shortest spanning subtree of a graph and the travelling salesman problem, *Proc. Amer. Math. Soc.* **7**, 48–50 (1956).

12. Moore, E. F., The firing squad synchronization problem, *in* "Sequential Machines Selected Papers," pp. 213–214 Addison-Wesley, Reading, Massachusetts, 1964.

13. Moore, E. F., The shortest path through a maze *in* "Proceedings of International Symposium in the Theory of Switching," p. II, pp. 285–292. Harvard Univ. Press, Cambridge, Massachusetts, 1959.

14. Moore, F. R., and Langdon, G. G., A generalized firing squad problem, *Information and Control* **12**, 212–220 (1968).

15. Rosenstiehl, P., Existence d'automates finis capables de s'accorder bien qu'arbitrairement connectés et nombreux, *Internat. Comp. Centre* **5**, 245–261 (1966).

16. Rosenstiehl, P., L'arbre minimum d'un graphe, *in* "Theory of Graphs, International Symposium," pp. 357–368. Gordon & Breach, New York, 1967.

17. Rosenstiehl, P., Graph problems solved by finite automata networks, Calgary International Conference on Combinatorial Structures and Their Applications (1969) (unpublished).

18. Rosenstiehl, P., Labyrinthologie mathématique *Math. Sci. Humaines* **33** (1971).

19. Tarry, G., Parcours d'un labyrinthe rentrant, *Assoc. Francais pour l'Avan. des Sciences* 49–53 (1886).

20. Tarry, G., Le problème des labyrinthes, *Nouvelles Annales de Math.* **14** (1895).

21. Varshavsky, V. J., Marakhovsky, V. B., and Peschansky, V. A., Synchronization of interacting automata, *in* "Mathematical Systems Theory," Vol. 4, No. 3, pp. 212–230. Springer-Verlag, Berlin and New York, 1970.

22. Waksman, A., An optimum solution to the firing squad synchronization problem, *Information and Control* **9**, 66–78 (1966).

23. Whitney, H., Non-separable and planar graphs, *Trans. Amer. Math. Soc.* **34**, 339–362 (1932).

24. Yamada, H., and Amoroso, S., Tessellation automata, *Information and Control* **14**, 299–317 (1969).

25. Arbib, M. A., Self-reproducing automata—Some implications for theoretical biology, *in* "Towards a Theoretical Biology" (C. M. Waddington ed.). Edinburgh University Press, Edinburgh, 1969.

AN ALGORITHM FOR A GENERAL CONSTRAINED SET COVERING PROBLEM

B. Roy

Groupe METRA
Université Paris-Dauphine
Paris, France

Let $B = \{b_j | j \in N, b_j \subset E\}$, E being, for instance, the set of vertices of a given graph H and the blocks b_j being particular subsets of E (edges, paths, and circuits of H). A *cover* is a family of blocks, the union of which is E. A *configuration* is a cover, which is an independent set of a graph G satisfying some additional constraints.

267

This contribution concerns the search for an optimal configuration for a broad class of objective functions and additional constraints.

A separation and evaluation procedure is proposed to solve this combinatorial programming problem. The problem is described and illustrated in Section 1. The main concepts are introduced in Section 2. Section 3 is devoted to the description of the algorithm.

1. The General Constrained Set Covering Problem

1.1. Statement of the Problem

1.1.1. INTRODUCTION

Let us consider a set E of m elements, where

$$E = \{e_i \mid i \in M\}, \qquad M = \{1,\dots,m\},$$

and a family B of n subsets of E called *blocks*, where

$$B = \{b_j \mid b_j \subset E, j \in N\}, \qquad N = \{1,\dots,n\}.$$

A cover C is a subfamily of B such that

$$(1) \qquad\qquad \bigcup_{b_j \in C} b_j = E.$$

The classical *set covering problem* (CP) is to find a cover C which minimizes

$$(2) \qquad\qquad \sum_{b_j \in C} p_j,$$

where p_j is a nonnegative real number assigned to b_j.

An important special case of the CP is the *partitioning problem* (PP) where C cannot be just any cover, but must be a partition of E. Lemke, Salkin, and Spielberg [7] have shown that, given any PP which has a feasible solution, the CP defined with the same blocks but with new weights p_j' such that

$$(3) \qquad\qquad p_j' = p_j + |b_j| P,^\dagger \qquad P = \sum_{b_j \in B} p_j,$$

has the same set of optimal solutions as the original PP.

Different algorithms have been proposed to solve the CP and the PP. Good syntheses are presented in Garfinkel [5] and Thiriez [14]. For many practical purposes we need

   (i)   to restrict the set of feasible solutions to covers which possess additional properties, less specific than to be a partition;

   (ii)   to use a more general objective function than (2).

† $|b|$ = cardinality of b.

This double generalization of the classical CP leads to the general *constrained set covering problem* (CCP) formulated in Section 1.1.3. The end of this first section will be devoted to making explicit this combinatorial programming formulation, particularly in connection with graph theory.

In spite of the interest of this problem, it seems that it has been considered only through applications or particular cases, for example, Salkin [13]. No general algorithm seems to have been proposed yet. The purpose of this paper is to present an algorithm which works without restricting too much the possible analytical forms for the additional conditions and the objective function.

This algorithm, described in Section 3, uses the approach introduced in Roy [11, Chapter VI, Section A and B]. The general concepts and notations are given in Section 2.

1.1.2. CONFIGURATIONS

Let us consider a subfamily C of B, and denote by X_C its characteristic vector. X_C is an $n \times 1$ column vector $(x_1, \ldots, x_n)$ with $x_j = 1$ if $b_j \in C$, and $x_j = 0$ if $b_j \notin C$. Conversely, for any binary $n \times 1$ column vector X, the above relations define a subfamily C_X of B.

All the additional properties of (1) we need to introduce may be expressed by conditions of the type

$$(4) \qquad \Phi_k(X) \leqslant \phi_k, \qquad k = 1, \ldots,$$

where $\Phi_k : \{0, 1\}^n \to R$ and $\phi_k \in R$. Various examples of such additional conditions are given in Section 1.3 together with the nature of the property they are able to formalize.

Frequently, the presence of a given block b_j in a cover C must be considered as incompatible with the presence in C of some other blocks b_h with $b_h \in \Gamma(b_j)$. For example, each block b_h having at least α elements of E in common with b_j may have to be excluded from any feasible cover to which b_j belongs ($\alpha = 1$ leads to PP). The importance of such a property will appear more clearly in Section 1.2.

All such additional properties, which are required in a problem, instead of being formalized by a series of conditions of type (4), can always be synthetized in an *incompatibility graph G*

$$G = (B, \Gamma)$$

in which two blocks are connected by an edge, if and only if they cannot occur simultaneously in a feasible cover. To be feasible, a cover must therefore be an independent set of G. By definition, a subset C of B is an *independent set* of G, if and only if two elements of C are never connected by an edge of G.

Let a *configuration* be any cover which is an independent set of the incompatibility graph G, and which satisfies conditions (4) imposed by the problem.

The general constrained set covering problem may then be stated as follows: find one configuration which minimizes an objective function $\Phi_0(X)$, where $\Phi_0:\{0,1\}^n \to R$.

1.1.3. COMBINATORIAL PROGRAMMING FORMULATION

(5)   Minimize $\Phi_0(X)$

subject to

(6) $x_j = 0 \text{ or } 1, \quad j = 1,...,n,$

(7) $A \cdot X \geqslant d,$

and

(8)† $X \in I_G,$

(9) $\Phi_k(X) \leqslant \phi_k, \quad k = 1,...,q,$

where d is an $m \times 1$ column vector of 1's, A is an $m \times n$ matrix ($A = \{a_{ij}\}$ with $a_{ij} = 1$ if $e_i \in b_j$ and $a_{ij} = 0$ if $e_i \notin b_j$), and I_G is the set of characteristic vectors of the independent sets of G.

1.2. Connections with Graph Theory

Let us consider a directed graph $H = (V, U)$, V being the set of vertices, and $U \subset V \times V$ being the set of arcs. Recall that, by definition, $W \subset V$ is an *externally stable set* of H, if and only if

$$\forall\, v \in V - W \quad \exists\, w \in W \qquad \text{such that} \quad (w,v) \in U.$$

We will say that such a vertex w covers vertex v. This terminology emphasizes the relationship between coverings and the externally stable property that we shall now clarify.

Consider first the particular directed graph $\bar{H} = (\bar{V}, \bar{U})$ derived from a covering problem by defining $\bar{V}$

$$\bar{V} = E \cup B \cup \{z\}, \quad z \text{ is an additional vertex,}$$

and $\bar{U}$ (see Fig. 1)

$$(z, b_j) \in \bar{U}, \qquad b_j \in B,$$

$$(b_j, e_i) \in \bar{U}, \qquad \text{iff} \quad e_i \in b_j.$$

$\dagger$ This constraint implicitly contains constraint (6), but we prefer to make the latter explicit.

(2) Δ_{ij} is a marginal cost for element e_i with respect to block b_j, and the ith term of the second sum gives the marginal cost of e_i with respect to configuration C_X, which is equal to the smallest Δ_{ij} over all the blocks in C_X which cover e_i.

This function may be used as the objective function in various problems where each block is defined by the assignment of a definite machine, truck, ship, etc., to perform a given subset of tasks or activities. When each subset of a block also appears itself as a block, it allows us to reduce the size of the problem by introducing in B only the maximal blocks. A maximal block is a block contained in no other block.

2. Notation and Main Concepts

The algorithm described in Section 3 belongs to the general family of what we called *separation and evaluation procedures*, discussed in Roy [10]. In such a procedure, the set of feasible solutions is divided into smaller and smaller subsets on the basis of a *separation principle* (see Section 2.1). Each subset so generated must be examined for emptiness and then for the possibility of existence of an optimal solution. These examination rules extensively use the concept of *optimistic evaluation* (see Section 2.2). To accelerate the exploration of the set of all configurations, a third concept is introduced in Section 2.3, the concept of *forced blocks*.

2.1. Separation Principle

2.1.1. PARTIAL SOLUTION

A *partial solution* ω is, by definition, a pair of disjoint subsets of B

$$\omega = (B_\omega^{\ 1}, B_\omega^{\ 0}), \qquad B_\omega^{\ 1} \subset B, \quad B_\omega^{\ 0} \subset B, \quad B_\omega^{\ 1} \cap B_\omega^{\ 0} = \varnothing.$$

We will denote by Ω the set of all partial solutions.

Consider now the set $\mathscr{C}$ of all configurations, that is, of all subsets C of B, with a characteristic vector X_C satisfying constraints (7), (8), and (9). To each $\omega \in \Omega$ is associated the subset $\mathscr{C}_\omega$ of $\mathscr{C}$ defined by

(10) $$\mathscr{C}_\omega = \{C \mid C \in \mathscr{C}, B_\omega^{\ 1} \subset C \subset B - B_\omega^{\ 0}\}.$$

Let $B_\omega^* = B_\omega^{\ 1} \cup B_\omega^{\ 0}$. The position of every block $b_j \in B_\omega^*$ with respect to configurations of $\mathscr{C}_\omega$ is completely determined. On the contrary, a block $b_j \notin B_\omega^*$ may be either in or out of a configuration $C \in \mathscr{C}_\omega$. Such a block will be called a *free block* in the partial solution ω.

2.1.2. VECTORS ASSOCIATED WITH A PARTIAL SOLUTION

The manipulations of partial solutions in the algorithm necessitate the introduction of three $n \times 1$ column vectors.

(11) $X_\omega = \{x_j | x_j = 1 \text{ if } b_j \in B_\omega{}^1 \text{ and } 0 \text{ if not}\}$,

(12) $\hat{X}_\omega = \{\hat{x}_j | \hat{x}_j = 0 \text{ if } b_j \in B_\omega{}^0 \text{ and } 1 \text{ if not}\}$,

(13) $Y_\omega = X_\omega + \hat{X}_\omega$.

Notice that a $n \times 1$ binary column vector X can be the characteristic vector of a configuration $C \in \mathscr{C}_\omega$ only if $X_\omega \leqslant X \leqslant \hat{X}_\omega$, with $X \leqslant Y$ if and only if the inequality holds for each component.

2.1.3. SEPARATION PRINCIPLE

Let ω be any partial solution and $\mathscr{C}_\omega$ the subset defined in (10). When a particular free block b_{j_ω} has been chosen by some rule, $\mathscr{C}_\omega$ is partitioned into two disjoint subsets $\mathscr{C}_{\omega_1}$ and $\mathscr{C}_{\omega_2}$ defined by

(14) $\omega_1 = (B_\omega{}^1 \cup \{b_{j_\omega}\}, B_\omega{}^0), \qquad \omega_2 = (B_\omega{}^1, B_\omega{}^0 \cup \{b_{j_\omega}\})$

in which b_{j_ω} belongs to all the configurations of the first one and is excluded from all the configurations of the second one. Equation (14) together with a rule for the selection of the *separating block* b_{j_ω} defines what we will call the *separation principle* for the problem.

Different selection rules may be considered (see Roy [11, Chapter VI, p. 31, first Remark]), but this aspect of the procedure will not be discussed in this paper. A particular selection method for the choice of j_ω has been adopted for the algorithm of Section 3. To explain it, let us introduce the subset $E_\omega{}^0$ of the elements of E, which do not belong to any block of $B_\omega{}^1$.

(15) $E_\omega{}^0 = \{e | e \in E, e \notin b \ \forall \ b \in B_\omega{}^1\}$.

The separating block b_{j_ω} is the first block, the smallest value of the index, such that

(1) b_j is a free block in ω, if $E_\omega{}^0 = \varnothing$;
(2) b_j is a free block containing the first element, the smallest index, of $E_\omega{}^0$, if this subset is not empty.

Obviously, other selection rules might be considered.

2.2. Optimistic Evaluation

2.2.1. DEFINITION

Consider one of the Φ functions either used for the objective function we want to minimize, see (5), or entering into one of the inequalities (9). An *optimistic evaluation* for Φ_k is a function v_k of the two vectors X_ω and $\hat{X}_\omega$ associated with a given partial solution ω such that[†]

$$(16) \qquad v_k(X_\omega, \hat{X}_\omega) \leqslant \Phi_k(X) \qquad \forall\, \omega \in \Omega,$$

for all the characteristic vectors X of a configuration of $\mathscr{C}_\omega$, that is, for all X's satisfying (6), (7), (8), (9), and

$$(17) \qquad X_\omega \leqslant X \leqslant \hat{X}_\omega.$$

In the building of an optimistic evaluation, we must try to obtain as good an approximation as possible of the exact value of the minimum of Φ_k on $\mathscr{C}_\omega$ without increasing too much the complexity, that is, the computational time of this evaluation. Nevertheless, it is more appropriate if the following coincidence property is satisfied:

$$v_k(X_\omega, \hat{X}_\omega) = \Phi_k(X_\omega), \qquad X_\omega = \hat{X}_\omega.$$

Before giving examples of how to make up optimistic evaluations (see Section 2.2.3), we will show why such functions are considered in our problem.

2.2.2. ASSOCIATED TEST

Suppose such an optimistic evaluation for Φ_k, $k = 1, \ldots, q$ has been built. By the test θ_k we will refer to the comparison between $v_k(X_\omega, \hat{X}_\omega)$ and ϕ_k. The test will be positive if and only if we obtain

$$(18) \qquad v_k(X_\omega, \hat{X}_\omega) \leqslant \phi_k.$$

In effect, if this inequality does not hold, it is the proof that $\mathscr{C}_\omega$ is empty.

Consider now the case $k = 0$. Suppose an optimistic evaluation v_0 has been built. Denote by ϕ_0 the value of the best configuration that is known. We can do the test θ_0, and if (18) does not hold, it is proof that the configurations of $\mathscr{C}_\omega$, if $\mathscr{C}_\omega \neq \varnothing$, are worse than those already known.

2.2.3. EXAMPLES

If we go back to the examples of Section 1.3, we see that it is relatively easy to obtain optimistic evaluation functions possessing the coincidence property.

The functions in Examples 1.3.1. and 1.3.2 are monotonic, that is, for all the pairs of vectors X and X' such that $X \leqslant X'$ either $\Phi(X) \leqslant \Phi(X')$ (nondecreasing function) or $\Phi(X) \geqslant \Phi(X')$ (nonincreasing function). In such cases, the optimistic evaluation may be $v(X_\omega, \hat{X}_\omega) = \Phi(X_\omega)$, if Φ is a nondecreasing function, and $v(X_\omega, X_\omega) = \Phi(\hat{X}_\omega)$, if Φ is a nonincreasing function.

If we now consider the functions of Examples 1.3.3, 1.3.4, and 1.3.5, it is

---

[†] When we want to maximize, or when inequality (9) is in the reverse orientation, $\leqslant$ must be changed into $\geqslant$ in (16) and (18).

easy to see that they are not monotonic. Nevertheless, as has been shown by Bendahan and Fayein [2], it is possible to define, on the same basis as before, an optimistic evaluation for those functions, and for many others as far as they are quasi-monotonic in the following sense: A Φ function is *quasi-monotonic* if it is possible to exhibit

(1)   some monotonic functions $\rho_p(X)$, $p = 1,...$;

(2)   a function, $\psi(\rho_1,...,\rho_p,...)$, which is monotonic with respect to each variable ρ_p,

such that

(19) $$\Phi(X) = \psi[\rho_1(X),..., \rho_p(X),...].$$

Then, if we substitute $\rho_p(X_\omega)$ or $\rho_p(\hat{X}_\omega)$ for $\rho_p(X)$ in the right-hand side of (19), according to the sense of variation of ψ with regard to ρ_p and of ρ_p with regard to X, we obtain an optimistic evaluation for $\Phi(X)$.

Another general approach, which is very useful in building an optimistic evaluation, consists in relaxing some of the constraints defining $\mathscr{C}_\omega$ so that the exact value of the minimum of Φ may be easily computable on the broader subset thus introduced. The relaxation, for instance, of constraint (6) together with some appropriate transformations of (8) and (9) may allow the use of classical linear programming to find the exact value of the minimum. Many suggestions have been made in this direction, especially for function (2). Rougerie and Viviant [9] give a very extensive list of possible optimistic evaluations for this classical objective function of the CP.

2.3.   Forced Blocks

2.3.1.   DEFINITION

Let us consider a partial solution ω together with its associated subset $\mathscr{C}_\omega$, see (10), and a block $b_j \in B - B_\omega{}^*$. For such a free block in ω, it may happen that either

(20) $b_j \in C \quad \forall C \in \mathscr{C}_\omega$

or

(21) $b_j \notin C \quad \forall C \in \mathscr{C}_\omega.$

We then say that b_j is a *forced block* in ω.

Suppose that, by studying constraints (7), (8), or (9), it has been possible to exhibit

(1)   a subset $B_\omega{}^+$ made of blocks for which (20) holds;

(2)   a subset $B_\omega{}^-$ made of blocks for which (21) holds.

It is possible to assert that

(1) $\mathscr{C}_\omega$ is empty if $B_\omega{}^+ \cap B_\omega{}^- \neq \varnothing$;
(2) $\mathscr{C}_\omega = \mathscr{C}_{\bar\omega}$, $\bar\omega$ being defined by

$$B_{\bar\omega}{}^1 = B_\omega{}^1 \cup B_\omega{}^+ \qquad \text{and} \qquad B_{\bar\omega}{}^0 = B_\omega{}^0 \cup B_\omega{}^-.$$

The end of this section will be devoted to showing how, by elementary considerations dealing with covering constraints (7), graph constraints (8), and Φ constraints (9), subsets like $B_\omega{}^+$ and $B_\omega{}^-$ can easily be exhibited.

2.3.2. FORCED BLOCKS BY COVERING CONSTRAINTS

Consider the vector Y_ω defined in (13), and compute

$$Z = A \cdot Y_\omega = \{z_i | i \in M\}.$$

Suppose that for some i, $z_i = 1$. This is proof that the element e_i can be covered by one and only one free block. Such a block must be included in $B_\omega{}^+$.

The detection of such forced blocks by covering constraints is developed in Step C of the algorithm in Section 3.

2.3.3. FORCED BLOCKS BY GRAPH CONSTRAINTS

From the definition of constraints (8), each free block of $\Gamma(B_\omega{}^1)$ is obviously an element of $B_\omega{}^-$. Step G of the algorithm is based on this remark.

2.3.4. FORCED BLOCKS BY Φ CONSTRAINTS

The detection of such forced blocks depends on the nature of the particular Φ_k function considered. We will give here only two elementary examples. Let us consider first the Φ function of Example 1.3.2. Since it is a monotonic nondecreasing function, the optimistic evaluation associated with it can be

$$v(X_\omega, \hat{X}_\omega) = \Phi(X_\omega) = \gamma_\omega{}^+ - \gamma_\omega{}^-$$

with

$$\gamma_\omega{}^+ = \max_{b_j \in B_\omega{}^1} \gamma_j, \qquad \gamma_\omega{}^- = \min_{b_j \in B_\omega{}^1} \gamma_j.$$

Suppose for a given ω we have

$$\gamma_\omega{}^+ - \gamma_\omega{}^- = \phi, \qquad \phi \text{ the right-hand side of (9)}.$$

Every block b_j for which either $\gamma_j > \gamma_\omega{}^+$ or $\gamma_j < \gamma_\omega{}^-$ can be included in $B_\omega{}^-$.

Consider now case (2) of Example 1.3.3. Let

$$n_1(t, \omega) = |B_t \cap B_\omega{}^1| \qquad \text{and} \qquad n(t, \omega) = |B_t \cap (B - B_\omega{}^*)|.$$

Suppose ω is such that

$$n_1(t, \omega) + n(t, \omega) = \lambda_t',$$

then each block of $B_t \cap (B - B_\omega^*)$ belongs to B_ω^+. If for another ω we have $n_1(t, \omega) = \lambda_t''$, then each free block of B_t belongs to B_ω^-.

Whatever the Φ_k function considered, the search for forced blocks involves the difference

$$\phi_k - v_k(X_\omega, \hat{X}_\omega).$$

That is the reason why this search must be performed in the algorithm together with the θ_k tests in Step E.

3. The Algorithm

Since it represents a separation and evaluation, or branch and bound, procedure, this algorithm consists of the exploration of a tree, the nodes of which are subsets generated by iterating the separation principle. Two fundamental types of exploration may be distinguished (see Roy [10]). With the first one, called *progressive* or *parallel*, we iterate from subset to subset according to the value of an optimistic evaluation for the objective function, so as to make it monotonically nondecreasing. Though very good results have been obtained with progressive separation and evaluation procedures in mixed integer programming (see Roy *et al.* [12]), we will resort to the other type of exploration which seems yet more appropriate to our particular problem. This second type of exploration, called *sequential* or *serial*, is related to implicit enumeration methods (see Balas [1] or Geoffrion [6]). It uses a complete order associated with the tree. Subsets generated are examined according to the corresponding *a priori* definite sequence. For the classical PP (see Section 1.1.1), efficient algorithms of this type have already been proposed by Pierce [8] and by Garfinkel and Nemhauser [4]. The following algorithm differs mainly in the role attributed to optimistic evaluations and in the introduction of the concept of forced blocks.

3.1. The Basic Tree and Its Exploration

Let us consider any given ω and its $\mathscr{C}_\omega$, its separating block $b_{j\omega}$ and corresponding partial solutions ω_1 and ω_2 defined by (14). Suppose that the considerations developed in Sections 2.3.2, 2.3.3, and 2.3.4 allow the exhibition of some forced blocks in ω_1 and in ω_2. With the support of assertion (2) of Section 2.3.1, it becomes possible to substitute new partial solutions ω' and ω'' respectively for ω_1 and ω_2, such that $\mathscr{C}_{\omega'} = \mathscr{C}_{\omega_1}$ and $\mathscr{C}_{\omega''} = \mathscr{C}_{\omega_2}$.

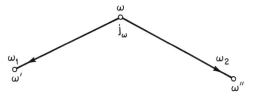

Fig. 2

If no forced blocks were found, ω' and ω'' can always be defined by $\omega' = \omega_1$ and $\omega'' = \omega_2$.

Fig. 2 shows the elementary module used to build the basic tree T. We will now complete its definition. Start with the partial solution ω_0, the root of T, defined by

$$B^1_{\omega_0} = \emptyset, \qquad B^0_{\omega_0} = \emptyset.$$

Define j_{ω_0} as explained in Section 2.1.3 and build the two corresponding partial solutions ω' and ω'' introduced above. Then restart from ω' (see Fig. 3), and

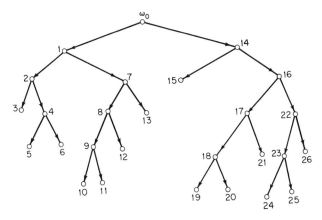

Fig. 3

iterate the separation process until each generated subset either has been separated or cannot be separated because no separating blocks were found (no free block remains in the partial solution considered). We obtain in that manner a finite rooted binary tree T (see Fig. 3).

Let us introduce, for each pair of nodes with the same predecessor in Γ (see Fig. 2), a *transverse order* and with this order, nodes derived from ω_1 are ranked before, on the left in Fig. 3 nodes derived from ω_2 [see (14)]. With this transverse order, the nodes of T may be completely ordered in a classical way:

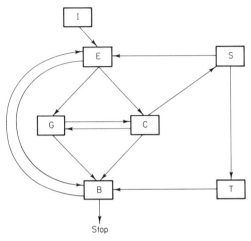

Fig. 4

the *Tarry order* (see Roy [11, Chapter V, p. 345]). This order is given by the natural order of integers on Fig. 3.

In the following algorithm, subsets $\mathscr{C}_\omega$ are examined according to the Tarry order introduced above, but as in all branch and bound procedures, the tree is not completely generated. In effect, tests θ_k for $k = 0, 1, ..., q$ (see Section 2.2.2) and assertion (1) a of Section 2.3.1, allow us to avoid the separation of many $\mathscr{C}_\omega$'s, because it is proved that they are empty, or that they do not include any optimal solution.

3.2. *Description of the Algorithm*

A simplified flow chart is given in Fig. 4.

Step I. Initialization: Introduce a stack which is empty at the beginning of the algorithm. Let $\phi_0 = +\infty$. Define ω by $B_\omega{}^1 = \varnothing$ and $B_\omega{}^0 = \varnothing$ and go to Step E.

Step E. Evaluation (see Section 2.3.4): Execute the following tests and go to Step B as soon as one of them is negative.

$$\theta_k \colon v_k(X_\omega, \hat{X}_\omega) \leqslant \phi_k, \qquad k = 0, 1, ..., q.$$

Let B_ω^{+k} and B_ω^{-k} be the two possibly empty subsets of forced blocks by constraint Φ_k, and

$$B_\omega{}^+ = \bigcup_{k=0}^{k=q} B_\omega^{+k}, \quad B_\omega{}^- = \bigcup_{k=0}^{k=q} B_\omega^{-k}.$$

If $B_\omega^+ \cap B_\omega^- \neq \emptyset$, go to Step B; if not, modify[†] ω by

$$B_\omega^1 \leftarrow B_\omega^1, \qquad B_\omega^0 \leftarrow B_\omega^0 \cup B_\omega^-,$$

and

(1) if $B_\omega^+ \neq \emptyset$, go to Step G, B_ω^+ being memorized for this next step;
(2) if $B_\omega^+ = \emptyset$, go to Step C.

Step G. *Graph exclusion* (see Section 2.3.3): Determine $B^- = \Gamma(B_\omega^+) \cap (B - B_\omega^*)$ with $\Gamma(B_\omega^+) = \bigcup_{b \in B_\omega^+} \Gamma(b)$, and

(1) if $B^- \cap B_\omega^+ \neq \emptyset$, go to Step B;
(2) if $B^- \cap B_\omega^+ = \emptyset$, go to Step C after having modified ω by

$$B_\omega^1 \leftarrow B_\omega^1 \cup B_\omega^+, \qquad B_\omega^0 \leftarrow B_\omega^0 \cup B^-.$$

Step C. *Covering* (see Section 2.3.2): Compute $A \cdot Y_\omega = Z$ and

(1) if for one $i \in M, z_i = 0$, go to Step B;
(2) if for all $i \in M, z_i \geq 2$, go to Step S;
(3) if $Z > 0$ and $M_\omega^+ = \{i \mid i \in M \text{ and } z_i = 1\} \neq \emptyset$, determine

$$N_\omega^+ = \{j \mid j \in N, y_j = 1, \sum_{i \in M_\omega^+} a_{ij} \geq 1\}, \qquad y_j \text{ the } j\text{th component of } Y_\omega,$$

$$B_\omega^+ = \{b_j \mid j \in N_\omega^+\},$$

and go back to Step G, B_ω^+ being memorized for this next step.

Step S. *Separation:* Test $B - B_\omega^* = \emptyset$. If the test is positive, go to Step T. Otherwise, determine the smallest value $i_\omega \in M$ such that the corresponding component of vector $A \cdot X_\omega$ equals 0, if there is one. Let j_ω be the smallest value $j \in N$ such that $b_j \in B - B_\omega^*$ and $a_{i_\omega j} = 1$, if i_ω is defined. Put on the top of the stack the pair $(B_\omega^1, B_\omega^0 \cup \{b_{j_\omega}\})$ and go to Step E after having modified ω by

$$B_\omega^1 \leftarrow B_\omega^1 \cup \{b_{j_\omega}\}, \qquad B_\omega^0 \leftarrow B_\omega^0 \cup \Gamma(b_{j_\omega}).$$

Step T. *Terminal nodes:* Execute the following tests, and go to Step B as soon as one of them is negative:

$$v_k(X_\omega, \hat{X}_\omega) \leq \phi_k, \qquad k = 0, 1, \ldots, q.$$

If all are positive, a better solution, characterized by X_ω, has been found. Memorize it. Modify ϕ_0 by $\phi_0 \leftarrow \Phi_0(X_\omega)$ and go to Step B.

Step B. *Backtracking:* If the stack is empty, let ω be the couple on its top, take it out of the stack, and go to Step E. If it is empty, terminate:

(1) If $\phi_0 = +\infty$, the problem has no solution.
(2) If $\phi_0 \neq +\infty$, it is the value of the optimal solution, which was found in Step T.

† The symbol $\leftarrow$ means that the left-hand side must be changed into the right-hand side.

3.3. Remarks on the Algorithm

(a)   It is very easy to prove that this algorithm works the way announced in Section 3.1. For this we have to remark that

(1)   each modification of ω in Steps E, G, C, and S leads effectively to another partial solution, the two subsets of the couple are disjoint;

(2)   whatever the ω with which we start Step E, C, G, or S, the subset $B_\omega^1 \cup \{b\}$ where b is any free block in ω, is an independent set of the incompatibility graph G.

The assertions of Step B then become evident in the light of the standard separation and evaluation procedures.

(b)   In order to quickly unearth the forced blocks corresponding to covering constraints, the rows of the matrix A may be renumbered so as to have

$$i < i' \qquad \text{iff} \quad \sum_{j \in N} a_{ij} \leqslant \sum_{j \in N} a_{i'j}.$$

The columns may also be arranged so as to facilitate the construction of sets like $\Gamma(b_{j_\omega})$ and $\Gamma(B_\omega^+)$. For this purpose, the n blocks may be partitioned into subsets $L_1, \ldots, L_p, \ldots$, each L_p corresponding to a complete subgraph in G. Thus, as soon as a block $b \in L_p$ must be included in B_ω^1, we know, without any search, that all the others blocks of L_p will be incompatible. The column order used by Pierce [8] and Garfinkel and Nemhauser [4] in their algorithm for the PP, resorts to this idea, although no incompatibility graph is explicitly introduced to formalize the partitioning constraint.

(c)   In practice it may be useful to compute in Step S the value of $v_0(X_\omega, \hat{X}_\omega)$ for the ω corresponding to the pair going on the stack, and to memorize this value, say $w(\omega)$, together with the pair. In Step B, it is then easy to compute w_0 which equals the smallest value of $w(\omega)$'s which remains in the stack. Now if Step T is modified in order to stop the algorithm as soon as

$$\Phi_0(X_\omega) - w_0 \leqslant \varepsilon, \qquad \varepsilon \quad \text{an } a \text{ priori given number,}$$

we may assert that, when the algorithm is interrupted by that additional rule, the best solution found to date differs from the optimal solution by at most ε.

(d)   Some better-hidden forced blocks may easily be introduced to accelerate the procedure, according to each particular problem studied. An important practical situation must now be mentioned.

Suppose it is possible to prove that the optimal configuration is a minimum cover. By *minimum cover*, we mean a cover C_0 such that

$$C \subset C_0, \qquad C \neq C_0 \Rightarrow C \quad \text{does not cover} \quad E.$$

This is obviously the case in the classical CP, and in many other more sophisticated real problems.

Let us now consider a partial solution ω such that $AX_\omega \geq 1$. It is clear that each free block in ω, if there is any, can be considered as a forced block, included in B_ω^-. Then, Step S may be modified as follows:

(1) drop out test $B - B_\omega^* \neq \emptyset$;
(2) when i_ω is not defined, go to Step T after having modified ω by

$$B_\omega^1 \leftarrow B_\omega^1, \quad B_\omega^0 \leftarrow B - B_\omega^1.$$

Moreover, it may be fruitful to include an additional test in Step E. Is there a $b \in B_\omega^1$ which is included in the union of the other blocks of B_ω^1? If so, go to Step B.

References

1. Balas, E., An additive algorithm for solving linear programs with zero–one variables, *Operations Res.* **13**, 517–546 (1965).
2. Bendahan, S., and Fayein, V., Problèmes périodiques d'affectation avec réemploi, Thèse 3 Cycle, Université Paris-Dauphine, May 1971.
3. Berge, C., Alternating chain methods: A survey, *in* "Graph Theory and Computing" (R. C. Read, ed.). Academic Press, New York, 1972.
4. Garfinkel, R. S., and Nemhauser, G. L., The set partitioning problem: Set covering with equality constraints, *Operations Res.* **17**, No. 5 (1969).
5. Garfinkel, R. S., and Nemhauser, G. L., Set Covering: A survey. Paper presented at XVII International Conference of the Institute of Management Sciences, London, July 1970.
6. Geoffrion, A. M., Integer programming by implicit enumeration and Balas' method, *SIAM Rev.* **9**, 178–190 (1967).
7. Lemke, C., Salkin, H. M. and Spielberg, K., Set covering by single branch enumeration with linear programming subproblems, *IBM New York Sci. Center Rep.* No. 320–2979 (October, 1969).
8. Pierce, J. F., Application of combinatorial programming to a class of all-zero-one integer programming problems, *Management Sci.* **15**, No. 3 (1968).
9. Rougerie, A., and Viviant, J. P., P.S.E.S. et Partition, Thèse 3 cycle, Université Paris-Dauphine, 1971.
10. Roy, B., Procédure d'exploration par séparation et évaluation (P.S.E.P. et P.S.E.S.), *Rev. Française Informat. Recherche Opérationnelle* No. V-1 (1969).
11. Roy, B., "Algèbre moderne et théorie des graphes orientées vers les sciences économiques et sociales." Vol. 1, Dunod, Paris, 1969. Vol. 2, Dunod, Paris, 1970.
12. Roy, B., Benayoun, R., and Tergny, J., From S.E.P. procedure to the mixed Ophelie program, *in* "Integer and Nonlinear Programming" (J. Abadie, ed.). North-Holland Publ., Amsterdam, 1970.
13. Salkin, H. M., An algorithm for the base constrained set covering problem. *Comm. TIMS, 11th American Meeting, Los Angeles, October 19–21* (1970).
14. Thiriez, H., The set covering problem: A group theoretic approach, *Rev. Française Informat. Recherche Opérationnelle* No. V-3 (1971).

TRIPARTITE PATH NUMBERS

R. G. Stanton   L. O. James   D. D. Cowan

Department of Computer Science
University of Manitoba
Winnipeg, Manitoba
Canada

Applied Analysis
   and Computer Science Department
University of Waterloo
Waterloo, Ontario
Canada

1. Introduction

The path number $pn(G)$ of a graph G was introduced by Harary at the Jamaica conference on graph theory and computing. In view of this origin, it goes without saying that G is assumed to be a finite undirected graph without multiple joins and without loops. For such a graph G, we consider paths in G and write

$$(1) \qquad G = P_1 \cup P_2 \cup \cdots \cup P_k,$$

where $P_i \cap P_j = \varnothing$ for $i \neq j$. Clearly, there are many ways of writing G as such a union of disjoint paths. We define $pn(G)$ to be the minimal value of k, the minimum being taken over all possible decompositions of the form (1).

The path number was studied by the authors [1], where various results were obtained for $pn(G)$ in the cases when (1) G is a tree, (2) G has at least one circuit, and (3) G is regular. Algorithms were given to produce a minimal path set for the complete graph K_n and for the complete bipartite graph $K_{m,n}$. The result for the latter graph, assuming $m \leqslant n$, is given by

$$pn(K_{m,2n}) = n, \qquad m < 2n, \qquad pn(K_{m,2n+1}) = n + 1, \qquad m \text{ even,}$$

$$= n + 1, \quad m = 2n; \qquad\qquad = n + (m+1)/2, \quad m \text{ odd.}$$

In this paper, we consider the complete tripartite graph $K_{a,b,c}$, and denote its path number by $k(a,b,c)$. This function $k(a,b,c)$ is studied, and some generalizations for the complete n-partite graph are suggested.

2. Elementary Results

We start by assuming that $abc \neq 0$. With this restriction two obvious lemmas may be obtained.

LEMMA 1.

$$k(a,b,c) \geqslant \frac{ab + bc + ca}{a + b + c - 1}$$

Proof: $K_{a,b,c}$ possesses $ab + bc + ca$ edges. However, no path can contain more than $a + b + c - 1$ edges. The lemma follows.

LEMMA 2. For any a, b, c, we have

$$k(a,b,c) \geqslant (a+b)/2.$$

Proof: Let the a vertices form a set A, the b vertices form a set B, and the c vertices a set C. The vertices of C have $(a + b)c$ edges joined to them, and no path may contain more than $2c$ of these edges. The result follows.

We now combine these results.

THEOREM 1. If $a \geqslant b \geqslant c$, then

$$k(a,a,a) \geqslant a + 1;$$

otherwise

$$k(a,b,c) \geqslant (a+b)/2.$$

Proof: We now compute the difference between the lower bounds of Lemma 2 and Lemma 1 and find that

$$\frac{a+b}{2} - \frac{ab+ac+bc}{a+b+c-1} = \frac{a(a-1-c)+b(b-1-c)}{2(a+b+c-1)}.$$

This expression is greater than or equal to zero, unless $b=c$ and $a=b$ or $b+1$. If $a=b=c$, we use Lemma 1 to obtain

$$k(a,a,a) \geq a+1.$$

If $a=b+1$, $b=c$, the results of Lemmas 1 and 2 are identical. In all other cases, Lemma 2 is stronger. Thus, we have the theorem.

THEOREM 2. For $a \geq b \geq 1$, we have $k(1,1,1)=2$ with $k(a,c,1) = \{(a+b)/2\}$, where we use $\{x\}$ to denote x if x is an integer. The least integer above x if x is nonintegral.

Proof: Theorem 1 gives these results as lower bounds. Thus, one merely needs to exhibit an algorithm attaining these bounds. The algorithm for $k(1,1,1)$ is trivial. For $k(a,b,1)$, we display the algorithm diagrammatically.

The diagram shows three axes, and each cell indicates an edge joining the vertices which coordinatize the cell. Thus, the shaded cell in Fig. 1 indicates

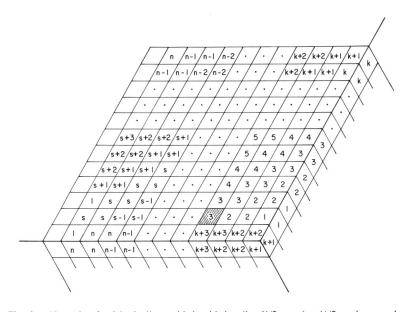

Fig. 1. Algorithm for $k(a, b, 1)$: a odd, b odd, $k=(b-1)/2$, $n=(a+b)/2$, and $s=n-k$.

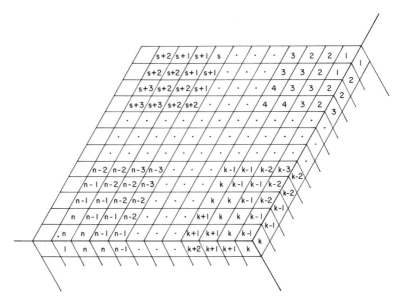

Fig. 2. Algorithm for $k(a, b, 1)$: a even, b even, $n = (a+b)/2$, $k = (b/2)+1$, and $s = n-k$.

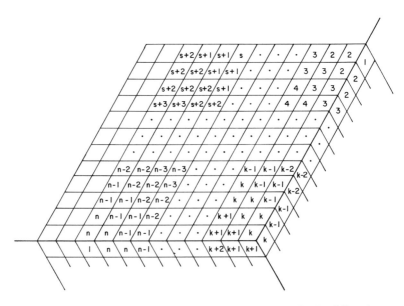

Fig. 3. Algorithm for $k(a, b, 1)$: a odd, b even, $k = (b/2)+1$, $n = (a+b+1)/2$, and $s = n-k$.

the edge joining point 4 in A to point 2 in B. Path number j is indicated by writing j in all the cells which form part of the path. With this convention, Figs. 1, 2, and 3, complete the proof of Theorem 2.

3. Extensions of Previous Algorithms

We establish three further lemmas.

LEMMA 3.   If $a > b + c$, $b > c$, and any two of a, b, c, are even, then

$$k(a, b, c) = \{(a+b)/2\}.$$

Proof: We consider the tripartite graph as being made up of an $(a, b+c)$ bipartite graph and a (b, c) bipartite graph. For a and b even, these graphs use $a/2$ and $b/2$ paths. With Theorem 1, this proves the result. For a and c even, these graphs employ $a/2 + (b+1)/2$ paths. Again, we have the result. Finally, for b and c even, the 2 graphs employ $(a+1)/2 + b/2$ paths. This completes the lemma.

LEMMA 4.   If $a \geqslant b \geqslant c$, $k(a, b, c) = \{(a+b)/2\}$, and $2t > b + c$, then

$$k(a+2t, b, c) = \{(a+b)/2\} + c.$$

Proof: Decompose the graph into an (a, b, c) tripartite graph and a $(2t, b+c)$ bipartite graph. The result is then immediate.

LEMMA 5.   If $a \geqslant b \geqslant c$ and $k(a, b, c) = \{(a+b)/2\}$ and $2t > a + c$, then

$$k(a, b+2t, c) = \{(a+b)/2\} + t.$$

Proof: Use a decomposition similar to that in Lemma 4.

We can now prove Theorem 3.

THEOREM 3.   If $a > c$, then

$$k(a, a, c) = a.$$

Proof: From Theorem 1, $k(a, a, c) \geqslant a$. An algorithm for constructing a covering of the graph with a paths is indicated in Figs. 4 and 5.

4. The Exceptional Case

We now consider the exceptional case $a = b = c$, and determine $k(a, a, a)$. To do this, we divide our edges into $3a$ classes named $d(i, j)$, where $1 \leqslant i \leqslant 3$ and $0 \leqslant j \leqslant a - 1$. The edge set $d(i, j)$ is found as follows: name the vertices

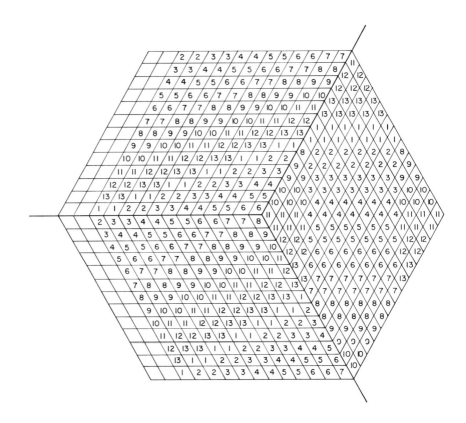

Fig. 4. Algorithm for $k(a,13,13)$, where $a \leqslant 12$.

of A as $(1,1),(1,2),\dots,(1,a)$, those of B as $(2,1),(2,2),\dots,(2,a)$, and those of C as $(3,1),(3,2),\dots,(3,a)$. Then $d(i,j)$ joins vertex $(i+1,\alpha)$ to vertex $(i+2,\alpha+j)$, as α ranges from 0 to $a-1$. Arithmetic in the first element of these pairs is modulo 3, in the second element is modulo a. This gives $3a$ sets of a elements, that is, all $3a^2$ elements of the edge set of G.

Lemma 6. If a is odd, then the complete (a,a,a) tripartite graph may be partitioned into a Hamiltonian circuits.

Proof: We merely display the circuits. Circuit i is given as the set $d(1,i)$, $d(2,i)$, $d(3,1-2i)$.

Lemma 7. If $a \equiv 0$ modulo 4, the complete (a,a,a) tripartite graph can be partitioned into a Hamiltonian circuits.

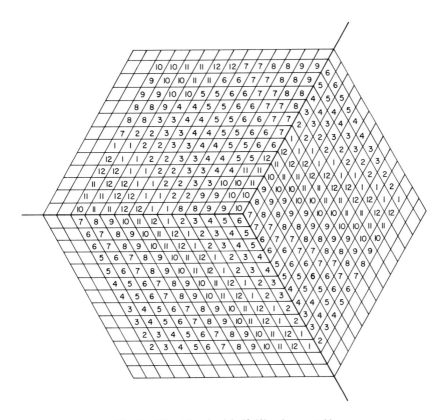

Fig. 5. Algorithm for $k(a,12,12)$, where $a \leqslant 11$.

Proof: There are three kinds of circuits. These are

$$d(1,i), \quad d(2,i), \qquad d(3,1-2i), \qquad 0 < i \leqslant a/2;$$

$$d(1,i), \quad d(2,i+1), \qquad d(3,-2i), \qquad a/2 < i < a;$$

$$d(1,0), \quad d(2,a/2+1), \quad d(3,0).$$

We may now deduce Theorem 4.

THEOREM 4.

$$k(a,a,a) = a + 1.$$

Proof: If $a \not\equiv 2$ modulo 4, we use Lemmas 6 and 7. If $a \equiv 2$ modulo 4, $a > 2$, the solution is a variant of that used in Lemma 7, and has a pattern indicated in Fig. 6 and Fig. 7. For $a = 2$, the result is trivial.

One can derive other similar results.

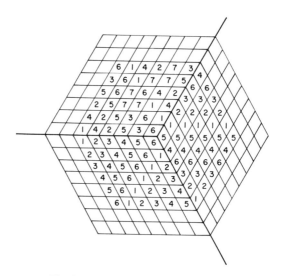

Fig. 6. Algorithm for $k(a, a, a) = a+1$.

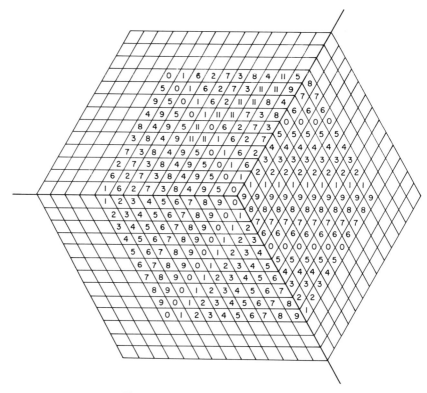

Fig. 7. Algorithm for $k(10, 10, 10) = 11$.

LEMMA 8.

$$k(a, 2, 2) = 1 + \{a/2\},$$

and we conjecture that Theorem 1 gives an exact bound.

5. The Complete n-Partite Graph

Let $v = (v_1, v_2, \ldots, v_n)$ be a vector of integers in nonincreasing order. We define $k(v)$ to be the path number of the complete n-partite graph on $(v_1, v_2, \ldots, v_n)$ vertices, and find that many of the previous results generalize. We indicate generalizations by an asterisk.

LEMMA 1*.

$$k(v) \geqslant \frac{\sum v_i v_j}{\sum v_i - 1}, \qquad i \neq j.$$

LEMMA 2*.   For n even and all v_i odd, we have

$$k(v) \geqslant \sum v_i/2.$$

LEMMA 3*.

$$k(v) \geqslant \sum v_i/2, \qquad i = 1, 2, \ldots, n - 1.$$

Theorem 1 generalizes to Theorem 1*.

THEOREM 1*.   If $\sum_{i=1}^{n-1}(v_i - v_n - 1)v_i < 0$, then

$$k(v) \geqslant \sum_{i \neq j} v_i v_j / (\sum v_i - 1);$$

otherwise

$$k(v) \geqslant \sum_{i=1}^{n-1} v_i/2.$$

For n even and all v_i odd, we have

$$k(v) \geqslant \sum v_i/2.$$

Proof: We use Lemmas 1*, 2*, and 3*, and denote $\sum v_i$ by S and $\sum v_i^2$ by T. Then we need to prove

(1) $S/2 \geqslant (S^2 - T)/2(S - 1)$;
(2)   if $\sum_{i=1}^{n-1}(v_i - v_n - 1)v_i < 0$, then $(S^2 - T)/2(S - 1) > (S - v_n)/2$;
(3)   if $\sum_{i=1}^{n-1}(v_i - v_n - 1)v_i \geqslant 0$, then $(S^2 - T)/2(S - 1) \leqslant (S - v_n)/2$.

The first result is equivalent to $T \geqslant S$, which is obvious. The second and third follow from writing

$$\sum_{i=1}^{n-1} (v_i - v_n - 1) v_i = (T - v_n^2) - v_n(S - v_n) - (S - v_n)$$

$$= T + v_n - S(v_n + 1)$$

$$= (S-1)(S-v_n) - (S^2 - T).$$

COROLLARY 1*.   If $v_n < v_{n-1}$, then Case 2 does not occur.

We can also state two further generalizations, which lend support to the conjecture that the bounds of Theorem 1* are exact.

LEMMA 4*.   Let $k(v) = (S - v_n)/2$. If w is a vector and $\sum w_i = S_1$, and if $k(w, S) = S_1/2$, then

$$k(w, v) = (S + S_1 - v_n)/2.$$

LEMMA 5*.   If $k(v) = (S - v_n)/2$ and $r > k(v)$, and $w = (0, 0, \ldots, 2r, \ldots, 0)$, where $2r$ may be in any position except the last, then

$$k(v + w) = k(v) + r.$$

Reference

1. Stanton, R. G., Cowan, D. D., and James, L. O., Some results on path numbers, *Proc. Louisiana Conference on Combinatorics, Graph Theory, and Computing, Baton Rouge*, 112–135 (1970).

NON-HAMILTONIAN PLANAR MAPS

W. T. Tutte

Faculty of Combinatorics and Optimization
University of Waterloo
Waterloo, Ontario
Canada

A *planar map* is a dissection of the sphere or closed plane into a finite number of simply connected polygonal regions called *faces* or *countries* by means of a graph drawn in the surface. It is assumed that this graph has no loop or isthmus. In this paper we shall use the term *map* as an abbreviation for *planar map*.

A *Hamiltonian circuit* in a map is a circuit in its graph passing through every vertex. A map is called *Hamiltonian* or *non-Hamiltonian* according as it does or does not have such a circuit. A map is said to be *cyclically n connected* if at least *n* edges must be removed in order to decompose the graph into two disjoint parts, each containing a circuit.

Special interest is attached to the *trivalent* or *cubic* maps, in which exactly three edges meet at each vertex. These are studied in connection with the four color conjecture, which asserts that the faces of a map can be colored in four colors so that no two of the same color have a common edge. Let us use the term *5 chromatic* for the hypothetical maps that do not satisfy this conjecture, and let a *minimal map* be defined as a trivalent 5-chromatic map with the least possible number of faces.

In the theory of the four-color problem it is shown that the conjecture is true for all maps if it is true for all trivalent ones. Various properties of minimal maps are determined. In particular it is shown that a minimal map must be cyclically 5 connected.

There is a connection between the four-color problem and the theory of Hamiltonian circuits. Let C be a Hamiltonian circuit in a map M, not necessarily trivalent. It separates the sphere into two regions that we may call the inside and the outside of C. The faces inside C can be colored alternately red and blue while those outside can be colored alternately green and yellow. Thus, all Hamiltonian maps can be 4 colored.

This fact suggests the possibility of verifying the four-color conjecture by showing that the established limitations on the structure of a minimal map permit the construction of a Hamiltonian circuit. This possibility was explored by Tait toward the end of the nineteenth century [2]. Tait observed that non-Hamiltonian trivalent maps exist, for example, see the map of Fig. 1, but conjectured that all cyclically 3-connected trivalent maps are Hamiltonian. Tait's conjecture was eventually shown to be false. The cyclically 3-connected non-Hamiltonian map of Fig. 2 was published in 1946 [3], and a cyclically 4-connected non-Hamiltonian trivalent map was exhibited in 1960 [4]. In 1965 H. Walther [5] published a cyclically 5-connected non-Hamiltonian trivalent map. Walther's map has 83 faces, of which 67 are pentagons, 6 are heptagons, 5 are octagons, and 4 are dodecagons. The remaining face has 21 sides.

The existence of Walther's map is of course discouraging to students of Hamiltonian tetrachromatology. The known properties of minimal maps no longer seem inconsistent with their non-Hamiltonian character. With Walther's discovery our interest turns anew to the theory of non-Hamiltonian trivalent maps. How can we best determine whether or not a given map is

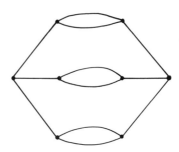

Fig. 1

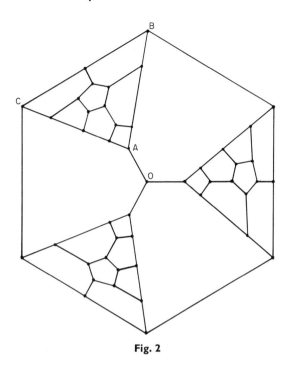

Fig. 2

Hamiltonian? How would we set about constructing a non-Hamiltonian trivalent map satisfying given conditions? It must be admitted that even up to this stage the progress of the theory seemed somewhat disappointing. By dint of much hard work a few highly complicated examples of special interest had been constructed, and that was all.

Soon after Walther's work a revolutionary discovery was made by two Russian mathematicians, V. Kozyrev and E. Grinberg. It was reported by Sachs in 1968 [1]. It shares one property with some other major advances: once explained it seems trivial. Every combinatorialist interested in this aspect of his discipline must cry, "Why didn't I think of this myself? How was it possible for Tait, or even Hamilton, to miss it?" The work of Kozyrev and Grinberg gives some useful sufficient conditions for a map to be non-Hamiltonian, and makes it reasonably easy to construct non-Hamiltonian trivalent maps, even cyclically 5-connected ones.

Suppose that we are given a map M, not necessarily trivalent, and that we assume it to have a Hamiltonian circuit C. Then the edges of M fall naturally into three sets, the edges of C, the diagonal edges crossing the inside of C, and the diagonal edges crossing the outside of C. Let there be c edges in the first set, d' in the second and d'' in the third.

The d' diagonal edges crossing the inside of C decompose that Jordan

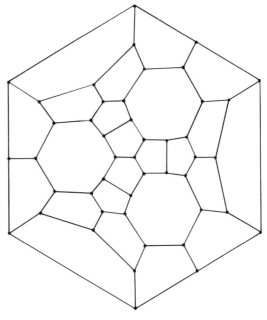

Fig. 3

domain into $d'+1$ faces of the map. Similarly the outside of C is decomposed into $d''+1$ faces by its d'' diagonals. Let us denote the number of i-sided faces of M by f_i. We assume that f_i' of these are inside C and f_i'' are outside.

Let us sum the numbers of sides of the faces inside C. The result can be obtained also by counting 1 for each edge of C and 2 for each diagonal edge crossing the inside of C. Thus,

$$\sum_{i=2}^{\infty} if_i' = c + 2d'$$

$$= c - 2 + 2\sum_{i=2}^{\infty} f_i',$$

$$\sum_{i=2}^{\infty} (i-2)f_i' = c - 2.$$

The same reasoning applied to the outside of C gives us the equation

$$\sum_{i=2}^{\infty} (i-2)f_i'' = c - 2.$$

Hence, by subtraction,

(1) $$\sum_{i=2}^{\infty} (i-2)(f_i' - f_i'') = 0.$$

Kozyrev and Grinberg pointed out that for some finite sequences $(f_2, f_3, \ldots, f_m)$ of nonnegative integers (1) must always be false however we partition each f_i into two non-negative integers f_i' and f_i''. Suppose for example that $f_i = 0$ whenever $i-2$ does not divide by 3, with the single exception that $f_j = 1$ for one particular j not congruent to 2 modulo 3. Then it is impossible to make the expression on the left of (1) divide by 3, and hence (1) is necessarily false. We conclude from this that any map corresponding to such a sequence must be non-Hamiltonian. Such maps exist; Kozyrev and Grinberg gave the trivalent and cyclically 5-connected example, Fig. 3. This has 25 faces, with $f_5 = 21$, $f_8 = 3$, and $f_9 = 1$. A slightly simpler example of a cyclically 5-connected non-Hamiltonian trivalent map was obtained at the Calgary Conference in June 1969. I was asked by H. V. Kronk if I knew of any Hamiltonian trivalent map in which there was one edge not belonging to any Hamiltonian circuit. I was able to reply in the affirmative, mentioning the map obtained from that of Fig. 2 by contracting the triangle ABC and the faces inside it into a single vertex. The resulting map is Hamiltonian but in it the edge OA belongs to no Hamiltonian circuit. I tried however to use the theory of Kozyrev and Grinberg to construct a more impressive example. Because of erroneous reasoning I stumbled instead upon the cyclically 5-connected non-Hamiltonian map of Fig. 4.

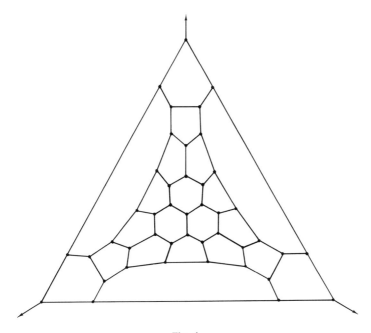

Fig. 4

The map of Fig. 4 has 24 faces, with $f_5 = 18$, $f_6 = 3$, and $f_8 = 3$. Now this sequence of numbers f_i does not make (1) impossible. We observe however that whenever that equation is satisfied one of the numbers f_6' and f_6'' is 3 and the other is 0. We deduce that for any Hamiltonian circuit of this map the three hexagons must be on the same side of the circuit. But this is impossible since any Hamiltonian circuit must pass through the common vertex of the three hexagons.

One wonders how far the Kozyrev–Grinberg theory could be extended. Can all the known non-Hamiltonian maps be simply explained by this theory or by refinements of it? It is amusing to note that the non-Hamiltonian character of the map of Fig. 1 can be demonstrated by the same three-hexagon argument that we have used for the map of Fig. 4. It does not seem that the map of Fig. 2 can be dealt with directly by the theory. However this map is obtained [3] by a simple construction based on the map shown below in Fig. 5. The essential fact is that no Hamiltonian circuit of the latter map passes through both of the edges A and B. This can be regarded as a consequence of the Kozyrev–Grinberg theory, as (1) can be satisfied only by a Hamiltonian circuit separating one of the quadrilaterals from the other four.

Perhaps a converse form of the Kozyrev–Grinberg theory could be found. One wonders for example about those trivalent maps in which the number of sides of each face is congruent to 2 modulo 3. For these (1) is trivially true,

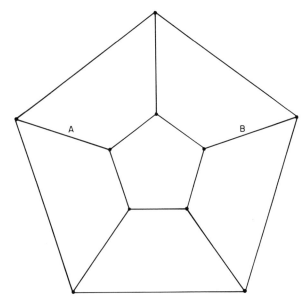

Fig. 5

whatever the numbers f_i' and f_i'' may be. Is Tait's conjecture valid for maps of this kind?

In conclusion let us note that the Kozyrev–Grinberg theory can be expressed in dual form as a theory of Hamiltonian bonds. A Hamiltonian bond in a graph G is a set H of edges such that the rest of the graph consists of two disjoint trees, and each edge of H has one end in each tree. Let us denote the number of vertices of G of valency i by f_i, and suppose f_i' of these to be in the first tree and f_i'' in the second. Then we can establish (1), much as before. This form of the theory applies to all graphs, planar or nonplanar.

References

1. Sachs, H., "Beiträge zur Graphentheorie," pp. 127–130. Barth, Leipzig, 1968.
2. Tait, P. G., *Phil. Mag.* (5) **17**, 30–46 (1884); *Scientific Papers*, Vol. II, 85–98.
3. Tutte, W. T., On Hamiltonian circuits, *J. London Math. Soc.* **21**, 98–101 (1946).
4. Tutte, W. T., A non-Hamilton planar graph, *Acta Math. Acad. Sci. Hungar.* **11**, 371–375 (1960).
5. Walther, H., Ein kubischer, planarer, zyklisch fünffach zusammenhangender Graph, der keinen Hamiltonkreis besitzt, *Wiss. Z. Techn. Hochsch. Ilmenau* **11**, 163–166 (1965).

A TOP-DOWN ALGORITHM
FOR CONSTRUCTING
NEARLY OPTIMAL LEXICOGRAPHIC TREES†

W. A. Walker‡

C. C. Gotlieb

Department of Computer Science
University of Toronto
Toronto, Ontario
Canada

1. Introduction

The binary search tree has been proposed as a data structure for lists of names which must be both searched and updated frequently. A *binary search tree* is a rooted, ordered tree such that the out-degree of every node is two for

† This research was supported in part by the National Research Council of Canada.
‡ Present address: Ontario Hydro, Toronto, Ontario, Canada.

303

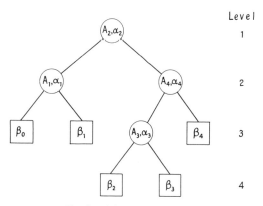

Fig. I. A lexicographic tree.

an internal node or zero for a leaf. If there are N internal nodes, there are $2N$ edges and $N+1$ leaves. Each internal node is associated with one name in a set of lexicographically ordered names, $A_1 < A_2 < \cdots < A_N$. If the internal nodes are labeled with the associated names, the post-order listing[†] of the internal nodes is $A_1, A_2, ..., A_N$. Windley [8], Booth and Colin [1], Hibbard [4], and Clampett [3] have given the average search time required to locate names in binary trees if all the names are equally likely to be used in a search, and are entered in the tree in a random order. When all the names are equally likely, a best possible tree, that is, the one with minimum average search time, is one in which the average path length from the root to a leaf is minimized.

A *lexicographic tree* is a binary search tree, such that:

(1)   there is a set of N frequencies $\alpha_1, ..., \alpha_N$, α_i being associated with A_i;

(2)   there is a set of $N+1$ frequencies $\beta_0, ..., \beta_N$, associated with the leaves, β_i being the frequency of encountering names which lie between A_i and A_{i+1}, β_0 being the frequency of names preceding A_1, and β_N being the frequency of names following A_N;

(3)   the post-order listing of the lexicographic tree, with the nodes labeled with the associated frequencies, is $\beta_0, \alpha_1, \beta_1, ..., \alpha_N, \beta_N$.

A special case of the lexicographic tree occurs when all the β_i are zero. In this case, only the names $A_1, A_2, ..., A_N$ would be used to search the tree and all searches would be successful, that is, terminate at an internal node.

Figure 1 illustrates a lexicographic tree with four names $A_1, ..., A_4$ and the nine associated frequencies. The weighted path length P of a lexicographic tree is given by

$$P = \sum_{i=1}^{N} L(\alpha_i)\,\alpha_i + \sum_{i=0}^{N} L(\beta_i)\,\beta_i,$$

† The post-order listing for a binary tree is left subtree, root, right subtree; see Knuth [5].

where $L(\alpha_i)$ is the level of the node with frequency α_i. For the tree shown in Fig. 1,

$$P = 3\beta_0 + 2\alpha_1 + 3\beta_1 + \alpha_2 + 4\beta_2 + 3\alpha_3 + 4\beta_3 + 2\alpha_4 + 3\beta_4.$$

The weight W of the tree is the sum of all the frequencies. The *average search length* is defined as the weighted path length divided by W and can be interpreted as being the average time to search the tree for a name. It is therefore of interest to construct a lexicographic tree with minimum weighted path length.

Knuth has given an algorithm for constructing an optimal lexicographic tree [6]. It is not necessarily unique. The algorithm requires a time proportional to N^2 and storage proportional to N^2. It constructs the tree from the bottom to top, that is, from the leaves to the root, and is practical when the number of names is small. However, for large values of N (in practice $N > 200$) it will require the use of secondary storage on all but the largest computers, and the time will be prohibitive. If N is very large, say 100,000, it is clear that the algorithm is impractical both with respect to time and space.

In this contribution, we will present a top-down algorithm which constructs nearly optimal lexicographic trees. The algorithm chooses a root for the tree and repeatedly chooses roots for the subtrees until a subtree is encountered which contains N_0 or fewer names, where N_0 is a parameter of the algorithm. Knuth's "Algorithm K" is then used to construct an optimal subtree. In a recent paper, Nievergelt and Wong [7] derive certain expressions for the expected search times for optimal trees, for balanced trees, and for random trees. From these they are led to suggest a heuristic method for constructing a nearly optimal tree. Our method corresponds in part to this suggestion.[†]

2. An Application

An application of a lexicographic search tree, where a large number of nodes is needed, is the author index of a library catalog, where author surnames are the nodes. In this case, the frequency associated with each name could be, for example, the number of entries for each author in the catalog, or, more usefully, the total number of references to an author's books in a given time interval. The author index could be maintained on a computer and accessed by librarians for updating and expanding the catalog, and by users for querying the catalog in a read-only manner. In addition to the frequency associated with each name, it would be possible to determine the frequencies for names which lie between the names in the lexicographic tree. Although the index would be continually updated, the frequency associated with most of the

---

† J. Bruno and E. G. Coffman, Jr., have also described a heuristic method for constructing a nearly optimal tree [2].

TABLE I

Average Search Length

	Optimal search length	$N_0 = 0$ $F = 1$	$N_0 = 0$ Root = centroid	$N_0 = 15$ $F = 4$	$N_0 = 15$ $F = 5$	$N_0 = 15$ $F = 6$	$N_0 = 15$ $F = 10^6$
Set 1: 1	4.2944	4.7271	4.9253	4.3635	4.3708	4.4103	4.4682
2	6.6060	7.6283	7.0881	6.6809	6.6812	6.6812	6.8741
3	5.8749	6.4392	6.1205	5.9439	5.9454	5.9454	5.9807
4	6.0650	6.7204	6.4168	6.1591	6.1254	6.1473	6.1640
5	6.6424	7.5619	6.8175	6.7192	6.7201	6.8078	6.7693
6	5.9633	7.0943	6.5883	6.0404	6.0777	6.0777	6.2380
7	5.8250	6.7102	6.3265	5.8256	5.8280	5.8560	5.9994
8	6.2576	7.8798	6.8535	6.3669	6.3793	6.3780	6.5602
9	7.0856	9.8991	7.4352	7.2746	7.3114	7.3512	7.2489
10	7.3376	8.2455	7.5238	7.3675	7.3487	7.3954	7.4175
Set 2: 1	4.6317	5.0299	5.6680	4.7003	4.6721	4.7795	4.7738
2	7.2555	8.1425	7.7977	7.2602	7.2685	7.2685	7.4535
3	6.2534	6.7129	6.8398	6.2854	6.3060	6.3175	6.3910
4	6.3976	6.8927	6.8801	6.4308	6.4539	6.4661	6.5360
5	7.0182	7.7191	7.2384	7.0570	7.0835	7.0870	7.0981
6	6.5382	7.6647	7.0262	6.5632	6.5793	6.6808	6.7689
7	6.4549	7.2480	6.9911	6.4638	6.4630	6.4831	6.6469
8	6.5810	8.4312	6.8701	6.8593	6.7082	6.7288	7.0154
9	7.1216	9.6699	7.3190	7.4623	7.4606	7.4161	7.3423
10	7.6958	8.3864	7.8933	7.7384	7.7054	7.7054	7.7128
Set 3: 1	4.0035	4.7786	4.5373	4.0547	4.0547	4.1393	4.0655
2	6.4230	8.1525	6.9852	6.4747	6.4759	6.5035	6.4868
3	5.5561	6.6162	5.8533	5.5814	5.5814	5.6105	5.7129
4	5.7628	6.8180	6.2072	5.8182	5.8089	5.8098	5.8764
5	6.2211	7.2739	6.3650	6.2596	6.2658	6.2681	6.3288
6	5.6524	7.1084	6.2465	5.7149	5.7569	5.7656	5.9927
7	5.5935	7.1817	6.0225	5.7259	5.7231	5.7313	5.7873
8	6.0850	9.2180	6.6197	6.3225	6.2860	6.2445	6.3284
9	7.0125	11.1363	7.2585	7.2805	7.2344	7.2344	7.2099
10	6.9738	8.3983	7.0934	7.0482	7.0626	7.0385	7.0921
Set 4: 1	5.0362	5.9324	5.8784	5.0572	5.0572	5.1860	5.1079
2	7.4164	8.8770	7.8493	7.4609	7.4606	7.4908	7.5341
3	6.5463	7.4664	6.9774	6.5611	6.5697	6.5697	6.6169
4	6.6505	7.5831	7.1837	6.6667	6.6758	6.6758	6.7062
5	7.2702	8.4620	7.4610	7.3371	7.3085	7.3173	7.3212
6	6.5494	7.6391	7.0804	6.5627	6.5627	6.5892	6.6194
7	6.7508	8.4449	7.1797	6.7954	6.8179	6.7669	6.9337
8	6.7320	8.7564	6.9632	6.9473	6.9595	6.9377	7.1285
9	7.1119	10.4205	7.3073	7.3833	7.3798	7.3798	7.3373
10	7.8730	8.9113	8.0264	7.8765	7.8863	7.8802	7.8880

TABLE I (continued)

		Optimal search length	$N_0 = 0$ $F = 1$	$N_0 = 0$ Root = centroid	$N_0 = 15$ $F = 4$	$N_0 = 15$ $F = 5$	$N_0 = 15$ $F = 6$	$N_0 = 15$ $F = 10^6$
Set 5:	1	4.1379	4.8235	5.3214	4.1815	4.2028	4.2060	4.2735
	2	5.9726	7.2296	6.7504	6.0523	6.1668	6.1718	6.1478
	3	5.7540	6.7917	6.2758	5.7765	5.8456	5.8301	5.8458
	4	5.9979	7.0442	6.5053	6.0469	6.0605	6.0479	6.0706
	5	6.6600	8.3690	6.9629	6.8071	6.7324	6.7515	6.8056
	6	5.3331	6.5523	5.9711	5.3672	5.3995	5.5638	5.5449
	7	5.4117	6.6474	6.3725	5.4284	5.4284	5.4522	5.7755
	8	5.8119	7.2683	6.3579	5.8935	5.9134	5.9896	5.9578
	9	6.9904	10.3917	7.2422	7.2023	7.1713	7.1620	7.1881
	10	7.2693	8.7128	7.4147	7.3624	7.2930	7.2949	7.3149

names would not change greatly over a short period. For the case where frequencies are determined by usage statistics, each access to the catalog would be recorded. The search tree should be reconstructed periodically to reflect the changing search patterns. The problem is to construct the binary tree which minimizes the time required to locate an author's name, or to ascertain that the name is not in the tree.

3. Basis of a Top-Down Algorithm

Knuth suggests two possible rules for structuring nearly optimal lexico-graphic trees, and points out that neither rule will produce an optimal tree in all cases. The first rule is to choose the A_i with the largest α_i as the root of the tree, then proceed similarly for the subtrees. A set of names whose β_i frequencies are 0 and whose α_i satisfy

$$\alpha_i > \sum_{j=1}^{i-1} \alpha_j, \qquad i = 2,\ldots,N$$

will be structured into an optimal tree if we use this rule. Tests on this rule were included with the tests on the algorithm described in Section 7. See Table I, Column $N_0 = 0$, $F = 1$, where the average search lengths obtained by this method are compared with the optimal lengths. The poor results using this rule are partially explained by noting that the β_i cannot influence the final structure of the tree. In general, the larger the sum of the β_i, compared to the sum of the α_i, the poorer is this rule. However, even the tests with all the

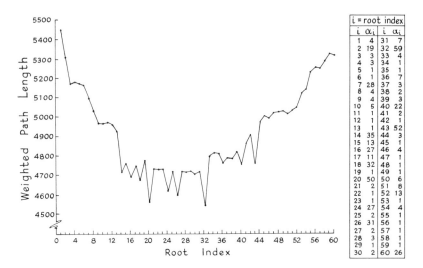

Fig. 2. Dependence of path length on root position.

β_i equal to zero, Sets 1–5, Case 1, do not produce acceptable nearly optimal trees.

The second rule suggested by Knuth is to choose as root a name whose left and right subtrees are most nearly equal in weight. There will be one or two such names. We will call the single name, or the lexicographically larger of the two names, the *centroid*. Choosing the centroid as root will result in an optimal tree when all the α_i and β_i are equal. This rule takes into account both the α_i and β_i. It nearly always is a significant improvement (see Table I, Column $N_0 = 0$, root = centroid) over choosing a name with largest α_i as the root. Since this second rule does not consider the individual α_i, a node with a very small α_i may be chosen as a root when there is an adjacent node with α_{i+1} very large which may, in fact, be the actual root of the optimal tree. Thus, neither of the rules suggested by Knuth is satisfactory for constructing a nearly optimal tree.

If the α_i and β_i are not all equal, we can regard the centroid as a first choice, and establish a rule for determining how far to move from this choice. To determine how far to move, let us examine how the minimum path length of a tree varies when different nodes are chosen as the root. From a set of names $A_1, \ldots, A_N$, construct trees $T_1, \ldots, T_N$ by choosing A_i as the root of T_i, and constructing optimal binary subtrees for A_i. As an example, Fig. 2 shows the weighted path length of the T_i for the first 60 names of data Set 1, Case 6. See Section 7 for a description of the test data. The minimum of this graph at 32 corresponds to the optimal tree.

Examining the weighted path length of many sets of T_i, such as that shown in Fig. 2, indicates that, in the majority of cases, the minimum weighted path length occurs when the frequency of the root of T_i is a local maximum, that is, $\alpha_{i-1} < \alpha_i > \alpha_{i+1}$. We would expect this relation, since if we choose α_{i-1} as the root in place of α_i, the increase in the weighted path length of the right subtree, due to adding α_i, will usually be larger than the decrease in the weighted path length of the left subtree, due to removing α_{i-1}. In addition, if the weighted path length of T_i, denoted by P_i, is a local minimum, that is, $P_{i-1} > P_i < P_{i+1}$, it is usually true that the associated α_i is a local maximum. In other words, a local minimum of P_i usually corresponds to a local maximum for the α_i. This is true for the central portion of the graph, that is, values of P_i for $10 < i < 50$, in Fig. 2. The relation should not be expected to hold when one of the subtrees contains very few nodes, since, in this case, the structure and weighted path length of a subtree can be significantly changed by the removal or addition of even a single name and its corresponding α_i frequency.

Based on the preceding discussion, our algorithm chooses as the root a name with the largest associated α_i in a neighborhood of the centroid. If this maximum is not unique, the name whose α_i is closest to the centroid is taken, enlarging if necessary the neighborhood which is being considered. The size of the neighborhood is determined by a parameter of the algorithm. The choice so determined may not be the root of the optimal tree, but it will usually be a name corresponding to a local minimum of the curve in Fig. 2. In practice, it is found that the value of the local minimum is not significantly larger than the minimum weighted path length. This rule for determining how far to move from the centroid does not consider the magnitude of the individual β_i, thus, for a small number of names, the tree can have a structure quite different from, and an average search length larger than, the optimal tree. Since Algorithm K is easily applied to small trees, it is used to determine the optimal subtree for subtrees containing fewer than a specified number of names N_0.

4. Algorithm for Nearly Optimal Lexicographic Trees

Given the ordered set $\{A\}$ of names, such that $A_1 < A_2 < \cdots < A_N$, $\alpha_1, \ldots, \alpha_N$, and $\beta_0, \ldots, \beta_N$, the steps of the algorithm to structure a nearly optimal lexicographic tree are as follows:

(1) If $N \leqslant N_0$, structure an optimal binary tree using Algorithm K.
(2) If $N > N_0$, let W_{k_1, k_2} be the weight of the subtree with frequencies

$\beta_{k_1}, \alpha_{k_1+1}, ..., \alpha_{k_2}, \beta_{k_2}$, F a parameter,[†] and A_c the centroid. Form the ordered set of names $\{A_F\} = \{A_L\} \cup A_c$, where the members of the set $\{A_L\}$ satisfy

$$|W_{0,L-1} - W_{L,N}| < W_{0,N}/F, \quad 1 \leqslant F \leqslant W_{0,N}.\text{[‡]}$$

(3) Find an index, max, such that $\alpha_{max} = \text{maximum}_i \alpha_i$, where $A_i \in \{A_F\}$.

(4) If in the set $\{A_F\}$ there is at least one name preceding or equal to A_c with associated frequency α_{max}, let p be the index such that A_p with $\alpha_p = \alpha_{max}$, is lexicographically closest to A_c. If there is no such p, let $\{A_Q\}$ be the null set and go to Step 6.

(5) If A_p is the first member of $\{A_F\}$ and $\alpha_{p-1} > \alpha_p$, form the set $\{A_Q\} = \{A_{p-1}, A_{p-2}, ..., A_u\}$, where $\alpha_{p-j-1} > \alpha_{p-j}$, $j = 0, ..., p-u-1$ and $\alpha_{u-1} \leqslant \alpha_u$ or $u - p = \lfloor \log_2 N \rfloor$; if A_p is not the first member of $\{A_F\}$, let $\{A_Q\}$ be the null set.

(6) If in the set $\{A_F\}$ there is at least one name following or equal to A_c with associated frequency α_{max} let r be the index such that A_r with $\alpha_r = \alpha_{max}$ is lexicographically closest to A_c; if there is no such r, let $\{A_S\}$ be the null set and go to Step 8.

(7) If A_r is the last member of $\{A_F\}$ and $\alpha_r < \alpha_{r+1}$, form the set $\{A_S\} = \{A_{r+1}, A_{r+2}, ..., A_v\}$, where $\alpha_{r+j} < \alpha_{r+j+1}$, $j = 0, 1, ..., v-r-1$, and $\alpha_v \geqslant \alpha_{v+1}$ or $v - r = \lfloor \log_2 N \rfloor$. If A_r is not the last member of $\{A_F\}$, let $\{A_S\}$ be the null set.

(8) Find an index, root, such that $\alpha_{root} = \text{maximum}_i \alpha_i$, where $A_i \in \{A_Q\} \cup \{A_F\} \cup \{A_S\}$ and $|W_{0,root-1} - W_{root,N}|$ is minimized; choose A_{root} as the root of the tree.

(9) Go to Step 1 and repeat the algorithm for the subtrees $A_1, ..., A_{root-1}$ and $A_{root+1}, ..., A_N$, where N is root -1 and N-root for the two cases.

5. Choosing Parameters of the Algorithm

The algorithm defined in the preceding section has two parameters, N_0 and F, which are influenced by the computer on which the algorithm is executed, the frequencies associated with the nodes, and the desired precision of the average search length. The parameter N_0 determines the maximum number of names which will be structured into an optimal subtree using Algorithm K. In practice, the available storage will usually determine an upper bound for N_0. The larger the value of N_0, the closer the average search length of the nearly optimal tree. We thus have a trade-off, which is examined further below, between the value of N_0 and the ratio of the average search length of the nearly optimal tree to the average search length of the optimal tree.

† F determines the neighborhood of the set $\{A_F\}$.

‡ $W_{0,N} = W$, the weight of the tree.

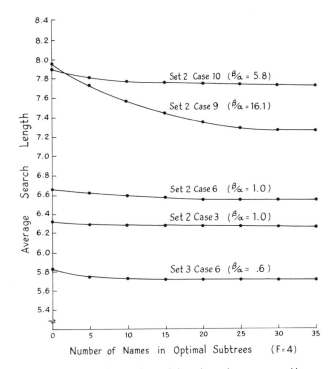

Fig. 3. Dependence of search length on the parameter N_0.

Let us examine Fig. 3 which shows 5 examples, from the tests described in Section 7, of how the average search length for a nearly optimal tree depends on N_0. As N_0 increases, the value of the average search length decreases. However, for each example, there is a value of N_0 beyond which the average search length decreases very slowly or remains constant. This value is largely determined by the sum of the β_i and the sum of the α_i, which we will refer to as the β and α frequency respectively. In general, if the β frequency is less than a few multiples of the α frequency, the value of N_0 beyond which it does not pay to go is small. In the examples shown, there is little advantage in choosing N_0 larger than 15. This is the value used in our tests described in Section 7. If the β frequency is many times greater than the α frequency, N_0 should be increased, perhaps, to 25 or 30.

The second parameter which can be varied is F, which determines the size of the neighborhood about the centroid from which the root will be chosen. With $F = 1$, the neighborhood is the entire tree. The name with the largest associated frequency becomes the new root.[†] Increasing F restricts the neigh-

---

[†] If $N_0 = 0$, this corresponds to the first rule of Knuth.

borhood. Choosing $F = W_{0,N}$ restricts the neighborhood so that little searching takes place, and a name near or equal to the centroid is chosen as the root.

The value of F also depends on the β frequency and the α frequency. Let us assume that the β frequency is many times greater than the α frequency, and all the α_i are small compared to the smallest β_i. If the smallest β_i is sufficiently large, the optimal binary tree will be the complete binary tree.[†] This tree is structured by our algorithm when $F = W_{0,N}$, that is, the centroid is chosen as the root. In practice a set of frequencies satisfying the above conditions will not occur often. However, it is found that having the β frequency many times greater than the α frequency is a sufficient condition for $F = W_{0,N}$ to produce good nearly optimal trees.

When the β frequency is less than a few multiples of the α frequency, the average search length of the nearly optimal tree is improved by increasing the neighborhood of the centroid which is searched for the root, as discussed in Section 3. If the β frequency is a few multiples of the α frequency, it is found that the choice of F is not critical and the average search length of the nearly optimal tree is almost constant for $F > 5$. If the α and β frequencies are nearly equal, the minimum average search length occurs for F near 4. If the β frequency is small compared to the α frequency, the individual α_i frequency will determine the best value of F.

Fig. 4 shows 5 examples, from tests described in Section 7, of how the average search length for a nearly optimal tree depends on F. Unless the β frequency is many times greater than the α frequency, a value of 4 for F seems to be acceptable.

6. Time to Construct the Nearly Optimal Tree

Before considering general timing formulas, let us determine the time required to construct a tree from N names with all frequencies uniform, that is, with $\alpha_i = \beta_i = $ constant. As usual, the parameters of our algorithm will be denoted by N_0 and F. Our algorithm consists of two steps, one for locating A_{root} in a subtree and the other for constructing an optimal subtree using Algorithm K.

The subtree roots, A_{root}, located by our algorithm will form the first L levels of the nearly optimal N name tree whenever N satisfies

(1) $2^{L-1}(N_0 + 1) + 2^{L-1} \doteq 1 \leqslant N \leqslant 2^L(N_0) + 2^L - 1$

The time required to locate A_{root} for a subtree with N_1 names and weight W_1 is proportional to W_1/F or equivalently, since the weights are all equal,

---

† A complete binary tree has leaves only on level L or $L+1$; see Knuth [5].

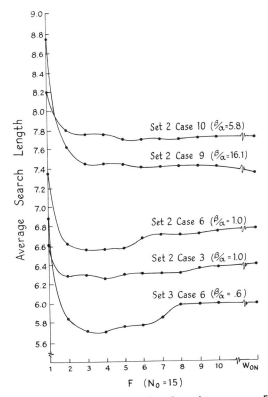

Fig. 4. Dependence of search length on the parameter F.

to N_1/F. On any level $L_1 \leqslant L$, there are 2^{L_1-1} subtree roots to be located, each requiring a time proportional to $N/F(2^{L_1-1})$. Hence, locating the subtree roots on each level requires a time proportional to N/F. The total time will be proportional to LN/F.

For any value of N satisfying (1) for some L, 2^L subtrees will be constructed by Algorithm K. The average number of names in each subtree varies from $N_0/2$ to N_0, as N varies from the lower bound to the upper bound of (1). If a subtree containing N_0 names is constructed in time $K(N_0)$ by Algorithm K, the total time required by our algorithm to construct the nearly optimal tree when N is equal to the upper bound in (1) is given by

$$(2) \qquad \text{Time} = K'LN/F + K(N_0)2^L$$

$$= K' \log_2\left(\frac{N+1}{N_0+1}\right)\frac{N}{F} + K(N_0)\left(\frac{N+1}{N_0+1}\right).$$

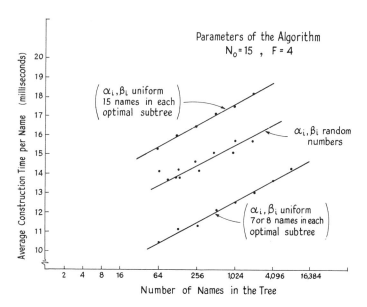

Fig. 5.   Time to construct a nearly optimal tree.

Hence, the time required to construct this tree is given by an equation of the form

(3) $$\text{Time} = K_1 N \log_2 N + K_2 N.$$

Similarly, when N equals the lower bound in (1) the time required to construct the tree will be given by

(4) $$\text{Time} = K_1 N \log_2 N + K_3 N,$$

where K_3 is determined by the time required by Algorithm K to construct an optimal subtree with $N_0/2$ names.

The average time per node required to construct the tree can be obtained by dividing (3) and (4) by N. Figure 5 shows the time per node required by our algorithm to construct the nearly optimal tree when the number of names in the optimal subtrees is $N_0/2$ and N_0. In both cases, the form of our normalized equation is verified, since the plotted times lie on a straight line on the semilogarithmic graph paper.

The nearly optimal trees of N names, with N not satisfying (1), have names chosen as subtree roots accounting for the names on the first L levels and some of the names on the $L+1$ level. The time required to locate A_{root} in an N_0+1 name subtree, and construct two optimal subtrees of $N_0/2$ names, is less than the time required to construct an optimal N_0+1 name subtree. Hence, the time required to construct any N name nearly optimal tree with uniform frequencies is given by the expression

$$K_1 N \log_2 N + K_4 N,$$

where $K_3 \leqslant K_4 \leqslant K_2$, and K_4 is determined by the number of optimal subtrees and the number of names in each optimal subtree.

The shape of the nearly optimal tree depends on the individual α_i and β_i frequencies. In general, the subtree roots will not form the first L levels as in the case of uniform frequencies. If most of the subtree roots occur on the first L levels, an equation of the form (3) may be used to estimate the time required to construct the tree. In practical applications, such as the library index, this will usually be the case. For example, Fig. 5 shows the time per node required to construct a tree in which the α_i and β_i frequencies are random numbers between 1 and 200. The slopes of the straight lines fitted to the plotted construction times for the random and uniform frequencies are almost equal. Since the expected average search lengths of the balanced tree and the optimal tree satisfy (see Nievergelt and Wong [7]).

$$|\text{expected average search length} - \log_2 N| < \text{constant},$$

we would expect the construction time per node of the nearly optimal trees to differ by a constant. That is, the slopes of the lines in Fig. 5 would be expected to be equal. The deviation of the construction times from the straight line for random frequencies results from the dependence of the time on the number of names in each optimal subtree. These numbers are determined by the random α_i and β_i frequencies.

The times in Fig. 5 were obtained from an ALGOL implementation of our algorithm executing on an IBM/360 Model 65, and should not be regarded as a measure of the minimum time required to construct a tree of N names. A much more efficient program could be written using assembler language for some sections. However, the form of the equation estimating the time should remain unchanged.

The algorithm requires $4N$ words of storage for the data. From the nature of our top down algorithm, it would be possible to use secondary storage for the data, when constructing trees for large values of N, without greatly increasing the running time beyond that achieved with a very large core store.

7. Tests of the Algorithm

Using a value of $N = 200$, the algorithm was tested for five sets of α_i frequencies, and for 10 sets of β_i frequencies with each α_i set. The sets of α_i were chosen in the following manner; Sets 1, 2, and 3 were obtained by

TABLE II

Test Data[a]

α_i Frequencies

Set 1

4	19	3	3	1	1	28	4	4	5	1	1	1	35	13	27	11	32	1	50
2	1	1	27	2	31	2	3	1	2	7	59	4	1	1	7	3	2	3	22
2	1	52	3	1	4	1	1	1	6	8	13	5	4	1	1	1	9	6	26
33	350	1	8	7	1	2	1	8	2	4	10	63	10	23	14	150	1	6	2
128	2	1	2	1	87	2	1	3	3	4	1	9	2	3	2	10	4	15	1
1	24	6	2	8	1	1	1	3	1	21	3	14	4	1	2	3	1	2	1
1	4	43	2	2	4	1	10	18	1	1	7	1	45	15	9	4	1	5	2
2	8	1	5	10	1	2	2	2	1	24	29	1	16	1	1	1	1	2	4
10	1	1	1	1	4	2	2	6	4	1	14	5	5	1	53	1	1	2	1
1																			4

Set 2

7	4	1	1	20	1	2	1	3	1	1	1	30	1	1	1	1	1	3	30
2	1	1	4	2	2	1	3	1	2	1	7	1	2	4	2	6	6	7	1
90	3	1	4	1	4	29	1	1	1	1	2	6	2	2	1	5	5	2	40
1	3	5	4	62	7	1	6	1	3	11	4	2	1	1	2	2	6	1	1
1	1	1	24	1	1	29	2	100	15	1	2	7	1	2	8	10	3	1	2
3	1	3	2	1	1	1	1	13	3	1	20	2	1	3	3	1	10	1	9
1	6	19	2	1	3	1	1	1	4	3	39	1	1	15	11	6	2	4	3
2	4	1	1	1	1	29	3	7	1	2	1	1	1	3	1	2	1	5	6
2	1	1	1	1	2	1	9	10	24	8	1	1	1	1	30	1	1	14	1
3	6	2	2	1	23			1	1			6				2			90

Set 3

1	1	1	2	4	1	12	4	4	8
2	1	5	12	9	1	5	4	1	1
1	1	1	7	1	3	1	2	7	1
2	1	1	1	1	1	30	1	107	11
2	1	2	7	227	1	37	1	2	1
3	1	4	1	227	5	7	173	10	2
2	3	1	4	54	1	84	15	1	5
2	12	2	1	11	3	1	1	2	2
2	2	1	1	1	1	3	4	3	2
1	4	1	2	2	1	2	1	4	1
10	1	42	1	51	7	1	2	1	1
1	1	2	1	2	1	1	7	1	1
1	1	5	2	40	2	3	2	2	6
13	1	3	11	1	40	183	3	2	
1	5	1	1	2	1	2	12	2	
12	1	1	1	2	3	33	17	2	
1	2	24	1	2	2	4	1	6	3
306	1	7	10	1	24	78	1	1	1
5	1	1	10	1	1	2	1	1	
8	1	2	1	4	3	1	1	4	1

Set 4

30	1	9	3	1	50	22	2	1	4
3	7	1	1	5	1	3	6	15	5
1	6	10	2	1	32	2	9	4	1
1	6	1	6	1	11	3	1	10	3
1	2	8	3	1	27	7	1	2	9
1	4	2	3	3	13	1	23	1	15
1	2	1	1	1	35	1	10	1	45
30	1	2	1	1	1	4	5	9	1
1	7	1	20	1	1	59	10	1	7
1	1	1	1	1	2	1	7	4	21	1
1	2	3	4	24	5	2	2	4	1
3	1	13	1	10	4	1	1	3	18
1	3	2	1	3	4	3	1	1	1
2	1	29	1	29	28	2	2	2	1
1	2	1	3	2	1	31	1	1	4
20	2	1	1	1	2	7	8	1	
1	4	2	2	1	3	27	8	1	2
1	1	3	19	1	3	1	1	1	43
4	1	1	6	1	19	1	1	2	4
7	2	3	1	2	4	2	33	1	2

Set 5

40	4	2	12	26	4	4	1	1	
2	4	1	5	1	5	2	9	2	1
5	2	3	1	1	1	1	1	1	
5	1	10	30	1	3	2	1	4	150
1	1	1	37	1	9	2	227	1	14
2	173	1	7	1	15	3	227	1	3
2	15	1	84	4	45	2	54	16	2
6	1	7	1	1	1	2	11	1	63
2	4	2	3	13	7	2	1	29	1
1	1	1	2	8	1	1	2	24	4
1	2	15	1	6	1	10	51	1	3
1	7	100	1	1	18	1	2	2	8
1	2	6	3	1	1	40	10	1	
29	183	1	40	1	13	11	2	2	
4	2	1	1	4	4	1	1	1	87
1	33	1	3	1	12	1	10	1	
4	1	24	4	3	2	1	2	5	2
1	1	1	78	52	43	306	1	1	1
3	2	1	1	1	4	5	10	8	350
90	1	1	1	2	2	8	4	10	128

TABLE II (continued)

β_i Frequencies

Case 6

3	1	3	2	1	1	29	2	13	3	1	1	2	1	2	1	8	1	10	1	9
1	6	19	2	1	3	1	1	1	4	20	1	1	1	1	6	3	6	2	1	3
2	1	1	1	2	3	29	3	10	24	2	7	1	1	1	1	1	1	1	5	1
2	1	1	4	20	2	1	3	1	2	1	1	30	2	1	2	2	6	6	7	1
7	4	3	1	1	1	28	4	4	1	35	1	1	1	13	27	11	32	3	30
4	19	1	3	1	31	2	1	1	5	1	59	4	1	1	7	7	3	1	2	50
2	1	1	8	2	1	2	1	1	2	10	10	5	10	23	1	1	10	2	9	22
33	1	1	1	7	1	2	1	3	2	4	1	9	1	1	2	3	6	9	2
1	2	1	2	8	4	1	1	18	4	21	7	1	45	15	9	3	10	4	15	1
2	4	43	2	1	4	1	1	3	1	1	1	1	1	1	2	5	1	4
4								18												

Case 7

2	4	43	2	1	4	1	1	18	1	1	7	1	45	15	9	3	1	5	4
1	2	1	1	8	1	2	1	3	4	21	1	9	1	1	2	10	4	15	1
33	1	1	8	7	1	2	1	1	2	4	10	5	10	23	1	1	9	6	2	2
2	1	1	27	2	31	2	3	1	2	7	59	4	1	1	7	3	3	2	22
4	19	3	3	1	1	28	4	4	5	35	1	1	35	13	27	11	32	1	50
3	1	3	2	1	3	29	2	13	3	1	1	2	1	2	8	1	10	5	9
1	6	19	2	1	2	1	1	1	4	20	1	1	3	6	1	2	1	3
2	1	1	2	1	2	29	3	10	24	2	7	1	1	3	1	6	6	5	1
2	1	4	1	2	1	1	3	1	2	1	1	1	2	3	2	1	6	7	1
7	4	1	20	1	1	2	1	3	1	1	1	30	1	4	1	1	3	30
4																		

Case 8

2	1	52	3	1	4	1	2	1	8
128	350	1	2	1	87	1	1	8	4
2	4	43	2	1	4	1	1	18	1
10	8	1	5	10	1	10	2	2	24
90	3	1	4	1	4	6	29	1	1
1	1	1	24	1	1	1	1	100	1
8	5	306	1	12	1	40	13	2	2
4	10	1	2	1	1	3	11	1	2
1	1	78	4	3	1	2	40	7	3
1	2	1	1	33	2		183		4

Case 9

12	8	315	18	7	24	6	8	6	49
900	6	14	12	623	12	8	19	19	6
24	248	13	7	28	6	8	6	6	42
49	24	30	68	6	12	60	9	9	183
6	14	37	11	9	4	34	14	14	6
7	11	41	8	7	7	211	18	18	14
70	7	15	15	121	23	18	7	7	6
307	14	6	66	323	999	960	6	6	4
40	73	7	12	6	6	195	12	12	43
25	22	11	18	7	497	45	221	190	30

ª The α_i and β_i vectors are arranged in the following matrices as

$$\alpha_1, \quad \alpha_2, \quad \alpha_3, \ldots, \quad \alpha_{19}, \quad \alpha_{20}$$
$$\alpha_{21}, \quad \alpha_{22}, \quad \ldots, \quad \alpha_{39}, \quad \alpha_{40}$$
$$\alpha_{181}, \quad \alpha_{182}, \quad \ldots, \quad \alpha_{199}, \quad \alpha_{200}$$

examining three sections[†] of the University of Toronto Library author–title catalog, and defining a name A_i to be a surname under which a card was filed. The surname could be a main or added entry, an author or title, as long as the card was filed under the surname. The frequency α_i associated with A_i was the number of cards on which the A_i appeared. If several copies of a book were in various departmental libraries, each copy might have a separate card, and each card was counted in determining the α_i. For each of the three sections, an α_i distribution was found. The fourth set of α_i was chosen by eliminating the most frequently occurring names from Sets 1 and 2, and then selecting 200 names from those remaining. Set 5 was obtained by including all of the larger frequencies found in Sets 1, 2, and 3 in the 200 frequencies.

For each set of α_i, 10 sets of β_i, divided in 4 classes were obtained as follows:

Class 1:   the β_i were all equal.

Case 1: $\beta_i = 0,\ 0 \leqslant i \leqslant N$;
Case 2: $\beta_i = 10,\ 0 \leqslant i \leqslant N$.

Class 2:   the β_i were calculated as a function of neighboring α_i, where α_i is taken as 0 for $i \leqslant 0$ or $i > N$.

Case 3: $\beta_i = (\alpha_i + \alpha_{i+1})/2,\ 0 \leqslant i \leqslant N$;
Case 4: $\beta_i = (\alpha_{i-1} + \alpha_i + \alpha_{i+1} + \alpha_{i+2})/4,\ 0 \leqslant i \leqslant N$;
Case 5: $\beta_i = |3\alpha_{i+1} - 2\alpha_{i+3}|,\ 0 \leqslant i \leqslant N$.

Class 3:   the β_i were chosen so the sum of the β_i would be equal to or less than the sum of the α_i.

Case 6: β_i chosen from Set 1, 2, and 3, Case 3 and 4;
Case 7: β_i chosen from Set 1, 2, and 3, Case 3 and 4.

Class 4:   the β_i were chosen so the sum of the β_i would be larger than the sum of the α_i.

Case 8: β_i chosen from Sets 1, 2, and 3, Case 3 and 4;
Case 9: β_i chosen randomly;
Case 10: $\beta_i = 1.5(\alpha_{i-1} + \alpha_i + \alpha_{i+1} + \alpha_{i+2}),\ 0 \leqslant i \leqslant N$.

The 5 sets of α_i frequencies and Cases 6, 7, 8, and 9 of the β_i frequencies are listed in Table II. In Table I the average search length of the optimal tree is compared with the results of our top down algorithm for different values of N_0 and F.

The best nearly optimal trees are obtained when the β frequency and the α frequency are approximately equal in value. In these cases the difference

---

[†] The third section, for example, corresponds to surnames between Newstead and Niedermayer.

between the average search length of the nearly optimal tree and the optimal tree is usually less than 1%. When the β frequency is many times greater than the α frequency, the poorest nearly optimal trees are constructed. This is expected, since the algorithm uses the β_i frequencies in choosing the centroid, but does not consider them when choosing another name for the actual root. Even for the poorest nearly optimal trees, the difference between the average search length of the optimal and the nearly optimal tree is less than 3.5%.

The algorithm was also tested over a sequence of values of N. The sets of names used for this test were drawn from a list of 144,486 distinct surnames, each having an associated frequency α_k.[†] The names were first ordered so that $\alpha_{i_1} \geqslant \alpha_{i_2} \geqslant \cdots \geqslant \alpha_{i_{144,486}}$. For each value of N in the sequence, the N most frequently occurring names $A_{i_1}, \ldots, A_{i_N}$ were selected and reordered lexicographically, that is, so that $A_{i_1} < \cdots < A_{i_N}$. The frequencies used to construct the tree were as follows: α_j was the frequency associated with A_{i_j} and β_j was the sum of the frequencies of names between A_{i_j} and $A_{i_{j+1}}$. Note that as the number of names in the tree increases, the α frequency increases by an amount equal to the β frequency decrease, that is, the total weight of the tree remains constant. As previously, the sum of the α's is called the α frequency, and the sum of the β's the β frequency.

The results of the test are given in Table III and shown in Fig. 6. For $N \leqslant 150$, optimal trees could be easily constructed and the difference in the average search length of the optimal and nearly optimal tree is always less than 2%. For $N \leqslant 150$, the β frequency is several times the α frequency, and from arguments given previously, the difference in the average search length of the optimal and nearly optimal tree is expected to decrease as N and α/β increases. Thus, it is reasonable to expect that the dependence of the nearly optimal tree on N will be the same as that of the optimal tree. From Fig. 6, we observe that for small values of N, the average search length behaves approximately like $\log_2 N$. However, for large values of N (>6000), the average search length of the nearly optimal tree is almost constant. This is in contrast to the average search length of approximately $\log_2 N$ for an optimal tree of N names, and $2N+1$ nodes, with uniform α_i and β_i frequencies.

The observed result is to be expected. When N is small, the large β_i account for most of the weight of the tree and the leaf nodes largely determine the average search length. As N increases, the β_i decrease, and the A_i with largest associated frequencies become nodes in the tree. Eventually the β_i become small enough, and the levels on which they occur deep enough, so that the contribution to the weighted path length of a subtree formed by splitting a β_i into α_k's and β_k's hardly changes. An interpretation of the above results is

† The original list consisted of over one million names, of which 144,486 were distinct. The frequencies are the frequencies of occurrences in the original list.

TABLE III

Average Search Length

N	α Frequency	β Frequency	Optimal tree	Nearly optimal tree $N_0 = 15, F = 5$
5	19,846	982,497	3.4114	3.4114
15	42,653	959,690	4.2864	4.2864
25	60,087	942,256	5.0638	5.1033
50	92,117	910,226	6.0483	6.1461
100	138,975	863,368	7.0007	7.0437
150	173,157	829,186	7.4885	7.5503
200	200,412	801,931		7.8795
500	305,266	697,077		8.9606
1000	401,288	601,055		9.6490
3000	561,956	440,387		10.6220
6000	655,538	346,805		11.1177
12,000	740,022	262,321		11.1592
144,486	1002,343	0		

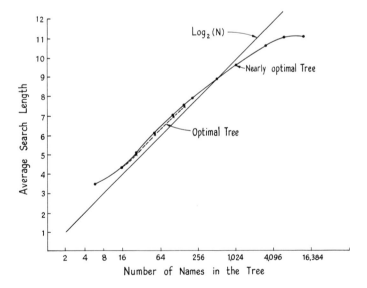

Fig. 6. Dependence of search length on the number of names in the tree.

that, if we know all the names which could be used to search a binary tree, and the frequencies with which they occur, then there is a value of N, 6000 in our example, such that, if more than the N most frequently occurring names are placed in the tree, the average search length will not be appreciably increased.

8. Summary

Our top-down algorithm for constructing nearly optimal lexicographic trees required a time proportional to $N \log_2 N$ for all N name trees constructed, and requires storage proportional to N. As the α_i and β_i frequencies change over a period of time, it would be necessary to completely restructure the tree to keep it in a nearly optimal form. In a practical application such as the library index discussed above, the frequencies associated with the nodes would not change greatly over a short period of time. It would be feasible to use our algorithm to occasionally restructure the tree. In addition, the β frequency would usually be less than the α frequency in such an application, and the nearly optimal trees can be expected to have an average search length within 1% of that of the optimal tree.

References

1. Booth, A. D., and Colin, A. J. T., On the efficiency of a new method of dictionary construction, *Information and Control* 3, 327–334 (1960).
2. Bruno, J., and Coffman, E. G., Jr., Nearly optimal binary search trees, IFIP Congress 71, TA-2, 29–32 (1971).
3. Clampett, H. A., Jr., Randomized binary searching with tree structures, *Comm. ACM* 9, 163–165 (1964).
4. Hibbard, T. N., Some combinatorial properties of certain trees with applications to searching and sorting, *J. Assoc. Comput. Mach.* 9, 13–28 (1962).
5. Knuth, D. E., "The Art of Computer Programming," Vol. 1. Addison-Wesley, Reading, Massachusetts, 1968.
6. Knuth, D. E., Optimum binary search trees, *Stanford Univ. Dept. Computer Sci. Tech. Rep. CS 149* (1970).
7. Nievergelt, J., and Wong, C. K., On binary search trees, IFIP Congress 71, TA-2, 23–28 (1971).
8. Windley, P. F., Trees, forests and rearranging, *Comput. J.* 3, 84–95 (1960).

INDEX